Nonlinear Systems -Theoretical Aspects and Recent Applications

Edited by Walter Legnani and Terry E. Moschandreou

Published in London, United Kingdom

IntechOpen

Supporting open minds since 2005

Nonlinear Systems - Theoretical Aspects and Recent Applications
http://dx.doi.org/10.5772/intechopen.77467
Edited by Walter Legnani and Terry E. Moschandreou

Part of IntechOpen Book Series: Nonlinear Systems, Volume 2
Book Series Editor: Mahmut Reyhanoglu

Contributors
Masoumeh Pourabd Rokne Saraie, Maria Ines Troparevsky, Silvia Alejandra Seminara, Marcela Antonieta Fabio, Osvaldo Rosso, Felipe Olivares, Lindiane Souza, Walter Edgardo Legnani, Vladimir Stepanovich Korolev, Elena Polyakhova, Irina Pototskaya, Barun Pratiher, Ali Al-Zughaibi, Vishakha Jadaun, Nitin Raja Singh, Parag Chatterjee, Leandro J. Cymberknop, Ricardo Armentano, Abderrahman Ait Aadi, El Hassan Zerrik, Mustafa Mamat, Ibrahim Sulaiman, Boubacar Diao, Khalil Ezzinbi, Anil Bajaj, Astitva Tripathi, Terry E. Moschandreou, Kamal Shah, Adnan Khan, Danfeng Luo

First published in London, United Kingdom, 2020 by IntechOpen
IntechOpen is the global imprint of INTECHOPEN LIMITED, registered in England and Wales, registration number: 11086078, 7th floor, 10 Lower Thames Street, London, EC3R 6AF, United Kingdom
Printed in Croatia

British Library Cataloguing-in-Publication Data
A catalogue record for this book is available from the British Library

Additional hard and PDF copies can be obtained from orders@intechopen.com

Nonlinear Systems - Theoretical Aspects and Recent Applications
Edited by Walter Legnani and Terry E. Moschandreou
p. cm.
Print ISBN 978-1-78985-471-8
Online ISBN 978-1-78985-472-5
eBook (PDF) ISBN 978-1-83880-308-7

For EU product safety concerns:
IN TECH d.o.o., Prolaz Marije Krucifikse Kozulić 3, 51000 Rijeka, Croatia, info@intechopen.com or visit our website at intechopen.com.

IntechOpen Book Series

Nonlinear Systems

Volume 2

Walter E. Legnani has a degree in Physics and a PhD in the area of Physics from the University of Buenos Aires (Argentina), and a thesis developed at the University of Colorado and the National Centre for Atmospheric Research (USA). He also has a postdoc from the Department of Applied Mathematics and Theoretical Physics, Cambridge University and Reading University (UK). He is a full professor in signal and system analysis and has run several postgraduate courses. He is also Director of the Signals and Image Processing Centre (Universidad Tecnologica Nacional-Facultad Regional Buenos Aires), a director of several postgraduate theses, a board scientific member of the Binational Centre Argentina Italia (CAIMAR), a former full professor at Favaloro University, a former secretary of science, technology and postgraduate studies (UTN), and a member of several international scientific societies and evaluation committees. He is a peer reviewer and guest editor of prestigious journals of high impact. www.researchgate.net/profile/Walter_Legnani

Dr. Terry E. Moschandreou is a professor in applied mathematics at the University of Western Ontario in the School of Mathematical and Statistical Sciences where he has taught for several years. He received his PhD degree in Applied Mathematics from the University of Western Ontario in 1996. The greater part of his professional life has been spent at the University of Western Ontario and Fanshawe College in London, Ontario, Canada. Dr. Moschandreou is also currently working for Goode Educational Services where he teaches students advanced calculus and linear algebra. For a short period, he worked at the National Technical University of Athens, Greece. Dr. Moschandreou is the author of several research articles in blood flow and oxygen transport in the microcirculation, general fluid dynamics, and theory of differential equations. Also, he has contributed to the field of finite element modeling of the upper airways in sleep apnea as well as surgical brain deformation modeling. More recently, he has been working with the partial differential equations of multiphase flow and level set methods as used in fluid dynamics.

Editors of Volume 2:
Walter E. Legnani
Signals and Images Processing Centre, Universidad Tecnologica Nacional
Facultad Regional Buenos Aires, Argentina

Terry E. Moschandreou
School of Mathematical and Statistical Sciences, University of Western Ontario, Canada

Book Series Editor:
Mahmut Reyhanoglu, PhD
University of North Carolina Asheville, Department of Engineering
Asheville, North Carolina, USA

Scope of the Series

The series will be both on classical materials, such as nonlinear dynamics, stability and optimality, and more modern topics such as differential geometry, nonlinear control theory and applications in robotics. The books can be used as a reference and guide in the active literature in these fields.

Topics will broadly include, but are not limited to:

- Nonlinear dynamical systems
- Lagrangian and Hamiltonian formulations
- Nonlinear analysis
- Differential geometry
- Nonlinear control theory
- Lyapunov methods
- Nonlinear observers
- Geometric mechanics
- Robotics applications

Contents

Preface

Nonlinear dynamical systems have been used in the most diverse areas of scientific knowledge. Along with this, differential equations of fractional order, whose theoretical formulation continues to grow and whose applications are increasingly diverse, have attracted outstanding interest. This book gathers clear examples of these fields along with the most recent knowledge. The contributions are from diverse authors from a remarkable variety of countries and show a diversity in fields of applications. This fact confirms the abundant current interest in these topics in the scientific and academic community.

Fractional calculus (FC) has roots that are deep in the theory of differential calculus. FC occurs in applications such as chaos and dynamical systems, modeling of memory-dependent theory, and complex media, for example in the study of porous media. Further applications are seen in the fields of digital circuits, heat diffusion, robotic theory, and controller tuning.

The development of FC is due to contributions from mathematicians like Euler, Liouville, Riemann, and Letnikov. Due to limitations in classical methods as applied to dynamical systems, FC has proven to be an efficient tool for this stream of study. Existence theory and hyperbolic differential equations are ery important parts of the study of FC. One of the most important contributions of FC is the Caputo fractional derivative. FC involves both derivatives and integrals up to an arbitrary order, which can be real or complex.

The Grunwald–Letnikov definition of fractional derivatives and the Riemann–Liouville definition are also important and use the gamma function. These operators associated with fractional derivatives are global operators defining memory events. The part of FC used in this book is new and has many of the features of FC that are important in the literature.

In addition to fractional theory, *Nonlinear Systems*, which is divided into theoretical and applied sections, has the following contributions.

In the theoretical section of the book, in the context of FC methods, Chapter 1, "A Review on Fractional Differential Equations and a Numerical Method to Solve Some Boundary Value Problems," is proposed.

In addition, related to FC, numerical methods is the content of Chapter 2, "Numerical Solutions to Some Families of Fractional Order Differential Equations."

Chapter 3, "A Shamanskii-Like Accelerated Method for Systems of Nonlinear Equation," starts with an initial iterate and moves through an intermediate sequence of iterates, which is a Newton iterate followed by several "cord" iterates. It is a generalization that encapsulates Newton's method.

Chapter 4 looks at the topic of "Modified Moving Least Squares Method for Two-Dimensional Linear and Nonlinear Systems of Integral Equations." In the moving least squares method an approximation value can be expressed as a linear combination of shape functions and known function values. The moving least squares method reconstructs continuous functions from a set of non-organized point samples of a biased weighted least squares indicator.

In the light of chaotic systems, a bi-dimensional and a causal plane are defined, in which different dynamical regimes appear very clear and give information on the process involved. This is the subject of Chapter 5, "Informational Time Causal Planes: A Tool for Chaotic Maps Dynamics Visualization."

Stability is an important area of nonlinear systems and is considered in Chapter 6, "On the Stabilization for Infinite Dimensional Semi-Linear Systems."

For Chapter 7, "Existence, Regularity and Compactness Properties in the Alpha-Norm for Some Partial Functional Integro-Differential Equations with Delay" is considered. The objective of this chapter is to study the alpha-norm, existence, continuity dependence in initial data, regularity, and compactness of solutions of mild solution for some semi-linear partial functional integro-differential equations in abstract Banach space.

In the applied section, "Recent Applications in Nonlinear Systems," Chapter 1 covers "Nonlinear Resonances in 3D Printed Structures." Here, nonlinear resonators are studied and the nonlinear behavior of such structures is analyzed. Computational methods are employed for structural design and the case of one to two internal resonances of hyperelastic materials.

The importance of health care cannot be understated and Chapter 2 considers the "Nonlinear Systems in Healthcare Towards Intelligent Disease Prediction." Here predictive analytics are considered with examples of intelligent systems toward disease prediction.

Shell structures are examined in Chapter 3, "Mathematical Modeling and Well-Posedness of Three-Dimensional Shell in Disorders of Human Vascular System" where a shell structure is a general three-dimensional structure that is elongated in two directions and thinned out in the other direction. In the human vascular system, the human anatomy develops cyst-related diseases with progressive severity. These cysts can be modeled as shells, albeit in higher dimensions.

Further applications are found in Chapter 4, "Features of Optimal Control in Photo-Gravitational Fields," and Chapter 5, "Nonlinear Friction Model Identification and Effectiveness." Chapter 6 of this section, "Electrostatically Driven MEMS Resonator: Pull-In Behavior and Nonlinear Phenomena," presents an interesting application to the stability and bifurcation analysis of highly nonlinear, electrically driven micro-electro-mechanical systems (MEMS).

Finally, in Chapter 7, "An Approximate Analytical Solution for Non-Linear Oxygen with Poiseuille Hemodynamic Flow in a Micro-Channel," an important nonlinear model of hemodynamic flow with oxygen transport is considered.

We hope that this book will be useful material for both graduate students and researchers in general.

It is not a point of arrival, only a starting point to be enriched with future developments that show the advances in the production of knowledge in these exciting fields of scientific work.

Walter E. Legnani
Signals and Images Processing Centre,
Universidad Tecnologica Nacional,
Facultad Regional Buenos Aires,
Argentina

Terry E. Moschandreou
School of Mathematical and Statistical Sciences,
University of Western,
Ontario, Canada

Section 1

Theoretical Aspects of Nonlinear Systems

Chapter 1

A Review on Fractional Differential Equations and a Numerical Method to Solve Some Boundary Value Problems

María I. Troparevsky, Silvia A. Seminara and Marcela A. Fabio

Abstract

Fractional differential equations can describe the dynamics of several complex and nonlocal systems with memory. They arise in many scientific and engineering areas such as physics, chemistry, biology, biophysics, economics, control theory, signal and image processing, etc. Particularly, nonlinear systems describing different phenomena can be modeled with fractional derivatives. Chaotic behavior has also been reported in some fractional models. There exist theoretical results related to existence and uniqueness of solutions to initial and boundary value problems with fractional differential equations; for the nonlinear case, there are still few of them. In this work we will present a summary of the different definitions of fractional derivatives and show models where they appear, including simple nonlinear systems with chaos. Existing results on the solvability of classical fractional differential equations and numerical approaches are summarized. Finally, we propose a numerical scheme to approximate the solution to linear fractional initial value problems and boundary value problems.

Keywords: fractional derivatives, fractional differential equations, wavelet decomposition, numerical approximation

1. Introduction

Fractional calculus is the theory of integrals and derivatives of arbitrary real (and even complex) order and was first suggested in works by mathematicians such as Leibniz, L'Hôpital, Abel, Liouville, Riemann, etc. The importance of fractional derivatives for modeling phenomena in different branches of science and engineering is due to their nonlocality nature, an intrinsic property of many complex systems. Unlike the derivative of integer order, fractional derivatives do not take into account only local characteristics of the dynamics but considers the global evolution of the system; for that reason, when dealing with certain phenomena, they provide more accurate models of real-world behavior than standard derivatives.

To illustrate this fact, we will retrieve an example from [1]. Recall the relationship between stress $\sigma(t)$ and strain $\varepsilon(t)$ in a material under the influence of external forces:

$$\sigma(t) = \eta \frac{d}{dt}\varepsilon(t) \tag{1}$$

is the Newton's law for a viscous liquid, with η the viscosity of the material, and

$$\sigma(t) = E\varepsilon(t) \tag{2}$$

is Hooke's law for an elastic solid, with E the modulus of elasticity. We can rewrite both Eqs. (1) and (2) as

$$\sigma(t) = \nu \frac{d^\alpha}{dt^\alpha}\varepsilon(t) \tag{3}$$

with $\alpha = 0$ for elastic solids and $\alpha = 1$ for a viscous liquid. But, in practice, there exist *viscoelastic* materials that have a behavior intermediate between an elastic solid and a viscous liquid, and it may be convenient to give sense to the operator $\frac{d^\alpha}{dt^\alpha}$ if $0 < \alpha < 1$.

There exist various definitions of fractional derivatives. All of them involve integral operators with different regularity properties, and some of them have singular kernels.

Next, we will briefly review the most frequently fractional derivatives cited in the bibliography (see [2] for a more complete review and [1, 3–6] for rigorous theoretical expositions and calculation methods).

The classical Cauchy formula for the n-fold iterated integral, with $n \in \mathbb{N}$, is

$${}_0I_t^n[f](t) = \frac{1}{(n-1)!}\int_0^t (t-s)^{n-1} f(s)ds. \tag{4}$$

Recalling that gamma function verifies $n\Gamma(n) = n!$, an immediate generalization of this formula for a real order α is

$${}_0I_t^\alpha[f](t) = \frac{1}{\Gamma(\alpha)}\int_0^t (t-s)^{\alpha-1} f(s)ds, \tag{5}$$

known as Riemann-Liouville fractional integral operator of order α (the term "fractional" is misleading but has a historical origin). From this, the Riemann-Liouville fractional derivative of order α, with $n - 1 < \alpha < n$, is defined as

$${}_0^{RL}D_t^\alpha[f](t) = \frac{1}{\Gamma(n-\alpha)}\frac{d^n}{dt^n}\int_0^t (t-s)^{n-\alpha-1} f(s)ds. \tag{6}$$

while the Caputo fractional derivative of order α is defined as

$${}_0^{C}D_t^\alpha[f](t) = \frac{1}{\Gamma(n-\alpha)}\int_0^t (t-s)^{n-1-\alpha} \frac{d^n}{ds^n}[f(s)]ds. \tag{7}$$

Both Eqs. (6) and (7) define left inverse operators for the integral operator of Riemann-Liouville of order α and are associated to the idea that "deriving α times may be thought as integrating $n - \alpha$ times and deriving n times." Of course these definitions aren't equivalent: clearly the domains of the operators ${}_0^{RL}D_t^\alpha[.]$ and ${}_0^{C}D_t^\alpha[.]$ are different; because of the different hypothesis about f, we need to impose to guarantee their existence. Besides that, with the appropriate conditions for f,

$$ {}_0^{C}D_t^{\alpha}[f](t) = {}_0^{RL}D_t^{\alpha}[f](t) - \sum_{k=0}^{\infty} \frac{t^{k-\alpha}}{\Gamma(k-\alpha+1)} f^{(k)}(0^{+}). \tag{8} $$

More recently, the *Caputo-Fabrizio fractional derivative of order α,* with $\alpha \in [0,1)$, was defined as

$$ {}_0^{CF}D_t^{\alpha}[f](t) = \frac{M(\alpha)}{1-\alpha} \int_0^t f'(s) e^{-\frac{\alpha(t-s)}{1-\alpha}} ds, \tag{9} $$

for $M(\alpha)$ is a normalizing factor verifying $M(0) = M(1) = 1$.

Let us point out that, both in Eqs. (7) and (9), the lower limit in the integral could be changed by any value $a \in [-\infty, t)$, i.e.,

$$ {}_a^{C}D_t^{\alpha}[f](t) = \frac{1}{\Gamma(n-\alpha)} \int_a^t (t-s)^{n-1-\alpha} \frac{d^n}{ds^n}[f(s)] ds \tag{10} $$

and

$$ {}_a^{CF}D_t^{\alpha}[f](t) = \frac{M(\alpha)}{1-\alpha} \int_a^t f'(s) e^{-\frac{\alpha(t-s)}{1-\alpha}} ds. \tag{11} $$

In [7] the authors prove that the operator defined in Eq. (9) verifies the following (convenient) properties:

${}_0^{CF}D_t^{\alpha}[k] = 0$, for any constant k.

$$ \lim_{\alpha \to 1} {}_0^{CF}D_t^{\alpha}[f](t) = \frac{df}{dt}. \tag{12} $$

$$ \lim_{\alpha \to 0} {}_0^{CF}D_t^{\alpha}[f](t) = f(t) - f(0). \tag{13} $$

Caputo-Fabrizio definition was then generalized by Atangana and Baleanu, who gave the following definition of the *Atangana-Baleanu fractional derivative in Riemann-Lioville sense*:

$$ {}_0^{ABR}D_t^{\alpha}[f](t) = \frac{M(\alpha)}{1-\alpha} \frac{d}{dt} \int_0^t f(s) E_{\alpha}\left(-\frac{\alpha(t-s)^{\alpha}}{1-\alpha}\right) ds \tag{14} $$

and the *Atangana-Baleanu fractional derivative in Caputo sense*

$$ {}_0^{ABC}D_t^{\alpha}[f](t) = \frac{M(\alpha)}{1-\alpha} \int_0^t f'(s) E_{\alpha}\left(-\frac{\alpha(t-s)^{\alpha}}{1-\alpha}\right) ds, \tag{15} $$

replacing the exponential by $E_{\alpha}(z) = \sum_{k=0}^{\infty} \frac{z^k}{\Gamma(\alpha k+1)}$, the generalized Mittag-Leffler function.

Other types of fractional derivatives are Grünwald-Letnikov's, Hadamard's, Weyl's, etc. In every definition it is clear that fractional derivative operators are not local, since they need the information of f in a whole interval of integration. When defined with 0 as lower limit of integration, as we did, function f is usually assumed to be causal (i.e., $f(t) \equiv 0$ for $t < 0$), but this limit can also be changed.

Authors choose one definition or the other depending on the real-world phenomena they need to model; the scope of application of each operator is still unknown, and, in relation to some of them, there is neither an agreement about whether it is appropriate or not to call them derivatives (see, for a discussion on this topic, [8, 9]) nor what are the criteria to decide on it ([10, 11]).

Caputo and Fabrizio ([12]) proposed the following terms to recognize if an integral operator merits to be called a fractional derivative:

1. The fractional derivative must be a linear operator.

2. The fractional derivative of an analytic function must be analytic.

3. If the order of the derivative is a positive integer, the derivative must be the classical one.

4. If the order is null, the original function must be recovered.

5. For $n \in \mathbb{N}$, $0 < \alpha < 1 : D_t^\alpha [D_t^n[f]](t) = D_t^n [D_t^\alpha[f]](t)$.

6. $D_t^\alpha[f](t)$ must depend on the past history of f.

Many applications of fractional calculus have been reported in areas as diverse as diffusion problems, hydraulics, potential theory, control theory, electrochemistry, viscoelasticity, and nanotechnology, among others (see, e.g., [13, 14], for a profuse listing of application areas). In Section 2 we will briefly exemplify a few of these applications, in quite different fields, and in Section 3, we will even mention some cases of fractional nonlinear systems which exhibit chaotic behavior.

Theoretical results concerning existence and uniqueness of solutions to fractional differential equations have been also developed.

In [15–17] the authors state conditions to guarantee the existence and uniqueness of solution to problems like

$$\begin{cases} {}_0^C D_t^\alpha [f](t) = F(t, f(t)) t \epsilon (0, T), T < \infty \\ \quad \textit{initial or boundary conditions} \end{cases} \tag{16}$$

or

$$\begin{cases} {}_0^{RL} D_t^\alpha [f](t) = F(t, f(t)) t \epsilon (0, T), T < \infty \\ \quad \textit{initial or boundary conditions} \end{cases} \tag{17}$$

for $0 < \alpha < 2$. After rewriting the equation as an integral equation with a kernel whose norm is bounded in a proper Banach space, they use generalizations of the fixed-point theorem. The function F, besides being continuous, must satisfy certain conditions that substitute the classical Lipschitz's one.

Similar results are stated in [18] for a coupled system of fractional differential equations involving Riemann-Liouville derivatives.

Analytical calculus of fractional operators is, in general, difficult. In [19–21] a few examples of quite different analytical methods are presented.

In [22, 23], existence and uniqueness, for the solution to a simple case,

$$\{ {}_0^{CF} D_t \alpha [f](t) + \beta f(t) = g(t) f(0) = 0 \tag{18}$$

are proved, and explicit formulae are presented when g is continuous, causal, and null at the origin. The case of Caputo derivative is also considered. In all cases, the computation of the primitive of the data function is required.

Numerical methods have also been proposed to obtain approximate solution to fractional differential equations.

In [6] some numerical approximations to solutions to different fractional differential equations are presented and experimentally verified on various examples, and in [24, 25] complete surveys on numerical methods are offered. But numerous articles appear continuously with new approximation methods: we finish this section commenting briefly some works on numerical methods of quite different nature.

In [26] a local fractional natural homotopy perturbation method is proposed to found the solution to partial fractional differential equations as a series.

A method based on a semi-discrete finite difference approximation in time, and Galerkin finite element method in space, is proposed in [27] to solve fractional partial differential equations arising in neuronal dynamics.

In [28] a new numerical approximation of Atangana-Baleanu integral, as the summation of the average of the given function and its fractional integral in Riemann-Liouville sense, is proposed.

Semi-discrete finite element methods are introduced in [29] to solve diffusion equations, and implicit numerical algorithms for the case of spatial and temporal fractional derivatives appeared in [30]. A high-speed numerical scheme for fractional differentiation and fractional integration is proposed in [31]. In [32], a new numerical method to solve partial differential equations involving Caputo derivatives of fractional variable order is obtained in terms of standard (integer order) derivatives.

In [33], a discrete form is proposed for solving time fractional convection-diffusion equation. The Laplace transform is used to solve fractional differential equations in [34].

Finally, in Section 4, we will present a numerical method we have developed, based on wavelets, to solve initial and boundary value problems with linear fractional differential equations.

2. Some mathematical models with fractional derivatives

The purpose of this section is to highlight the role of fractional derivatives for modeling certain real evolution processes. We enumerate several mathematical models of different fields, found in the recent literature. In each of them, the fractional order of derivation is justified by the nature of the phenomenon that is described. Usually, in the papers, both the symbols $\frac{\partial^\alpha}{\partial t^\alpha}$ and D_t^α are used to indistinctly represent any of the fractional derivatives, whose type is clarified in the text.

In [35] the authors review the evolution of the general fractional equation:

$$\frac{\partial^\beta u}{\partial t^\beta} = a\,\frac{\partial^2 u}{\partial x^2} \tag{19}$$

for $a > 0$, $0 < \beta \leq 2$, where $x \in S \subset \mathbb{R}$ and $t \in \mathbb{R} > 0$ denote the space and time variables. This equation is obtained from the classical D'Alembert wave equation by replacing the second-order time derivative with the Caputo fractional derivative of order $\beta \in (0, 2]$. The authors show that, for $1 < \beta < 2$, the behavior of the fundamental solutions turns out to be intermediate between diffusion (for a viscous fluid) and wave propagation (for an elastic solid), thus justifying the attribute of fractional diffusive waves.

In [36] another approach for time fractional wave (Eq. (19)) is proposed. It is solved, for $0<\beta<1$ and special initial conditions, by the method of separation of variables.

In [37] the authors study the particular linear fractional Klein-Gordon equation:

$$\begin{cases} \dfrac{\partial^\alpha u}{\partial t^\alpha}-\dfrac{\partial^2 u}{\partial x^2}+u=6x^3t+\left(x^3-6x\right)t^3 & t>0, x\in\mathbb{R} \\ u(x,0)=0 \\ u_t(x,0)=0 \end{cases} \tag{20}$$

considering the fractional Caputo derivative with $1<\alpha\leq 2$. They achieve a numerical solution by variational iteration method and multivariate Padé approximation.

In [38] the following mathematical model, using Fick's law of diffusion, is developed to study the effect of fractional advection diffusion equation (cross flow) for the calcium profile, considering the Caputo fractional derivative $(0<\alpha\leq 1)$:

$$\begin{cases} \dfrac{\partial^\alpha C}{\partial t^\alpha}=D\dfrac{\partial^2 C}{\partial x^2} & x\geq 0, t\geq 0 \\ C(x,0)=C_0 \\ \lim\limits_{x\to+\infty} C(x,t)=C_\infty \end{cases} \tag{21}$$

Here, the calcium concentration $C(x,t)$ varies in time and space, D is a diffusion constant, and C_∞ is calcium concentration at infinity and is assumed that, at initial state of time and at a long distance, calcium concentration vanishes or becomes zero. The authors note that the physical parameter α characterizes the cytosolic calcium ion in astrocytes.

In [39] the authors explain that arteries, like other soft tissues, exhibit visco-elastic behavior and part of the mechanical energy transferred to them is dissipative (viscosity) and the other part is stored in a reversible form (elasticity). They modified the standard model by a fractional-order one and test it in human arterial segments. They conclude that fractional derivatives, in Riemann-Liouville sense, are a good alternative to model arterial viscosity.

The generalized Voigt model consists of a spring in parallel with two *springpots* (a neologism for a model that is between a spring—purely elastic—and a dashpot, purely viscous) of fractional orders α and β. The governing fractional-order differential equation is

$$\sigma(t)=E_0\varepsilon(t)+\eta_1\frac{\partial^\alpha\varepsilon(t)}{\partial t^\alpha}+\eta_2\frac{\partial^\beta\varepsilon(t)}{\partial x^\beta} \tag{22}$$

where E_0 is the elastic constant for a spring and η_1 and η_2 represent the viscosities of two springpots in parallel with the spring.

In Ref. [40] the authors developed an accurate and efficient numerical method for the fractional-order standard model described by Eq. (22) with $\alpha=\beta$, combining a fast convolution method with the spectral element discretization based on a general Jacobi polynomial basis that can be used to generate 3D polymorphic high-order elements. In that way they model complicated arterial geometries, such as patient-specific aneurysms, and apply it to 3D fluid-structure interaction simulations.

In [12] the authors use fractional derivatives to model the magnetic hysteresis, a phenomenon where the "memory" of the ferromagnetic material is crucial. They use a nonlinear model for the constitutive law of an isotropic ferromagnetic material:

$$\lambda H(t) = \theta_c \left(M^2(t) + 1 \right) M(t) - \theta(t) M(t) - C_0 D_t^\alpha [M](t) \tag{23}$$

for $\lambda > 0$, $H(t)$ is the magnetic excitation field, $M(t)$ is the magnetization vector, $\theta(t)$ is the temperature, θ_c is the Curie temperature below which the hysteresis is observed, and C_0 is the tensor with the constitutive properties of the magnetic material. They compare the resulting behavior when D_t^α is the Caputo fractional derivative with the one that results when the derivative is the Caputo-Fabrizio one. By numerical simulations they obtain examples of the classical hysteresis cycles and conclude that Caputo derivative expresses a stronger memory than the Caputo-Fabrizio operator.

These are just a few examples of the huge variety of problems that can be modeled by means of fractional differential equations. The nonlocality of the associated operators is the key to the success in the description of these phenomena.

3. Some simple systems exhibiting chaos

Chaos theory is also an area where fractional derivatives play an important role.

In this section we comment on some nonlinear systems modeled with fractional derivative, recently published, that exhibit chaos.

In [41] the authors studied a system based on the classical Lorenz one, but described by the Atangana-Baleanu fractional derivative (in the Caputo sense) with $0 < \alpha < 1$:

$$\begin{cases} {}^{ABC}_{\ \ 0}D_t^{\ \alpha}[x](t) = & \sigma\, y(t) - \alpha\, x(t) \\ {}^{ABC}_{\ \ 0}D_t^{\ \alpha}[y](t) = & \rho\, x(t) - x(t)\, z(t) - y(t) \\ {}^{ABC}_{\ \ 0}D_t^{\ \alpha}[z](t) = & x(t)\, y(t) - \beta\, z(t) \end{cases} \tag{24}$$

Under certain assumptions on the physical problem, they proved existence of solutions, and, by means of an iterative algorithm, numerical evidence of chaos is shown when $0.25 < \alpha < 0.3$ and $0.4 < \alpha < 0.5$ for the usual set of parameters $\sigma = 10$, $\rho = \frac{8}{3}$, and $\beta = 28$.

In [42] a three-dimensional fractional-order dynamical system for cancer growth is proposed replacing the standard derivatives in the evolution equations:

$$\begin{cases} \dot{x}_1(t) = & x_1(t) - A\, x_1(t)\, x_2(t) - B\, x_1(t) x_3(t) \\ \dot{x}_2(t) = & C x_2(t)(1 - x_2(t)) - D\, x_1(t)\, x_2(t) \\ \dot{x}_3(t) = & E \dfrac{x_1(t) x_3(t)}{x_1(t) + F} - G\, x_1(t)\, x_2(t) - H\, x_3(t) \end{cases} \tag{25}$$

by the Caputo-Fabrizio and the Atangana-Baleanu (Caputo sense) derivatives. The system parameters are related to the rate of change in the population of the different cells: healthy and tumor ones. The authors prove that the system has a unique solution and show that the system exhibits chaos for a proper choice of the parameters values and initial conditions.

In [43] a fractional Lorenz system is studied considering the *generalized Caputo derivative*, defined, for $0 < \alpha < 1$, $\rho > 0$, as

$$ {}_{0}^{GC}D_t^{\alpha,\rho}[f](t) = \frac{\rho^\alpha}{\sigma(1-\alpha)} \int_0^t \frac{f'(s)}{(t^\rho - s^\rho)^\alpha} ds \tag{26} $$

A detailed analysis of the stability of the system is performed. Adomian method is used to find semi-analytical solution to the fractional nonlinear equations. Chaotic behavior and strange attractors are numerically found for some values of α and ρ.

4. A numerical approximation scheme to solve linear fractional differential equations

After having exemplified several applications of fractional differential equations to different real-world problems, including chaotic ones, we will show a method that we have developed to obtain numerical solutions to linear fractional initial value and boundary value problems modeled with Caputo or Caputo-Fabrizio derivatives. The idea of the approximation scheme is to transform the derivatives into integral operators acting on the Fourier transform and to perform a wavelet decomposition of the data. The wavelet coefficients of the unknown are then recovered from a linear system of algebraic equations, and the solution is built up from its coefficients. The properties of the chosen wavelet basis guarantee numerical stability and efficiency of the approximation scheme. In the case of singular kernel, this procedure enables us to handle the singularity.

We note that choosing $a = -\infty$ in definition of Eqs. (4) or (5) and being $f \in H^1(-\infty, b)$, the Sobolev space of functions with (weak) first derivative in $L^2(-\infty, b)$, both derivatives can be expressed as a convolution.

For the Caputo-Fabrizio fractional derivative of order $0 < \alpha < 1$, changing variables in Eq. (9), we have

$$ {}_{-\infty}^{CF}D_t^{\alpha}[f](t) = \frac{M(\alpha)}{1-\alpha} \int_0^\infty f'(t-s) e^{-\frac{\alpha}{1-\alpha}s} ds = \frac{M(\alpha)}{1-\alpha} (f' * k)(t) \tag{27} $$

where k is a causal function, $k(t) = e^{-\frac{\alpha t}{1-\alpha}}$ for $t \geq 0$, and $k(t) = 0$ for $t < 0$.

Consequently, since $\hat{k}(\omega) = \frac{1-\alpha}{\alpha + i\omega(1-\alpha)}$,

$$ {}_{-\infty}^{CF}D_t^{\alpha}[f](t) = \frac{M(\alpha)}{2\pi(1-\alpha)} \int_R \hat{f'}(\omega) \hat{k}(\omega) e^{i\omega t} d\omega. \tag{28} $$

Using the properties of the Fourier transform, we can rewrite the last equality:

$$ {}_{-\infty}^{CF}D_t^{\alpha}[f](t) = \frac{M(\alpha)}{1-\alpha} \int_R \hat{f}(\omega) h(\omega) e^{i\omega t} d\omega \tag{29} $$

where $h(\omega) = \frac{i\omega}{2\pi} \hat{k}(\omega)$.

Meanwhile, in the Caputo case, we have

$$ {}_{-\infty}^{CF}D_t^{\alpha}[f](t) = \frac{1}{\Gamma(1-\alpha)} \int_{-\infty}^t \frac{f'(s)}{(t-s)^\alpha} ds = \frac{1}{2\pi\Gamma(1-\alpha)} \int_R \hat{f}(\omega) \hat{k}(\omega) e^{i\omega t} d\omega \tag{30} $$

where $k(t) = \frac{1}{t^\alpha}$ and $\hat{k}(\omega) = \frac{\Gamma(1-\alpha)}{(i\omega)^{1-\alpha}}$.

4.1 An initial value problem

Let us consider the initial value problem (IVP):

$$\begin{cases} {}_{0}^{CF}D_t^{\alpha}[f](t) + \sigma_0 f(t) + \sigma_1 f'(t) = g(t) \\ f(0) = 0 \end{cases} \tag{31}$$

We look for f satisfying Eq. (31), where g is a causal and smooth function with $g(0) = 0$. Other situations where the initial condition is not null can also be faced adapting the following scheme. We will consider that the fractional derivative is the Caputo-Fabrizio one; the Caputo case can be solved similarly (see [44] for a detailed description).

First we choose a wavelet basis with special properties: well localized in both time and frequency domain, smooth, band limited and infinitely oscillating with fast decay. The mother wavelet $\psi \in S$ (the Schwartz space) and its Fourier transform satisfy supp $|\hat{\psi}_{jk}| = \Omega_j$ where $\Omega_j = \left\{\omega : 2^j(\pi - \beta) \leq |\omega| \leq 2^j(\pi + \beta)\right\}$, with $0 < \beta \leq \frac{\pi}{3}$ ([45]).

The family $\left\{\psi_{jk}/\psi_{jk} = 2^{\frac{j}{2}}\psi\left(2^j t - k\right), j, k \in \mathbb{Z}\right\}$ is an orthonormal basis (BON) of $L^2(\mathbb{R})$ associated to a multiresolution analysis (MRA). We denote, by $W_j = span\left\{\psi_{jk}, j, k \in \mathbb{Z}\right\}$ and $V_J = \oplus_{j<J} W_j$, the wavelet and scale subspaces, respectively, and decompose the space $L^2(\mathbb{R}) = \oplus_{j\in\mathbb{Z}} W_j = \oplus_{j>n} W_j + V_n, n \in \mathbb{Z}$. We have also a scale function $\varphi \in V_0$ so that $\{\varphi(t-k), k \in \mathbb{Z}\}$ is a BON of V_0 ([46]).

The sets Ω_{j-1}, Ω_j, Ω_{j+1} have little overlap, and W_j is nearly a basis for the set of functions whose Fourier transform has support in Ω_j. This property of the basis will be crucial in the procedure.

Now we decompose the data g as

$$g(t) = \sum_{n\in\mathbb{N}} \langle g, \varphi_{Jn}\rangle \varphi_{Jn}(t) + \sum_{j>J}\sum_{k\in\mathbb{Z}} \left\langle g, \psi_{jk}\right\rangle \psi_{jk}(t) \tag{32}$$

where the first and second terms are the projections of g in V_J and W_j, respectively.

The properties of localization of the wavelets guarantee absolute convergence in each W_j (see [47] for details).

Now we choose the levels where the energy of g is concentrated, $J_{min}, J_{max} \in \mathbb{Z}$ so that

$$g(t) = \sum_{J_{min}}^{J_{max}} g_j(\mathrm{t}) + r(t) = \sum_{J_{min}}^{J_{max}}\sum_{k\in\mathbb{Z}} \left\langle g, \psi_{jk}\right\rangle \psi_{jk}(t) + r(t), \ \|r\|_2 \leq \varepsilon \|g\|_2 \tag{33}$$

for small ε, and truncate the component in each level so that the following approximation of g_j arises:

$$\widetilde{g}_j(t) = \sum_{k\in K_j} c_{jk}\psi_{jk}(t), \ c_{jk} = \left\langle g, \psi_{jk}\right\rangle. \tag{34}$$

Afterwards we obtain the fractional Caputo-Fabrizio derivatives of the wavelet basis by means of Eq. (29):

$$v_{jk}(t) = {}_{-\infty}^{CF}D_t^{\alpha}\left[\psi_{jk}\right](\mathrm{t}) = \frac{M(\alpha)}{1-\alpha}\int_{\Omega_j} \hat{\psi}_{jk}(\omega) h(\omega) e^{i\omega t} d\omega \tag{35}$$

(recall supp$|\hat{\psi}_{jk}| = \Omega_j$).

Let us consider for a moment that in Eq. (31) we have ${}_{-\infty}^{CF}D_t^{\alpha}$, i.e., $a = -\infty$. Note that, since $supp\, |v_{jk}| \subset \Omega_j$, $v_{jk} \in W_{j-1} \bigcup W_j \bigcup W_{j+1}$, but, from the properties of the chosen basis, we can consider $v_{jk} \in W_j$. This fact enables us to work on each level j separately (details can be found in [44, 48]).

Then, since the unknown f can be expressed as $f(t) = \sum_{j \in \mathbb{Z}} f_j(t)$, where $f_j(t) = \sum_{k \in \mathbb{Z}} b_{jk} \psi_{jk}(t)$ and $b_{jk} = \left\langle f, \psi_{jk} \right\rangle$, we have

$$f(t) \approx \sum_{J_{min}}^{J_{max}} f_j(t) \approx \sum_{J_{min}}^{J_{max}} \sum_{k \in K_j} b_{jk} \psi_{jk}(t). \tag{36}$$

Finally, we replace this last expression in Eq. (31), where for simplicity we will first consider $\sigma_1 = 0$ and look for the wavelet coefficients b_{jk} that satisfy, for each $j \in [J_{min}, J_{max}]$:

$$\sum_{k \in K_j} b_{jk} v_{jk}(t) + \sigma_0 \sum_{k \in K_j} b_{jk} \psi_{jk}(t) = \sum_{k' \in K_j} c_{jk'} \psi_{jk'}(t) \tag{37}$$

or

$$\sum_{k \in K_j} b_{jk} \left\langle v_{jk}, \psi_{jm} \right\rangle + \sigma_0 \sum_{k \in K_j} b_{jk} \left\langle \psi_{jk}, \psi_{jm} \right\rangle = \sum_{k' \in K_j} c_{jk'} \left\langle \psi_{jk'}, \psi_{jm}, \right\rangle \tag{38}$$

that, in matrix form, results in

$$\mathcal{M}^j \boldsymbol{b}^j = \boldsymbol{c}^j \tag{39}$$

where $\boldsymbol{b}^j = (b_{jk})_{k \in K_j}$, $\boldsymbol{c}^j = (c_{jk'})_{k' \in K_j}$, $\mathcal{M}^j \in \mathbb{R}^{K_j \times K_j}$ and $(\mathcal{M}^j)_{kl} = \langle v_{kl}, \psi_{kl} \rangle + \sigma_0 \left\langle \psi_{kj}, \psi_{kl} \right\rangle$.

From the properties of the wavelet basis, and those of v_{jk}, it results that $\mathcal{M}^j$ is a diagonal dominant matrix and, consequently, the vector of coefficients $\boldsymbol{b}^j$ can be computed in a stable and accurate way. The solution f can be obtained from Eq. (36). Moreover, it can be shown that $f(0) = 0$. To correct the effect of having considered ${}_{-\infty}^{CF}D_t^{\alpha}$ instead of ${}_{0}^{CF}D_t^{\alpha}$, we set $\widetilde{f} = f \cdot \chi_{[0,T]}$, where $\chi_{[0,T]}$ is the characteristic function of the interval $[0, T]$. Finally, $\widetilde{f}$ is an approximate solution to Eq. (31).

The error introduced in the approximation can be controlled and reduced: a more accurate truncated projection of the data into the wavelet subspaces can be considered, and the elements of the matrix can be computed with good precision since they can be expressed as integrals over compact subsets; finally, the matrix of the resulting linear system is a diagonal dominant matrix, and the solution can be computed accurately. In summary, the good properties of the basis and the operator guarantee that the resulting approximation scheme is efficient and numerically stable and no additional conditions need to be imposed.

4.1.1 Example 1

We illustrate the performance of the proposed approximation scheme by solving the IVP described by Eq. (31) for $\sigma_0 = 0.9$, $\sigma_1 = 0$, and $\alpha = 0.5$ and g a causal function defined as $g(t) = v(t)\sin(2\pi t)\ \cos(0.5\pi t)$, where v is a smooth window in the interval $[0, 4]$. Wavelet analysis indicates that the energy of the data g is concentrated in the subspaces W_0 and W_1; thus, we consider levels $-1 \le j \le 2$ for the reconstruction.

For this case, being $\sigma_1 = 0$, $\sigma_0 \neq -\frac{1}{1-\alpha}$, $g \in C(0, \infty)$, and $g(0) = 0$, there exists a formula for the "exact" solution to Eq. (31) (see [22]). The approximate solution $\tilde{f}$ to the IVP is plotted (in green) in **Figure 1**, together with the exact solution (in blue).

If $\sigma_1 \neq 0$, since $f'(t) = \frac{1}{2\pi}\int_{\mathbb{R}} i\omega \hat{f}(\omega)e^{i\omega t}d\omega$, we obtain similar equations for the coefficients on each level:

$$\sum_{k\in K_j} b_{jk}\left\langle v_{jk}, \psi_{jm}\right\rangle + \sigma_0 \sum_{k\in K_j} b_{jk}\left\langle \psi_{jk}, \psi_{jm}\right\rangle + \sigma_1 \sum_{k\in K_j} b_{jk}\left\langle i\omega\psi_{jk}, \psi_{jm}\right\rangle = \sum_{k'\in K_j} c_{jk'}\left\langle \psi_{jk'}, \psi_{jm}\right\rangle \tag{40}$$

Once more the matrix of the resulting linear system is diagonal dominant, and the system can be solved efficiently (see [49] for details).

The following example shows the performance of the method for $\sigma_1 = 0.3$.

4.1.2 Example 2

Now we consider IVP described by Eq. (31) for $\sigma_0 = 0.9$, $\sigma_1 = 0.3$, $\alpha = 0.5$, and $f(t) = t^2 v(t)(2\sin(2.5\pi t) - 0.5\cos(6\pi t))$, with v as a smooth window over interval $[0, 7]$. Since $\sigma_1 \neq 0$ and we have no formula to test the performance of the approximation, for this example we will set f, calculate g from Eq. (31), and then apply the proposed method to recover f.

Choosing $-1 \le j \le 2$, it results in $\tilde{f} = \sum_{j=-1}^{2} \tilde{f}_j$. The plots of the f (blue) and $\tilde{f}$ (green) appear in **Figure 2**.

This scheme can be adapted to solve boundary value problems. We show the procedure finding the solution to the fractional heat equation.

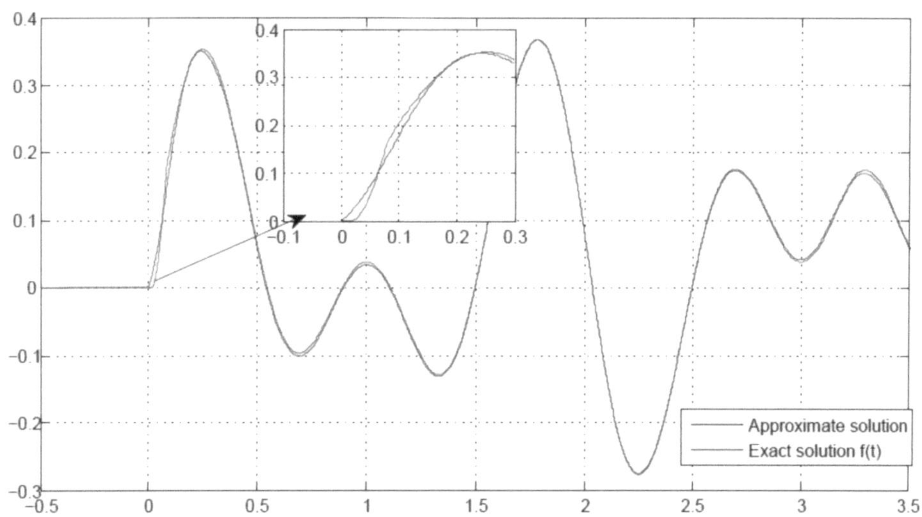

Figure 1.
The approximation and the exact solution for the Example 1.

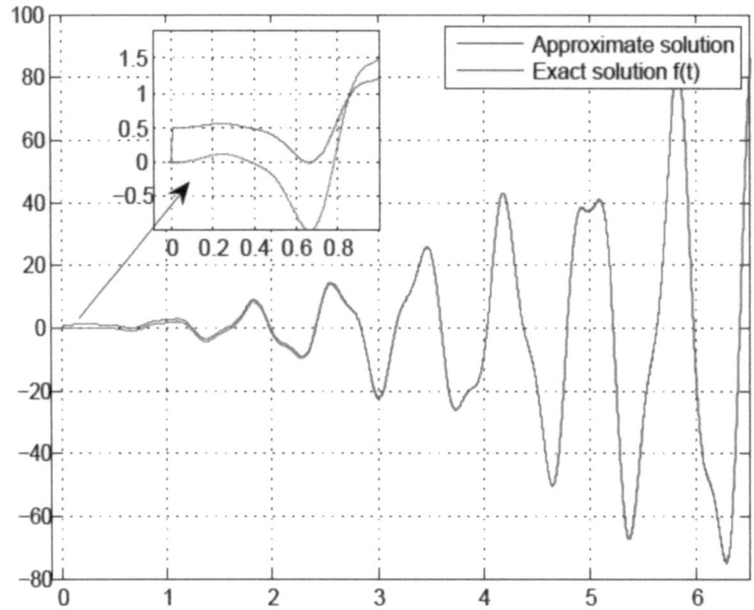

Figure 2.
f and $\tilde{f}$ of Example 2.

4.2 Boundary value problem

We show how to adapt the scheme used for initial value problem for solving boundary value problems with fractional partial differential equations in an example.

We will consider a fractional heat problem where we have replaced the classical time derivative by the Caputo-Fabrizio fractional derivative of order α:

$$\begin{cases} {}_{0}^{CF}D_{t}^{\alpha}[u](x,t) - u_{xx}(x,t) = g(x,t), & x \in [0,1], t \in (0,T) \\ u(x,0) = 0 & x \in [0,1] \\ u(0,t) = u(1,t) = 0 & t \in (0,T) \end{cases} \tag{41}$$

This equation models the evolution of temperatures in a bar of length 1, constituted by a heterogeneous material which has “memory,” due to the fluctuations introduced by elements at different dimension scales ([7]).

The smooth and causal function g represents an external source. We look for smooth solutions $u \in C^2(0,1) \times (0,T)$ by separating variables and pose

$$u(x,t) = \sum_{k \in \mathbb{Z}} u_k(t) \sin(k\pi x) \tag{42}$$

where $u_k(t)$ is the Fourier coefficients of $u(x,t)$ for each $t \in (0,T)$.

For the temporal part of the function, after replacing in Eq. (36), we obtain an initial value problem like that described by Eq. (31) for each coefficient $u_k(t)$:

$$\begin{cases} D_0^{\alpha}[u_k](t) + (k\pi)^2 u_k(t) = B_k(t), t \in (0,T) \\ u_k(0) = 0 \end{cases} \tag{43}$$

with $B_k(t) = 2\int_0^1 g(x,t)\sin(k\pi x)dx$ and the Fourier coefficients of $g(x,t)$ for each $t \in (0,T)$. The uniqueness of solution is guaranteed because $u(x,0) = 0$, so $B_k(0)$ is null.

We show the approximate solution to Eq. (41) for $\alpha = 0.5$, $T = 3$, and $g(x,t) = v(t)e^{-t/2}\sin(5\pi t)\sin(2\pi x)$, with v a smooth window in $[0,3]$.

In this case we only need to solve Eq. (43) for $k = 2$, with $B_2 = v(t)e^{-t/2}\sin(5\pi t)$. Wavelet analysis indicates that the 95% of the energy of B_2 is concentrated in subspaces W_1, W_2, and W_3, and we obtained the following condition numbers for

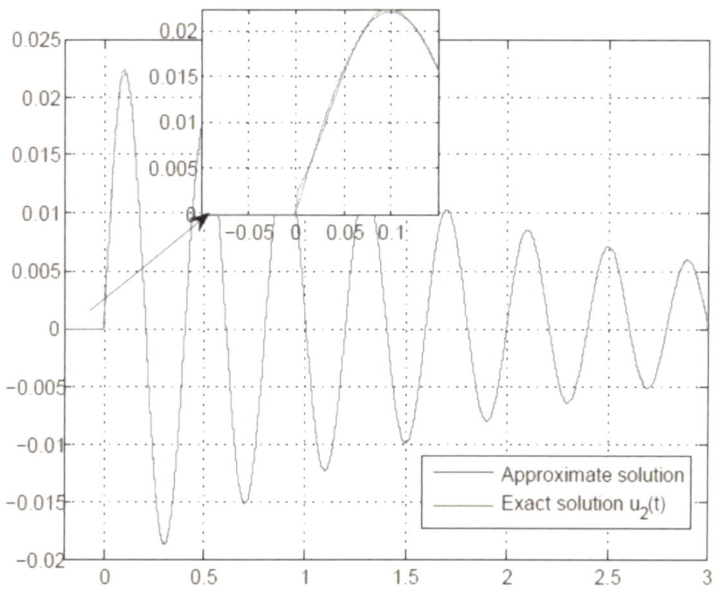

Figure 3.
The approximate and the exact solutions to Eq. (41) with $\alpha = 0.8$, $T = 3$, *and* $B_3 = v(t)e^{-\frac{t}{2}}\sin(5\pi t)$.

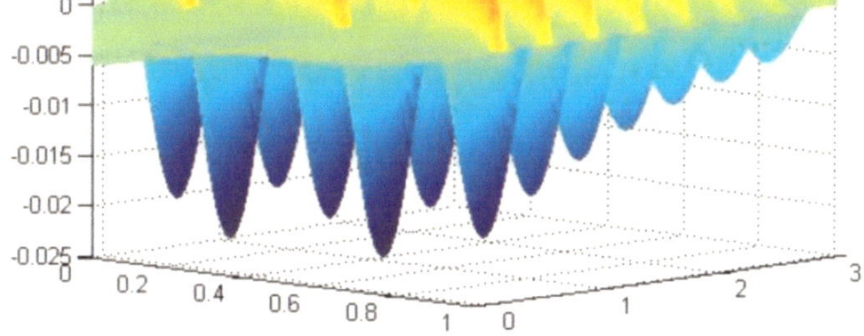

Figure 4.
The approximate solutions to Eq. (39) with $\alpha = 0.8$, $T = 3$, $g(x,t) = v(t)e^{-t/2}\sin(5\pi t)\sin(2\pi x)$, *and* v *a smooth window in* $[0,3]$.

the band matrices of Eq. (39): $cond_{\infty}(\mathcal{M}^0) = 1.1153$, $cond_{\infty}(\mathcal{M}^1) = 1.0663$, $cond_{\infty}(\mathcal{M}^2) = 1.0132$, $cond_{\infty}(\mathcal{M}^3) = 1.0098$, and $cond_{\infty}(\mathcal{M}^4) = 1.0076$. We consider levels $0 \leq j \leq 4$ for the reconstruction, and the mean square error in this case is $\|u_2 - \widetilde{u}_2\|_{L_2} = 3.5020\ 10^{-4}$.

Figure 3 shows the approximate and the exact solution to Eq. (43). In **Figure 4** we draw the approximate solution u of the heat problem described by Eq. (41), and in **Figure 5** the difference between the true solution $u(x,t)$ and its approximation is plotted. The mean square error obtained in this case is $4.0016\ 10^{-4}$.

Finally, in **Figure 6**, we show the approximate solutions u_2 for different orders of derivation α, $\alpha \to 1$, exhibiting the tendency to the solution of Eq. (43) with

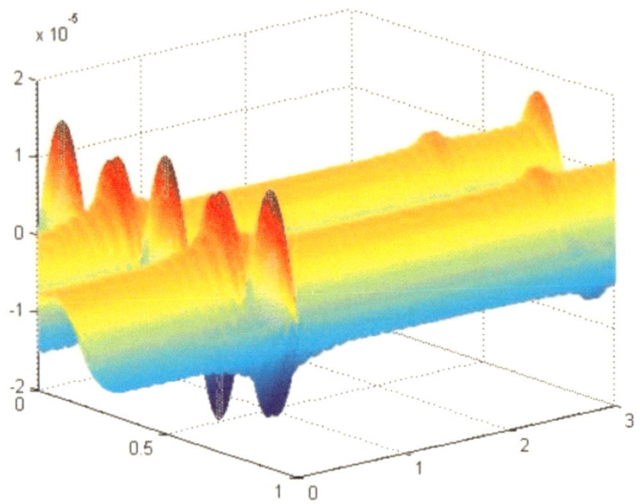

Figure 5.
The difference between the true solution $u(x,t)$ to Eq. (39) and its approximation by means of the wavelet scheme.

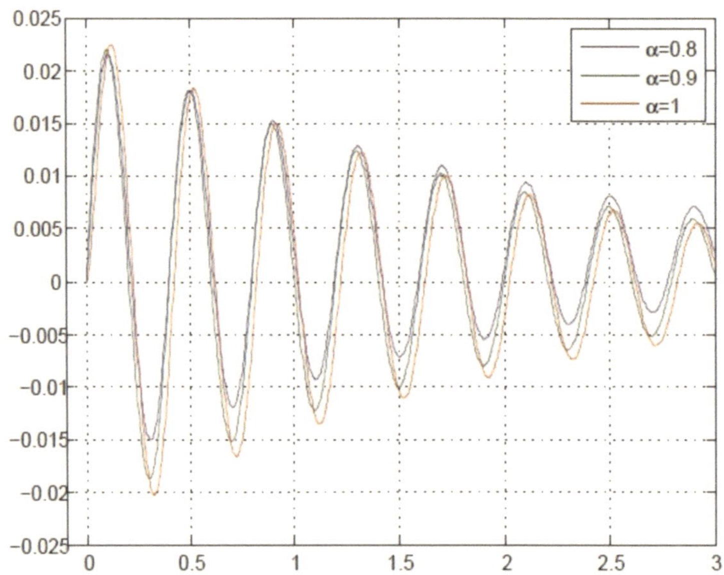

Figure 6.
The approximate solutions to Eq. (41) with $\alpha \to 1$.

$\alpha = 1$, as expected (for $\alpha = 1$, Eq. (41) describes classical heat problem, for a bar made of a homogeneous material).

5. Conclusions

In this chapter we have presented a summary of some recent works showing the relevance and the intense research work in the area of fractional calculus and its applications. We have focused on nonlinear models describing different phenomena where fractional differentiation plays an important role. In the last section we have presented an approximation scheme that we have developed to solve linear initial and boundary value problems based on wavelet decomposition, and the performance of the method is illustrated by examples. Possible extensions and adaptation to nonlinear equations are still under study.

Acknowledgements

This work was partially supported by University of Buenos Aires through UBACyT (2018–2021) 20020170100350BA.

Conflict of interest

The authors declare no conflicts of interest regarding this chapter.

Author details

María I. Troparevsky[1]*, Silvia A. Seminara[1] and Marcela A. Fabio[2]

1 Faculty of Engineering, University of Buenos Aires, Buenos Aires, Argentina

2 School of Science and Technology, University of San Martín, Buenos Aires, Argentina

*Address all correspondence to: mariainestro@gmail.com

References

[1] Diethelm K. The Analysis of Fractional Differential Equations. Berlin: Springer Verlag; 2010. 264p. ISBN: 9783642145735. DOI: 10.1007/978-3-642-14574-2

[2] Capelas de Oliveira E, Tenreiro Machado J. A review of definitions for fractional derivatives and integral. Mathematical Problems in Engineering. 2014;**2014**:238459. DOI: 10.1155/2014/238459. 6p

[3] Oldham K, Spanier J. The Fractional Calculus. New York-London: Academic Press Inc.; 1974. 322p. ISBN: 9780080956206

[4] Miller K, Ross B. An Introduction to the Fractional Calculus and Fractional Differential Equations. New York: John Wiley & Sons, Inc; 1993. 382p. ISBN: 0471588849

[5] Samko S, Kilbas A, Marichev O. Fractional Integrals and Derivatives. Theory and Applications. Amsterdam: Gordon and Breach Science Publishers; 1993. 1016p. ISBN: 2881248640

[6] Podlubny I. Fractional Differential Equations. San Diego: Academic Press; 1998. 340p. ISBN: 9780125588409

[7] Caputo M, Fabrizio M. A new definition of fractional derivative without singular kernel. Progress in Fractional Differentiation and Applications. 2015;**1**(2):73-85. DOI: 10.12785/pfda/010201

[8] Ortigueira M, Tenreiro Machado J. A critical analysis of the Caputo-Fabrizio operator. Communications in Nonlinear Science and Numerical Simulation. 2017;**59**:608-611. DOI: 10.1016/j.cnsns.2017.12.001

[9] Giusti A. A comment on some new definitions of fractional derivative. Nonlinear Dynamics. 2018;**93**(3):1757-1763. DOI: 10.1007/s11071-018-4289-8

[10] Sales Teodoro G, Capelas de Oliveira E. Derivadas fracionárias: criterios para classificação. Revista Eletrônica Paulista de Matemática. 2017; **10**:10-19. DOI: 10.21167/cqdvol10ermacic201723169664gsteco1019

[11] Sales Teodoro G, Oliveira D, Capelas de Oliveira E. Sobre derivadas fracionárias. Revista Brasileira de Ensino de Física. 2018;**40**(2):e2307. DOI: 10.1590/1806–9126-RBEF-2017-0213

[12] Caputo M, Fabrizio M. On the notion of fractional derivative and applications to the hysteresis phenomena. Meccanica. 2017;**52**(1):3043-3052. DOI: 10.1007/s11012-017-0652-y

[13] Mainardi F. Fractional calculus. In: Carpinteri A, Mainardi F, editors. Fractals and Fractional Calculus in Continuum Mechanics. International Centre for Mechanical Sciences (Courses and Lectures). Vol. 378. Vienna: Springer Verlag; 1997. DOI: 10.1007/978-3-7091-2664-6_7

[14] Tenreiro Machado J, Silva M, Barbosa R, Jesus I, Reis C, Marcos M, et al. Some applications of fractional calculus in engineering. Mathematical Problems in Engineering. 2010;**2010**:639801. DOI: 10.1155/2010/639801. 34p

[15] Baleanu D, Rezapour A, Mohammadi H. Some existence results on nonlinear fractional differential equations. Philosophical Transactions. Series A, Mathematical, Physical, and Engineering Sciences. 2013;**371**(1990):20120144. DOI: 10.1098/rsta.2012.0144

[16] Baleanu D, Agarwal R, Mohammadi H, Rezapour S. Some existence results

for a nonlinear fractional differential equation on partially ordered Banach spaces. Boundary Value Problems. 2013; **2013**:112. DOI: 10.1186/1687-2770-2013-112

[17] Abbas S. Existence of solutions to fractional order ordinary and delay differential equations and applications. Electronic Journal of Differential Equations. 2011;**2011**(9):1-11. ISSN: 1072-6691

[18] Shah K, Li Y. Existence theory of differential equations of arbitrary. In: Differential Equations—Theory and Current Research. London: IntechOpen; 2018. pp. 35-55. DOI: 10.5772/intechopen.75523. Ch 2

[19] Maitama S, Abdullahi I. New analytical method for solving linear and nonlinear fractional partial differential equations. Progress in Fractional Differentiation and Applications. 2016; **2**(4):247-256. DOI: 10.18576/pfda/020402

[20] Sabatier J, Agrawal O, Tenreiro Machado J, editors. Advances in Fractional Calculus. The Netherlands: Springer; 2007. 549p. ISBN: 978-1-4020-6041-0

[21] Odzijewicz T, Malinowska A, Torres D. Fractional calculus of variations in terms of a generalized fractional integral with applications to physics. Abstract and Applied Analysis. 2012;**2012**:871912. DOI: 10.1155/2012/871912. 24p

[22] Al Salti N, Karimov E, Kerbal S. Boundary value problems for fractional heat equation involving Caputo-Fabrizio derivative. New Trends in Mathematical Science. 2017;**4**(4):79-89. DOI: 10.20852/ntmsci.2016422308

[23] Losada J, Nieto J. Properties of a new fractional derivative without singular kernel. Progress in Fractional Differentiation and Applications. 2015; **1**(2):87-92. DOI: 10.12785/pfda/010202

[24] Baleanu D, Diethelm K, Scalas E, Trujillo J. Fractional Calculus: Models and Numerical Methods. Singapore: World Scientific Publishing; 2012. 400p. ISBN: 978-981-4355-20-9

[25] Landy A. Fractional differential equations and numerical methods [Thesis]. University of Chester; 2009

[26] Maitama S. Local fractional natural Homotopy perturbation method for solving partial differential equations with local fractional derivative. Progress in Fractional Differentiation and Applications. 2018;**4**(3):219-228. DOI: 10.18576/pfda/040306

[27] Zhuang P, Liu F, Turner I, Anh V. Galerkin finite element method and error analysis for the fractional cable equation. Numerical Algorithms. 2016; **72**(2):447-466. DOI: 10.1007/s11075-015-0055-x

[28] Djida J, Area I, Atangana A. New Numerical Scheme of Atangana-Baleanu Fractional Integral: An Application to Groundwater Flow within Leaky Aquifer. 2016. arXiv: 1610.08681

[29] Sun H, Chen W, Sze K. A semi-discrete finite element method for a class of time-fractional diffusion equations. Philosophical Transactions of the Royal Society A - Mathematical Physical and Engineering Sciences. 2013;**371**(1990):20120268. DOI: 10.1098/rsta.2012.0268

[30] Yu Q, Liu F, Turner I, Burrage K. Stability and convergence of an implicit numerical method for the space and time fractional Bloch–Torrey equation. Philosophical Transactions of the Royal Society A - Mathematical Physical and Engineering Sciences. 2013;**371**(1990): 20120150. DOI: 10.1098/rsta.2012.0150

[31] Fukunaga M, Shimizu N. A high-speed algorithm for computation of fractional differentiation and fractional integration. Philosophical Transactions

of the Royal Society A - Mathematical Physical and Engineering Sciences. 2013;**371**:20120152. DOI: 10.1098/rsta.2012.0152

[32] Tavares D, Almeida R, Torres D. Caputo derivatives of fractional variable order: numerical approximations. Communications in Nonlinear Science and Numerical Simulation. 2016;**35**:69-87. DOI: 10.1016/j.cnsns.2015.10.027

[33] Zhang J, Zhang X, Yang B. An approximation scheme for the time fractional convection–diffusion equation. Applied Mathematics and Computation. 2018;**335**:305-312. DOI: 10.1016/j.amc.2018.04.019

[34] Lin S, Lu C. Laplace transform for solving some families of fractional differential equations and its applications. Adv. Difference Equ. 2013; **2013**:137. DOI: 10.1186/1687-1847-2013-137. 9p

[35] Mainardi F, Paradisi P. Fractional diffusive waves. Journal of Computational Acoustics. 2001;**9**(4): 1417-1436. DOI: 10.1016/S0218-396X(01)00082-6

[36] Parsian H. Time fractional wave equation: Caputo sense. Advanced Studies in Theoretical Physics. 2012; **6**(2):95-100

[37] Turut V, Güzel N. On solving partial differential equations of fractional oder by using the variational iteration method and multivariate Padé approximations. European Journal of Pure and Applied Mathematics. 2013; **6**(2):147-171. ISSN: 1307-5543

[38] Agarwal R, Sonal J. Mathematical modeling and analysis of dynamics of cytosolic calcium ion in astrocytes using fractional calculus. Journal of Fractional Calculus and Applications. 2018;**9**(2): 1-12. ISSN: 2090-5858

[39] Craiem D, Rojo F, Atienza J, Guinea G, Armentano R. Fractional calculus applied to model arterial viscoelasticity. Latin American Applied Research. 2008; **38**(2):141-145. ISNN: 0327–0793

[40] Yu Y, Perdikaris P, Karniadakis G. Fractional modeling of viscoelasticity in 3D cerebral arteries and aneurysms. Journal of Computational Physics. 2016; **323**:219-242. DOI: 10.1016/j.jcp.2016.06.038

[41] Atangana A, Koca I. Chaos in a simple nonlinear system with Atangana–Baleanu derivatives with fractional order. Chaos, Solitons & Fractals. 2016;**89**:447-454. DOI: 10.1016/j.chaos.2016.02.012. 8p

[42] Gómez-Aguilar J, López-López M, Alvarado-Martínez V, Baleanu D, Khan H. Chaos in a cancer model via fractional derivatives with exponential decay and Mittag-Leffler law. Entropy. 2017;**19**:681. DOI: 10.3390/e19120681. 19p

[43] Baleanu D, Wu G, Zeng S. Chaos analysis and asymptotic stability of generalized caputo fractional differential equations. Chaos, Solitons and Fractals. 2017;**102**:99-105. DOI: 10.1016/j.chaos.2017.02.007

[44] Fabio M, Troparevsky M. An inverse problem for the caputo fractional derivative by means of the wavelet transform. Progress in Fractional Differentiation and Applications. 2018;**4**(1):15-26. DOI: 10.18576/pfda/040103

[45] Meyer Y. Ondelettes et Operateurs II. Operatteurs de Calderon Zygmund. Paris: Hermann et Cie; 1990. 381p. ISBN: 2705661263

[46] Mallat S. A Wavelet Tour of Signal Processing. Academic Press; 2009. 832p. ISBN: 978–0–12-374370-1. DOI: 10.1016/B978-0-12-374370-1.X0001-8

[47] Fabio M, Serrano E. Infinitely oscillating wavelets and an efficient implementation algorithm based on the FFT. Revista de Matemática: Teoría y Aplicaciones. 2015;**22**(1):61-69. ISSN: 1409–2433 (Print), 2215–3373 (Online)

[48] Fabio M, Troparevsky M. Numerical solution to initial value problems for fractional differential equations. Progress in Fractional Differentiation and Applications. 2019; **5**(3):1-12. DOI: 10.18576/pfda/INitialVP18NSP29nov. [in press]

[49] Troparevsky M, Fabio M. Approximate solutions to initial value problems with combined derivatives [in Spanish]. Mecánica Computacional. 2018;**XXXVI**(11):449-459. ISSN: 2591–3522

Chapter 2

Numerical Solutions to Some Families of Fractional Order Differential Equations by Laguerre Polynomials

Adnan Khan, Kamal Shah and Danfeng Luo

Abstract

This article is devoted to compute numerical solutions of some classes and families of fractional order differential equations (FODEs). For the required numerical analysis, we utilize Laguerre polynomials and establish some operational matrices regarding to fractional order derivatives and integrals without discretizing the data. Further corresponding to boundary value problems (BVPs), we establish a new operational matrix which is used to compute numerical solutions of boundary value problems (BVPs) of FODEs. Based on these operational matrices (OMs), we convert the proposed (FODEs) or their system to corresponding algebraic equation of Sylvester type or system of Sylvester type. The resulting algebraic equations are solved by MATLAB® using Gauss elimination method for the unknown coefficient matrix. To demonstrate the suggested scheme for numerical solution, many suitable examples are provided.

Keywords: FODEs, numerical solution, Laguerre polynomials, operational matrices

1. Introduction

The theory of integrals as well as derivatives of arbitrary order is known by the special name "fractional calculus." It has an old history just like classical calculus. The chronicle of fractional calculus and encyclopedic book can be studied in [1, 2]. Researchers have now necessitated the use of fractional calculus due to its diverse applications in different fields, specially in electrical networks, signal and image processing and optics, etc. For conspicuous work on FODEs in the fields of dynamical systems, electrochemistry, advanced techniques of microorganisms culturing, weather forecasting, as well as statistics, we refer to peruse [3, 4]. Fractional derivatives show valid results in most cases where ordinary derivatives do not. Also annotating that fractional order derivatives as well as fractional integrals are global operators, while ordinary derivatives are local operators. Fractional order derivative provides greater degree of freedom. Therefore from different aspects, the aforesaid areas were investigated. For instance, many researchers have provide understanding to existence and uniqueness results about FODEs, for few results, we refer [5–7], and many others have actualized the instinctive framework of fractional differential equations in various problems [8–19] with many references included in them.

Often it is very difficult to obtain the exact solution due to global nature of fractional derivatives in differential equations. Contrarily approximate solutions are obtained by numerical methods assorted in [20–22]. Various new numerical methods have been developed, among them is one famous method called "spectral method" which is used to solve problems in various realms [23]. In this method operational matrices are obtained by using orthogonal polynomials [24]. Many authors have successfully developed operational matrices by using Legender, Jacobi, and various other polynomials [25, 26]. For delay differential and various other related equations, Laguerre spectral methods have been used [27–32]. Bernstein polynomials and various classes of other polynomials were also used to obtain operational matrices corresponding to fractional integrals and derivatives [33–40]. Apart from them, operational matrices were also developed with the collocation method (see Refs. [41–43]). Since spectral methods are powerful tools to compute numerical solutions of both ODEs and FODEs. Therefore, we bring out numerical analysis via using Laguerre polynomials of some families and coupled systems of FODEs under initial as well as boundary conditions. In this regard we investigate the numerical solutions to the given families under initial conditions

$$\begin{cases} {}_0^cD_t^\gamma z(t) \pm z(t) = 0, \quad 0 < \gamma \leq 1, \\ z(0) = z_0, \quad z_0 \in R, \end{cases} \tag{1}$$

and subject to boundary conditions

$$\begin{cases} {}_0^cD_t^\gamma z(t) \pm z(t) = 0, \ 1 < \gamma \leq 2, \\ z(0) = z_0, z(1) = z_1, \quad z_0, z_1 \in R. \end{cases} \tag{2}$$

By similar numerical techniques, we also investigate the numerical solutions to the following systems with fractional order derivatives under initial and boundary conditions as

$$\begin{cases} {}_0^cD_t^\gamma z(t) + az(t) + by(t) = f(t), \\ {}_0^cD_t^\gamma y(t) + cy(t) + dz(t) = g(t), \end{cases} \\ z(0) = z_0, y(0) = y_0 \tag{3}$$

for $0 < \gamma \leq 1$ and

$$\begin{cases} {}_0^cD_t^\gamma z(t) az(t) + by(t) = f(t), \\ {}_0^cD_t^\gamma y(t) + cy(t) + dz(t) = g(t), \\ z(0) = z_0, y(0) = y_0, \quad z(1) = z_1, y(1) = y_1, \end{cases} \tag{4}$$

for $1 < \gamma \leq 2$ where $f, g : [0, 1] \times R^2 \to R$ and $z_0, y_0, z_1, y_1 \in R$. We first obtain OMs for fractional derivatives and integrals by using Laguerre polynomials. Also corresponding to boundary conditions, we construct an operational matrix which is needed in numerical analysis of BVPs. With the help of the OMs we convert the considered problem of FODEs under initial/boundary conditions to Sylvester-type algebraic equations. Solving the mentioned matrix equations by using MATLAB®, we compute the numerical solutions of the considered problems.

2. Preliminaries

Here we recall some basic definition results that are needed in this work onward, keeping in mind that throughout the paper we use fractional derivative in Caputo sense.

Definition 1. The fractional integral of order $\gamma > 0$ of a function $z : (0, \infty) \to R$ is defined by

$$ {}_0I_t^{\gamma} z(t) = \frac{1}{\Gamma(\gamma)} \int_0^t \frac{z(s)}{(t-s)^{1-\gamma}} ds, $$

provided the integral converges at the right sides. Further a simple and important property of ${}_0I_t^{\gamma}$ is given by

$$ {}_0I_t^{\gamma} t^{\delta} = \frac{\Gamma(\delta+1)}{\Gamma(\delta+\gamma+1)} t^{\gamma+\delta}. $$

Definition 2. Caputo fractional derivative is defined as

$$ {}_0^cD_t^{\gamma} f(t) = \frac{1}{\Gamma(n-\gamma)} \int_0^t (t-s)^{n-\gamma-1} f^{(n)}(s) ds, $$

where n is a positive integer with the property that $n-1 < \gamma \le n$. For example, if $0 < \gamma \le 1$, then Caputo fractional derivative becomes

$$ {}_0^cD_t^{\gamma} f(t) = \frac{1}{\Gamma(1-\gamma)} \int_0^t (t-s)^{-\gamma-1} f'(s) ds. $$

Theorem 1. The FODE given by

$$ {}_0^cD_t^{\gamma} f(t) = 0 $$

has a unique solution, such that

$$ f(t) = d_0 + d_1 t + d_2 t^2 + \dots + d_{n-1} t^{n-1}, \quad n = [\gamma] + 1. $$

Lemma 1. Therefore in view of this result, if $h \in L^n[0, T]$, then the unique solution of nonhomogenous FODE

$$ {}_0^cD_t^{\gamma} f(t) = h(t), \quad n-1 < \gamma \le n $$

is written as

$$ f(t) = d_0 + d_1 t + d_2 t^2 + \dots + d_{n-1} t^{n-1} + {}_0I_t^{\gamma} h(t), $$

where d_i for $i = 0, 1, 2, 3 \dots n-1$ are real constants.

The above lemma is also stated as

$$ f(t) = {}_0I_t^{\gamma} h(t) + \sum_{i=0}^{n-1} \frac{f^i(0)}{i!} t^i. $$

Definition 3. The famous Laguerre polynomials are represented by $L_i^{\gamma}(t)$ and defined as

$$ L_i^{\gamma}(t) = \sum_{k=0}^{i} \frac{(-1)^k \Gamma(i+\gamma+1)}{\Gamma(k+1+\gamma)\Gamma(i-k+1)\Gamma(k+1)} t^k. $$

They are orthogonal on $[0, \infty]$. If $L_i^\gamma(t)$ and $L_j^\gamma(t)$ are Laguerre polynomials, then the orthogonality condition is given as

$$\int_0^\infty L_i^\gamma(t) L_j^\gamma(t) W^\gamma(t) dt = \delta_{i,j} U_k,$$

where

$$W^\gamma(t) = t^\gamma e^{-t},$$

is the weight function and

$$U_k = \begin{cases} \dfrac{\Gamma(1+\gamma+k)}{\Gamma(1+k)}, & i = j \\ 0 & i \neq j. \end{cases}$$

Now let $Z(t)$ be any function, defined on the interval $[0, \infty]$. We express the function in terms of Laguerre polynomials as

$$\begin{aligned} Z(t) &= \sum_{i=0}^{n} c_i L_i^\gamma(t). \\ &= c_0 L_0^\gamma(t) + c_1 L_1^\gamma(t) + \dots + c_N L_N^\gamma(t) \\ &= [c_0 \ c_1 \ \dots \ c_N] \begin{bmatrix} L_0^\gamma(t) \\ \vdots \\ L_n^\gamma(t) \end{bmatrix}. \end{aligned} \tag{5}$$

We set the above two vectors into their inner product and represent the column matrix by $\Psi(t)$, so that

$$Z(t) = c^t \Psi(t).$$

Again as

$$Z(t) = \sum_{i=0}^{n} c_i L_i^\gamma(t),$$

$$\int_0^L Z(t) W^\gamma(t) L_j^\gamma(t) dt = \int_0^L \sum_{i=0}^{n} c_i L_i^\gamma(t) L_j^\gamma(t) W^\gamma(t) dt,$$

which is written as

$$\sum_{i=0}^{n} c_i \int_0^L L_i^\gamma(t) L_j^\gamma(t) W^\gamma(t) dt.$$

We call h_i to the general term of integration

$$\int_0^L Z(t) W^\gamma(t) L_j^\gamma(t) dt = \sum_{i=0}^{n} c_i h_i.$$

Hence the coefficient c_i is

$$c_i = \frac{1}{h_i}\int_0^L Z(t)W^{\gamma}(t)L_j^{\gamma}(t)dt.$$

In vector form we can write Eq. (5) as

$$Z(t) = c_M^t \Psi_M(t).$$

where $M = m + 1$, c_M is the M terms coefficient vector and $\Psi_M(t)$ is the M terms function vector.

2.1 Representation of Laguerre polynomial with Caputo fractional order derivative

If the Caputo fractional order derivative is applied to Laguerre polynomial, by considering whole function constant except t^k. We use the definition of Caputo fractional order derivative for t^k to obtain (6) as

$${}_0^cD_t^{\gamma}L_i^{\gamma}(t) = \sum_{k=0}^{i}\left(t^{k-\gamma}\right)\frac{(-1)^k\Gamma(i+\gamma+1)}{\Gamma(k+1+\gamma)\Gamma(i-k+1)\Gamma(1+k-\gamma)}. \tag{6}$$

2.2 Error analysis

The proof of the following results can be found with details in [20].

Lemma 2. Let $L_i^{\beta}(t)$ be given; then

$${}_0^cD_t^{\gamma}L_i^{\beta}(t) = 0, \qquad i = 0, 1, 2, \cdots, [\beta] - 1, \ \gamma > 0.$$

Theorem 2. For error analysis, we state the theorem such that, a be any integer and $0 \le s \le a$, and then

$$\|P_{M,az} - z(t)\|A_{\alpha}^s, \Lambda \le cM^{\frac{s-a}{2}}|z(t)|A_{\alpha}^a, \Lambda, \forall z(t)\epsilon A_{\alpha}^a(\Lambda),$$

where $A_{\alpha}^a = \{\, z/z$ is measurable on Λ and $\|z\|\ A_{\alpha}^a, (\Lambda) < \infty \,\}$ and

$$|z|A_{\alpha}^a, (\Lambda) = \|\partial_p^a z\|_{w\alpha+a,\Lambda},$$

$$\|z\|A_{\alpha}^a, (\Lambda) = \left(\sum_{k=0}^{a}|z|^2_{A_{\alpha}^a,(\Lambda)}\right)^{\frac{1}{2}}.$$

Now let $\Lambda = \varrho/0 < \varrho < \infty$ with $\chi(\varrho)$ be a weight function. Then
$L_{\chi}^2(\Lambda) = \{\kappa \,/\, \kappa$ is measurable on Λ and $\|u\|_{L_{\chi}^2}, \Lambda < \infty\}$.
with the following inner product and norm

$$(u, v)_{\chi,\Lambda} = \int_{\Lambda} u(\varrho)v(\varrho)d\varrho, \qquad \|v\|\chi, \Lambda = \sqrt{\langle u, v\rangle}_{\chi,\Lambda}.$$

3. Operational matrices corresponding to fractional derivatives and integrals

Here in this section, we provide the required OMs via Laguerre polynomials of fractional derivatives and integrals.

Lemma 3. Let $\Psi_M(t)$ be a function vector; the fractional integral of order γ for the function $\Psi_M(t)$ can be generalized as

$$ {}_0I_t^{\gamma}\Psi_M(t) \approx G_{N\times N}^{\gamma}\Psi_M(t), $$

where $G_{N\times N}^{\gamma}$ is the OM of integration of fractional order γ and given by

$$ \begin{bmatrix} \daleth_{0,0,k,r}^{\gamma} & \daleth_{0,1,k,r}^{\gamma} & \cdots & \daleth_{0,j,k,r}^{\gamma} & \cdots & \daleth_{0,m,k,r}^{\gamma} \\ \daleth_{1,0,k,r}^{\gamma} & \daleth_{1,i,k,r}^{\gamma} & \cdots & \daleth_{1,j,k,r}^{\gamma} & \cdots & \daleth_{1,m,k,r}^{\gamma} \\ \vdots & \vdots & \vdots & \vdots & \cdots & \vdots \\ \daleth_{i,0,k,r}^{\gamma} & \daleth_{i,1,k,r}^{\gamma} & \cdots & \daleth_{i,j,k,r}^{\gamma} & \vdots & \daleth_{i,m,k,r}^{\gamma} \\ \vdots & \vdots & \vdots & \vdots & \cdots & \vdots \\ \daleth_{m,0,k,r}^{\gamma} & \daleth_{m,1,k,r}^{\gamma} & \cdots & \daleth_{m,j,k,r}^{\gamma} & \cdots & \daleth_{m,m,k,r}^{\gamma} \end{bmatrix}, $$

where

$$ \daleth_{i,j,k,r}^{\gamma} = \sum_{k=0}^{i}\sum_{r=0}^{i} \frac{(-1)^{k+r}\Gamma(j+1)\Gamma(i+\gamma+1)\Gamma(k+\gamma+\alpha+r+1)}{\Gamma(j-r+1)\Gamma(i-k+1)\Gamma(r+1)\Gamma(k+\gamma+1)\Gamma(k+\alpha+1)\Gamma(\gamma+r+1)}. $$

Proof. We apply the fractional order integral of order γ to the Laguerre polynomials

$$ {}_0^cI_t^{\gamma} L_i^{\gamma}(t) = \sum_{k=0}^{i} \frac{\Gamma(i+\gamma+1)}{\Gamma(i-k+1)\Gamma(k+\gamma+1)\Gamma(k+1)} {}_0^cI_t^{\gamma} t^k. \tag{7} $$

Since from (7), we have

$$ {}_0^cI_t^{\gamma} t^k = \frac{\Gamma(k+1)}{\Gamma(1+k+\alpha)} t^{k+\gamma}. $$

Therefore Eq. (7) implies that

$$ {}_0^cI_t^{\gamma} L_i^{\gamma}(t) = \sum_{k=0}^{i} t^{k+\gamma} \frac{\Gamma(i+\gamma+1)}{\Gamma(i-k+1)\Gamma(k+\gamma+1)\Gamma(k+1)} \frac{\Gamma(k+1)}{\Gamma(1+k+\alpha)}, $$

which is equal to

$$ {}_0^cI_t^{\gamma} L_i^{\gamma}(t) = \sum_{k=0}^{i} (-1)^k \frac{\Gamma(i+\gamma+1)}{\Gamma(i-k+1)\Gamma(k+\gamma+1)\Gamma(1+k-\gamma)} t^{k+\gamma}. \tag{8} $$

We approximate $t^{k+\gamma}$ in (8) with Laguerre polynomials, i.e.

$$ t^{k+\gamma} \approx \sum_{j=0}^{n} H_j L_j^{\gamma}(t). $$

By using the relation of orthogonality, we can find coefficients

$$ H_j = \sum_{r=0}^{j} (-1)^k \frac{\Gamma(j+1)\Gamma(k+\alpha+r+\gamma+1)}{\Gamma(1+j-r)\Gamma(1+r)\Gamma(1+r+\gamma)}. $$

So Eq. (8) implies

$$ {}_0^cI_t^\gamma L_i^\gamma(t) = \sum_{k=0}^{i}(-1)^k \frac{\Gamma(i+\gamma+1)}{\Gamma(i-k+1)\Gamma(k+\gamma+1)\Gamma(1+k-\gamma)} \times \sum_{r=0}^{j}(-1)^r \frac{\Gamma(j+1)\Gamma(k+\alpha+r+\gamma+1)}{\Gamma(j-r+1)\Gamma(r+1)\Gamma(r+\gamma+1).} $$

$$ {}_0^cI_t^\gamma L_i^\gamma(t) = \sum_{k=0}^{i}\sum_{r=0}^{j}(-1)^{k+r}\frac{\Gamma(j+1)\Gamma(i+\gamma+1)\Gamma(k+\alpha+r+\gamma+1)}{\Gamma(1-k+i)\Gamma(j-\gamma+1)\Gamma(\gamma+1)\Gamma(k+\gamma+1)\Gamma(k+\alpha+1)\Gamma(\gamma+r+1)}. $$

which is the desired result.

Lemma 4. Let $\Psi_M(t)$ be a function vector; then the fractional derivative of order γ for $\Psi_M(t)$ is generalized as

$$ {}_0^cD_t^\gamma \Psi_M(t) \approx \mathbf{W}_{M\times M}^\gamma \Psi_M(t), $$

where $\mathbf{W}_{M\times M}^\gamma$ is the OM of derivative of order γ, defined as in (9)

$$ \mathbf{W}_{M\times M}^\gamma = \begin{bmatrix} 0 & 0 & 0 & 0 & \cdots 0 & \\ \vdots & \vdots & \vdots & \vdots & \vdots & \\ \Theta^\gamma_{\lceil\gamma\rceil,0,k,\alpha} & \Theta^\gamma_{\lceil\gamma\rceil,1,k,\alpha} & \cdots\Theta^\gamma_{\lceil\gamma\rceil,j,k,\alpha} & \cdots & \cdots & \Theta^\gamma_{\lceil\gamma\rceil,n,k,\alpha} \\ \Theta^\gamma_{i,0,k,\alpha} & \Theta^\gamma_{i,1,k,\alpha} & \Theta^\gamma_{i,j,k,\alpha} & \cdots & \cdots & \Theta^\gamma_{i,n,k,\alpha} \\ \vdots & \vdots & \vdots & \vdots & \vdots & \\ \Theta^\gamma_{n,0,k,\alpha} & \Theta^\gamma_{n,1,k,\alpha} & \Theta^\gamma_{n,j,k,\alpha} & \cdots & \cdots & \Theta^\gamma_{n,n,k,\alpha} \end{bmatrix}, \tag{9} $$

where

$$ \Theta^\gamma_{i,j,k,\alpha} = \sum_{k=\gamma}^{i}\sum_{r=0}^{i}\frac{(-1)^{\gamma+k}\Gamma(j+1)\Gamma(i+\alpha+1)\Gamma(k+\alpha-r+\gamma+1)}{\Gamma(j-r+1)\Gamma(i-k+1)\Gamma(r+1)\Gamma(k+\alpha+1)\Gamma(k-\gamma+1)\Gamma(\alpha+\gamma+1)}. $$

Proof. Leaving the proof as it is very similar to the proof of the above lemma.

Lemma 5. We consider a function $Z(t)$ defined on $[0, \infty]$ and $y(t) = K_M\Psi_M^T(t)$; then

$$ Z(t)[{}_0I_t^\gamma y(t)] = K_M Q_{M\times M}^\gamma \Psi_M(t), $$

where $Q_{M\times M}^\gamma$ is the operational matrix, given by

$$ \begin{bmatrix} C_{0,0}, & C_{0,1} & \cdots & C_{0,j} & \cdots & C_{0,m} \\ C_{1,0} & C_{1,1} & \cdots & C_{1,j} & \cdots & C_{1,m} \\ \vdots & \vdots & \vdots & \vdots & \cdots & \vdots \\ C_{i,0} & C_{i,1} & \cdots & C_{i,j} & \vdots & C_{i,m} \\ \vdots & \vdots & \vdots & \vdots & \cdots & \vdots \\ C_{m,0} & C_{m,1} & \cdots & C_{m,j} & \cdots & C_{m,m} \end{bmatrix}, $$

where

$$C_{i,j} = \frac{1}{h_i}\int_0^1 \Delta_{i,\gamma,k} Z(t) L_j^{\gamma}(t)dt,$$

with

$$w_i = \sum_{k=0}^{i} \frac{(-1)^{i+1}\Gamma(i+1+\gamma)}{\Gamma(k+\gamma+1)\Gamma(1-k+i)\Gamma(k+\gamma)}.$$

Proof. By considering the general term of $\Psi_M(t)$

$$_0I_1^{\gamma} L_i(t) = \frac{1}{\Gamma(\gamma)}\int_0^1 (1-s)^{\gamma-1} L_i(s)ds.$$

$$_0I_1^{\gamma} L_i(t) = \frac{1}{\Gamma(\gamma)}\int_0^1 (1-s)^{\gamma-1}\sum_{k=0}^{i}(s)^k \frac{(-1)^k\Gamma(i+1+\gamma)}{\Gamma(-k+1+i)\Gamma(k+1+\gamma)\Gamma(1+k)}ds.$$

$$_0I_1^{\gamma} L_i(t) = \sum_{k=0}^{i} \frac{(-1)^k\Gamma(i+1+\gamma)}{\Gamma(\gamma)\Gamma(-k+1+i)\Gamma(k+1+\gamma)\Gamma(1+k)}\int_0^1 (1-s)^{\gamma-1}(s)^k ds. \tag{10}$$

Using the famous Laplace transform, we have from (10)

$$\pounds\left(\int_0^1 (1-s)^{\gamma-1}s^k ds = \frac{\Gamma(\gamma)\Gamma(k+1)}{\Gamma(\gamma+k)}.\right.$$

$$_0I_1^{\gamma} L_i(t) = \sum_{k=0}^{i} \frac{(-1)^k\Gamma(i+1+\gamma)}{\Gamma(\gamma)\Gamma(-k+1+i)\Gamma(k+1+\gamma)\Gamma(1+k)}\frac{\Gamma(\gamma)\Gamma(k+1)}{\Gamma(\gamma+k)}.$$

$$\sum_{k=0}^{i} \frac{(-1)^k\Gamma(i+1+\gamma)}{\Gamma(-k+1+i)\Gamma(k+\gamma+1)\Gamma(1+k)} = \Delta_{i,\gamma,k}.$$

Now using Laguerre polynomials, we have

$$\Delta_{i,\gamma,k} z(t) = \sum_{j=0}^{m} C_{i,j} L_i(t),$$

where $C_{i,j}$ is calculated by using orthogonality as

$$C_{i,j} = \frac{1}{hi}\int_0^1 \Delta_{i,\gamma,k} z(t) L_j^{\gamma}(t)dt. \tag{11}$$

To get the desired result, we evaluate the above (11) relation for $i = 0, 1, \ldots, m$ and $j = 0, 1, \ldots, m$.

4. Main result

In this section, we discuss some cases of FODEs with initial condition as well as boundary conditions. The approximate solution obtained through desired method is compared with the exact solution. Similarly we investigate numerical solutions to various coupled systems under some initial conditions as well as boundary conditions.

4.1 Treatment of FODEs under initial and boundary conditions

Here we discuss different cases.

Case 1. In the first case, we consider the fractional order differential equation

$$\begin{cases} {}_0^cD_t^\gamma z(t) \pm z(t) = 0, & 0 < \gamma \leqslant 1, \\ z(0) = z_0, & z_0 \in R \end{cases} \tag{12}$$

we see that

$${}_0^cD_t^\gamma z(t) = Ł_M \psi_M^T(t).$$

and applying ${}_0I_t^\gamma$ by the Lemma 1, on (12) we write

$$z(t) = e_0 + \ {}_0I_t^\gamma \left[Ł_M \psi_M^T(t)\right],$$

Using the initial condition to get $e_0 = z_0$ and approximate z_0 as $z_0 \approx F_M \ \psi_M^T(t)$, Eq. (12) implies

$$Ł_M \ \psi_M^T(t) + Ł_M \ G_{M\times M}^\gamma \ \psi_M^T(t) + F_M \ \psi_M^T(t) = 0.$$

Finally the Sylvester-type algebraic equation is obtained as

$$Ł_M + Ł_M \ G_{M\times M}^\gamma \ \psi_M^T(t) + F_M = 0.$$

Solving the Sylvester matrix for $Ł_M$, we get the numerical value for $z(t)$.

Example 1.

$$\begin{cases} {}_0^cD_t^\gamma z(t) \pm z(t) = 0, & 0 < \gamma \leq 1, \\ z(0) = 1, & z_0 \in R. \end{cases}$$

Since the exact solution is given by

$$z(t) = E_\gamma(-t^\gamma),$$

where E_γ is the Mittag-Leffler representation, and at $\gamma = 1$, $z(t) = e^{-t}$.

Approximating the solution through the proposed method and plotting the exact as well as numerical solution by using scale $M = 8$ corresponding to $\gamma = 1$ in **Figure 1**, we see that the proposed method works very well.

Case 2.

$$\begin{cases} {}_0^cD_t^\gamma z(t) + z(t) = 0, & 1 < \gamma \leqslant 2, \\ z(0) = z_0, \ z(1) = z_1, & z_0, \ z_1 \in R. \end{cases} \tag{13}$$

We take

$${}_0^cD_t^\gamma z(t) = K_M \psi_M^T(t). \tag{14}$$

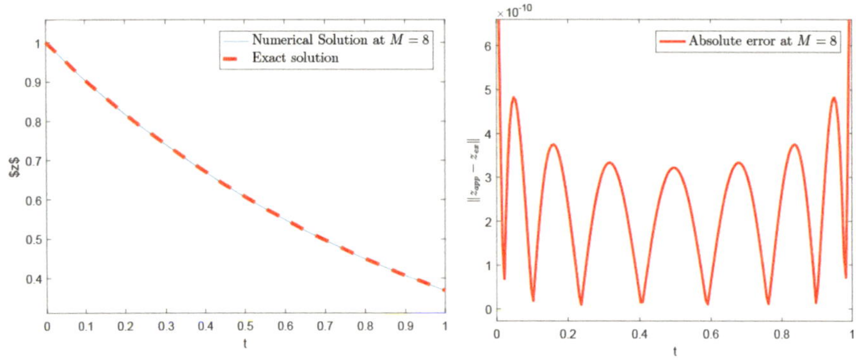

Figure 1.
Plots of both approximate and exact solution for the Example 1 for Case 1.

Applying Lemma 1 to Eq. (14), we get

$$z(t) = e_0 + e_1(t) + {}_0I_t^{\gamma} K_M \psi_M^T(t). \tag{15}$$

Using the conditions by putting $t = 0$ and $t = 1$ to get $e_0 = z_0$ and

$$e_1 = z_1 - z_0 - K_{M0} I_1^{\gamma} \psi_M^T(t)/_{t=1}.$$

Equation (15) implies

$$z(t) = z_0 + (z_1 - z_0)t - tK_{M0} I_1^{\gamma} \psi_M^T(t)/_{t=1} + {}_0I_t^{\gamma} K_M \psi_M^T(t),$$

where $z_0 + (z_1 - z_0)t$ is the smooth function of t and constants; we approximate it as

$$z_0 + (z_1 - z_0)t \approx G_{M\times M}^{\gamma} \psi_M^T(t)$$

and

$$tK_{M0} I_1^{\gamma} \psi_M^T(1) \approx K_M Q_{M\times M}^{\gamma} \psi_M^T(t).$$

Hence

$$z(t) = G_{M\times M}^{\gamma} \psi_M^T(t) - K_M Q_{M\times M}^{\gamma} \psi_M^T(t) + K_M G_{M\times M}^{\gamma} \psi_M^T(t)$$

So Eq. (13) implies

$$K_M \psi_M^T(t) + G_{M\times M}^{\gamma} \psi_M^T(t) - K_M Q_{M\times M}^{\gamma} \psi_M^T(t) + K_M G_{M\times M}^{\gamma} \psi_M^T(t) = 0$$

which is further solved for K_M to get the required numerical solution.
For Case 2, we give the following example.
Example 2.

$$\begin{cases} {}_0^cD_t^{\gamma} z(t) + z(t) = 0, & 0 < \gamma \leq 2, \\ z(0) = -1, \ z(1) = 1. \end{cases} \tag{16}$$

At $\gamma = 2$, we get the exact solution as of (16) as given by (17)

$$z(t) = 114.58 \sin(x) - \cos(x) \tag{17}$$

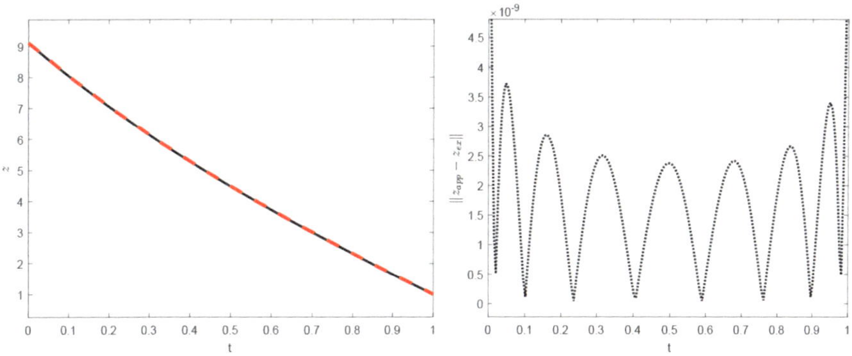

Figure 2.
The plot of exact and approximate solution for Example 2 for Case 2.

Upon using the suggested method, we see from the subplot at the left of **Figure 2** that exact and numerical solutions are very close to each other for very low scale level. Also, the absolute error is given in subplot at the right of **Figure 2**.

4.2 Coupled systems of linear FODEs under initial and boundary conditions

In this subsection, we consider different forms of coupled systems of FODEs with the initials as well as boundary conditions.

Case 1. First we take the coupled system of FODEs as

$$\begin{cases} {}_0^cD_t^\gamma z(t) + az(t) + by(t) = f(t) \\ {}_0^cD_t^\gamma y(t) + cy(t) + dz(t) = g(t), \end{cases} \tag{18}$$

with the conditions

$$z(0) = z_0, \qquad y(0) = y_0, z_0, y_0 \in R. \tag{19}$$

Let

$${}_0^cD_t^\gamma z(t) = Ł_M\psi_M^T(t), \ {}_0^cD_t^\gamma y(t) = K_M\psi_M^T(t). \tag{20}$$

Applying Lemma 1 to Eq. (20), we get

$$\begin{cases} z(t) = e_0 + Ł_M G_{M\times M}^\gamma \psi_M^T(t), \\ y(t) = d_0 + K_M G_{M\times M}^\gamma \psi_M^T(t). \end{cases} \tag{21}$$

Using the initial conditions given in Eq. (19), from Eq. (21), we get

$$\begin{cases} z(t) = F_M^1\psi_M^T(t) + Ł_M G_{M\times M}^\gamma \psi_M^T(t), \\ y(t) = y_0 \approx F_M^2\psi_M^T(t) + K_M G_{M\times M}^\gamma \psi_M^T(t). \end{cases} \tag{22}$$

We take approximation as

$$z_0 \approx F_M^1\psi_M^T(t),$$

and

$$y_0 \approx F_M^2 \psi_M^T(t),$$

while source functions are approximated as

$$f(t) \approx F_M^3 \Psi_M^T(t),$$

and

$$g(t) \approx F_M^4 \Psi_M^T(t).$$

Therefore the consider system on using (19)–(22), (18) becomes

$$\begin{cases} Ł_M \psi_M^T + a\left(F_M^1 \psi_M^T(t) + Ł_M G_{M\times M}^{\gamma} \psi_M^T(t)\right) \\ +b(F_M^2 \psi_M^T(t) + K_M G_{M\times M}^{\gamma} \psi_M^T(t) = F_M^3 \psi_M^T(t). \end{cases}$$
$$\begin{cases} K_M \psi_M^T + c\left(F_M^2 \psi_M^T(t) + K_M G_{M\times M}^{\gamma} \psi_M^T(t)\right) \\ +d\left(F_M^1 \psi_M^T(t) + Ł_M G_{M\times M}^{\gamma} \psi_M^T(t) = F_M^4 \psi_M^T(t). \end{cases}$$

On further rearrangement we have

$$\begin{cases} Ł_M + a\left(F_M^1 + Ł_M G_{M\times M}^{\gamma}\right) + b\left(F_M^2 + K_M G_{M\times M}^{\gamma} = F_M^3 \right. \\ K_M + c\left(F_M^2 + K_M G_{M\times M}^{\gamma}\right) + d\left(F_M^1 + Ł_M G_{M\times M}^{\gamma} = F_M^4. \right. \end{cases}$$

which further can be written as

$$\begin{cases} Ł_M\left(I_{M\times M} + aG_{M\times M}^{\gamma}\right) + K_M\left(bG_{M\times M}^{\gamma}\right) + \left(aF_M^1 + bF_M^2 - F_M^3\right) = 0 \\ K_M\left(I_{M\times M} + cG_{M\times M}^{\gamma}\right) + Ł_M\left(dG_{M\times M}^{\gamma}\right) + \left(cF_M^2 + dF_M^1 - F_M^4\right) = 0. \end{cases}$$

In matrix form we write as

$$[Ł_M \ K_M]\begin{bmatrix} I_{M\times M} + aG_{M\times M}^{\gamma} & 0 \\ 0 & I_{M\times M} + cG_{M\times M}^{\gamma} \end{bmatrix} + \begin{bmatrix} Ł_M & K_M \end{bmatrix}\begin{bmatrix} 0 & dG_{M\times M}^{\gamma} \\ bG_{M\times M}^{\gamma} & 0 \end{bmatrix} + \begin{bmatrix} aF_M^1 + bF_M^2 - F_M^3 \\ cF_M^2 + dF_M^1 - F_M^4 \end{bmatrix} = 0.$$

We solve this system of matrix equation for $[Ł_M \ K_M]$ by using Gaussian's elimination method. The considered system is in the form of $X\overline{A} + X\overline{B} + \overline{C} = 0,.$

where $X = [Ł_M \ K_M]$ $\overline{A} = \begin{bmatrix} I_{M\times M} + aG_{M\times M}^{\gamma} & 0 \\ 0 & I_{M\times M} + cG_{M\times M}^{\gamma} \end{bmatrix},.$

$\overline{B} = \begin{bmatrix} 0 & dG_{M\times M}^{\gamma} \\ bG_{M\times M}^{\gamma} & 0 \end{bmatrix}$ and $\overline{C} = \begin{bmatrix} aF_M^1 + bF_M^2 - F_M^3 \\ cF_M^2 + dF_M^1 - F_M^4. \end{bmatrix}.$

Upon computation of matrices $Ł_M, K_M$ by using MATLAB®, we put these matrices in Eq. (22) to find z_{app} and y_{app}, respectively.

Example 3. We now provide its example by considering the system of FODEs:

$$\begin{cases} {}_0^cD_t^{\gamma} z(t) + z(t) + y(t) = f(t) \\ {}_0^cD_t^{\gamma} y(t) + y(t) + z(t) = g(t), \\ z(0) = 2, \ y(0) = 1. \end{cases}$$

By taking $\gamma = 1$, the exact solution is obtained as

$$z(t) = \cos(t) + e^t, \quad y = \sin(t) + e^{-t},$$

where the external source functions are given by $f(t) = \cos(t) + e^{-t} + 2e^t$ and $g(t) = e^{-t} + \sin(t) + 2\cos(t)$. The exact solution z_{ex}, y_{ex} can be computed by any method of ODEs. Approximating the problem by the considered method, we see that the computed numerical and exact solutions have close agreement at very small-scale level. The corresponding accuracy has been recorded in **Table 1**. Further the comparison between exact and numerical solution and the results about absolute error have been demonstrated in **Figures 3** and **4**, respectively. In **Figure 3** we are given the comparison between exact solution and approximate solutions by using proposed method. Similarly the absolute errors have been described in **Figure 4**.

By comparing the exact and numerical solution through the proposed method, we observe that our numerical solution does not show any disagreement with the exact solution as can be seen in **Figure 3**. The absolute errors $\|z_{app} - z_{ex}\|$ and $\|y_{app} - y_{ex}\|$ plotted at the scale $M = 5$ are very low as given in **Figure 4**, which describes the efficiency of the proposed method.

Case 2. Similarly for the coupled system of FODEs with boundary conditions, we consider

$$\begin{cases} {}_0^cD_t^\gamma z(t) + az(t) + by(t) = f(t), \\ {}_0^cD_t^\gamma y(t) + cy(t) + dz(t) = g(t), \end{cases} \tag{23}$$
$$z(0) = z_0, y(0) = y_0, z(1) = z_1, y(1) = y_1.$$

t	CPU time (s)	Absolute error $\|z_{app} - z_{ex}\|$	Absolute error $\|y_{app} - y_{ex}\|$	CPU time (s)
0	30.5	0.00003	0.000006	32.5
0.15	32.7	0.000016	0.000034	33.3
o.35	35.8	0.000013	0.00003	33.9
0.65	33.6	0.000012	0.00003	35.6
0.87	34.8	0.000018	0.000036	36.5
1	35.9	0.00003	0.000006	36.8

Table 1.
Absolute error at $M = 5$, $\gamma = 0.9$, for different values of t in Example 3.

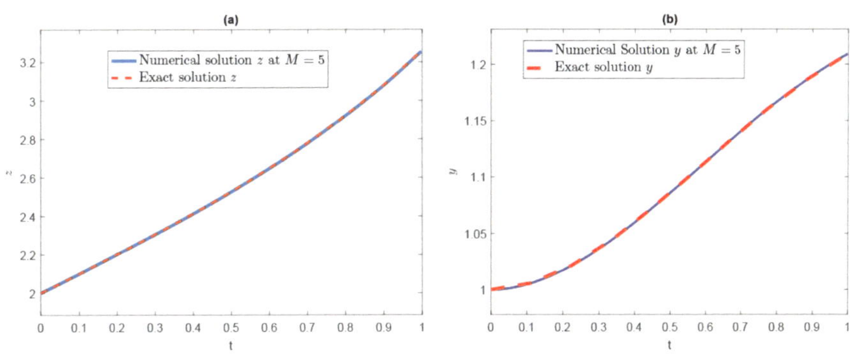

Figure 3.
Plots of exact and approximate solution of Example 3.

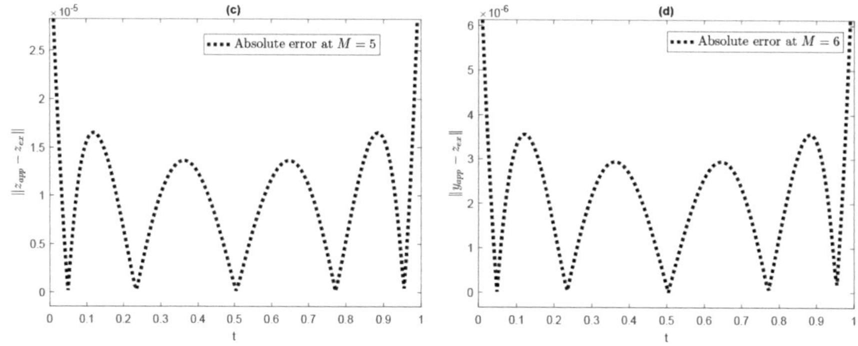

Figure 4.
Plots of absolute error of Example 3.

Let us assume

$$\begin{cases} {}_0^c D_t^\gamma z(t) = Ł_M \psi_M^T(t), \\ {}_0^c D_t^\gamma y(t) = K_M \psi_M^T(t). \end{cases} \tag{24}$$

Applying Lemma 1 to Eq. (24), we get

$$\begin{cases} z(t) = e_0 + e_1(t) + Ł_M G_{M\times M}^\gamma \Psi_M^T(t) \\ y(t) = d_0 + d_1(t) + K_M G_{M\times M}^{\star\gamma} \Psi_M^T(t), \end{cases} \tag{25}$$

where $d_0, d_1, e_0, e_1 \in R$. Using the initial conditions in Eq. (25), we have $e_0 = z_0$, $d_0 = y_0$. On using boundary conditions, we have from Eq. (25)

$$z(1) = z_0 + e_1 + Ł_M G_{M\times M}^\gamma \Psi_M^T(t)\big|_{t=1},$$
$$z(1) - z_0 - Ł_M G_{M\times M}^\gamma \Psi_M^T(t)\big|_{t=1} = e_1.$$

Similarly

$$y(1) = y_0 + d_1 + K_M G_{M\times M}^{\star\gamma} \Psi_M^T(t)\big|_{t=1},$$
$$y(1) - y_0 - K_M G_{M\times M}^{\star\gamma} \Psi_M^T(t)\big|_{t=1} = d_1.$$

Equation (25) implies that

$$\begin{cases} z(t) = z_0 + t(z_1 - z_0) - t(L_M G_{M\times M}^\gamma \Psi_M^T(t)\big|_{t=1}) + L_M G_{M\times M}^\gamma \Psi_M^T(t) \\ y(t) = y_0 + t(y_1 - y_0) - t(K_M G_{M\times M}^{\star\gamma} \Psi_M^T(t)\big|_{t=1}) + K_M G_{M\times M}^{\star\gamma} \Psi_M^T(t). \end{cases} \tag{26}$$

Let $z_0 + t(z_1 - z_0) \approx F_M^1 \Psi_M^T(t)$ and $y_0 + t(y_1 - y_0) \approx F_M^2 \psi_M^T(t)$, with

$$\begin{aligned} Ł_M G_{M\times M}^\gamma \Psi_M^T(t) &= Ł_M Q_{M\times M}^{\gamma,z} \Psi_M^T(t) \\ t K_M G_{M\times M}^{\star\gamma} \Psi_M^T(t) &= K_M Q_{M\times M}^{\gamma,y} \Psi_M^T(t). \end{aligned} \tag{27}$$

Hence Eq. (26) implies

$$\begin{cases} z(t) = F_M^1 \Psi_M^T(t) - L_M Q_{M\times M}^{\gamma,z} \Psi_M^T(t) + L_M G_{M\times M}^\gamma \Psi_M^T(t) \\ y(t) = F_M^2 \Psi_M^T(t) - K_M Q_{M\times M}^{\gamma,y} \Psi_M^T(t) + K_M G_{M\times M}^{\star\gamma} \Psi_M^T(t). \end{cases} \tag{28}$$

approximating $f(t)$ and $g(t)$ such that

$$\begin{cases} f(t) \approx F_M^3 \Psi_M^T(t) \\ g(t) \approx F_M^4 \Psi_M^T(t). \end{cases} \tag{29}$$

On using (24)–(29), system (23) can be written as

$$\begin{cases} L_M\Psi_M^T(t) + a\left(F_M^1\Psi_M^T(t) - L_M Q_{M\times M}^{\gamma,z}\Psi_M^T(t) + L_M G_{M\times M}^{\gamma}\Psi_M^T(t)\right) \\ +b\left(F_M^2\Psi_M^T(t) - K_M Q_{M\times M}^{\gamma}\Psi_M^T(t) + K_M G_{M\times M}^{\gamma}\Psi_M^T(t)\right) - F_M^3\Psi_M^T(t) = 0 \\ K_M\Psi_M^T(t) + c\left(F_M^2\Psi_M^T(t) - K_M Q_{M\times M}^{\gamma,y}\Psi_M^T(t) + K_M G_{M\times M}^{\gamma}\Psi_M^T(t)\right) \\ +d\left(F_M^1\Psi_M^T(t) - Ł_M Q_{M\times M}^{\gamma,z}\Psi_M^T(t) + Ł_M G_{M\times M}^{\gamma}\Psi_M^T(t)\right) - F_M^4\Psi_M^T(t) = 0. \end{cases}$$

On rearrangement of terms, the above equations give

$$\begin{cases} Ł_M\left(I_{M\times M} - aQ_{M\times M}^{\gamma,z} + aG_{M\times M}^{\gamma}\right) + K_M\left(I_{M\times M} - bQ_{M\times M}^{\gamma,y} + bG_{M\times M}^{\gamma}\right) \\ +aF_M^1 + bF_M^2 - F_M^3 = 0 \\ K_M\left(I_{M\times M} - cQ_{M\times M}^{\gamma,y} + cG_{M\times M}^{\gamma}\right) + Ł_M\left(I_{M\times M} - dQ_{M\times M}^{\gamma,z} + dG_{M\times M}^{\gamma}\right) \\ +cF_M^2 + dF_M^1 - F_M^4 = 0. \end{cases}$$

In matrix form, we can write

$$\begin{aligned} &[Ł_M\ K_M]\begin{bmatrix} I_{M\times M} - aQ_{M\times M}^{\gamma,z} + aG_{M\times M}^{\gamma} & 0 \\ 0 & I_{M\times M} - cQ_{M\times M}^{\gamma,y} + cG_{M\times M}^{\gamma} \end{bmatrix} \\ &+[L_M K_M]\begin{bmatrix} 0 & I_{M\times M} - dQ_{M\times M}^{\gamma,z} + dG_{M\times M}^{\gamma} \\ I_{M\times M} - bQ_{M\times M}^{\gamma,y} + bG_{M\times M}^{\gamma} & 0 \end{bmatrix} \\ &+\begin{bmatrix} aF_M^1 + bF_M^2 - F_M^3 \\ cF_M^2 + dF_M^1 - F_M^4 \end{bmatrix} = 0. \end{aligned}$$

We convert the system to algebraic equation by considering

$$\overline{L} = \begin{bmatrix} I_{M\times M} - aQ_{M\times M}^{\gamma,z} + aG_{M\times M}^{\gamma} & 0 \\ 0 & I_{M\times M} - cQ_{M\times M}^{\gamma,y} + cG_{M\times M}^{\gamma} \end{bmatrix}$$

$$\overline{M} = \begin{bmatrix} 0 & I_{M\times M} - dQ_{M\times M}^{\gamma,z} + dG_{M\times M}^{\gamma} \\ I_{M\times M} - bQ_{M\times M}^{\gamma,y} + bG_{M\times M}^{\gamma} & 0 \end{bmatrix}$$

$$\text{and } \overline{N} = \begin{bmatrix} aF_M^1 + bF_M^2 - F_M^3 \\ cF_M^2 + dF_M^1 - F_M^4 \end{bmatrix}.$$

so that the system is of the form

$$X\overline{L} + X\overline{M} + \overline{N} = 0,$$

and solving the given equation for the unknown matrix $X = [L_M K_M]$, we get the required solution.

Example 4. As an example, we consider the Caputo fractional differential equation for the coupled system with the boundary conditions as

$$
\begin{cases}
{}_0^cD_t^\gamma z(t) + 2z(t) - 2y(t) - f(t) = 0, \\
{}_0^cD_t^\gamma y(t) - 3y(t) + 2z(t) - g(t) = 0, \\
z(0) = 4 \quad z(1) = -4, \\
y(0) = 2, \quad y(1) = -2.
\end{cases}
$$

At $\gamma = 2$, the exact solutions are

$$
\begin{cases}
z(t) = t^6 + t^5 + t^4 - t^3 + t + 1, \\
y(t) = t^7 - t^6 + t^5 + t^4 + t^3 - t^2 - t + 1.
\end{cases}
$$

where the source functions are given by

$$
\begin{cases}
f(t) = -2t^7 + 4t^6 + 30t^4 + 16t^3 + 12t^2 - 2t + 2 \\
g(t) = -3t^7 + 12t^6 + 35t^5 - 27t^4 - 19t^3 + 20t^2 + 9t - 4.
\end{cases}
$$

We approximate the solution at the considered method by taking scale level $M = 5$. One can see that numerical plot and exact solution plot coincide very well as shown in **Figure 5**. Similarly the absolute error has been plotted at the given scale $M = 5$ in **Figure 6**, which is very low. The lowest value of absolute error $\|z_{app} - z_{ex}\|$ and $\|y_{app} - y_{ex}\|$ indicates efficiency of the proposed method. The table shows the

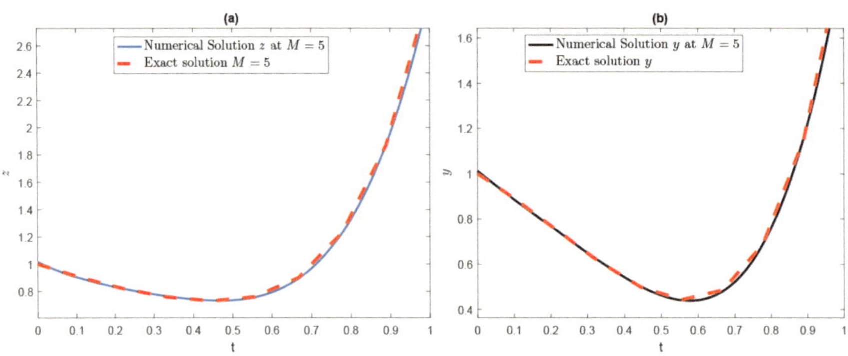

Figure 5.
Plots of exact and approximate solution for Case 4, boundary value problem.

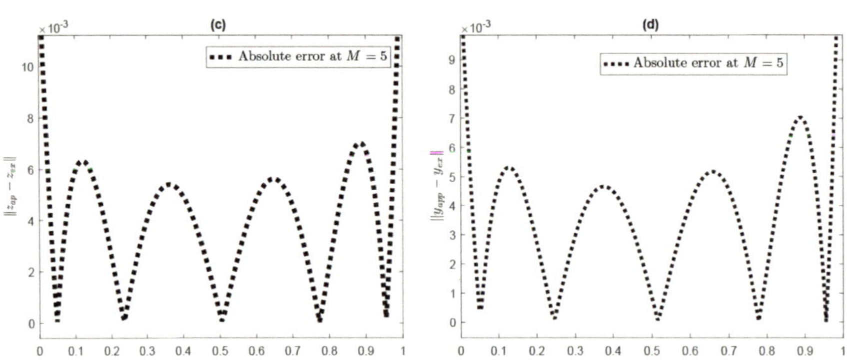

Figure 6.
Plots of absolute error for Case 4, boundary value problem.

t	Absolute error $\|z_{app} - z_{ex}\|$	CPU time (s)	Absolute error $\|y_{app} - y_{ex}\|$	CPU time (s)
0	0.011	49.4	0.010	50.0
0.15	0.0062	50.3	0.0052	52.5
0.35	0.0058	51.2	0.0047	54.6
0.65	0.006	51.5	0.005	55.5
0.85	0.0075	52.6	0.007	56.4
1	0.011	53.8	0.010	56.2

Table 2.
Absolute error at different values of t for Example 4.

comparison of errors for exact and approximate solutions for fixed scale level $M = 5$ and order $\gamma = 1.9$. Further the absolute error has been recorded at different values of space variable in **Table 2** which provides the information about efficiency of the proposed method.

5. Conclusion

We have successfully used the class of orthogonal polynomials of Laguerre polynomials to establish a numerical method to compute the numerical solution of FODEs and their coupled systems under some initial and boundary conditions. By using these polynomials, we have obtained some operational matrices corresponding to fractional order derivatives and integration. Also we have computed a new matrix corresponding to boundary conditions for boundary value problems of FODEs. Using the aforementioned matrices, we have converted the considered problem of FODEs to Sylvester-type algebraic equations. To obtain the numerical solution, we easily solved the desired algebraic equations by taking help from MATLAB®. Corresponding to the established procedure, we have provided numbers of examples to demonstrate our results. Also some error analyses have been provided along with graphical representations. By increasing the scale level, the accuracy is increased and vice versa. On the other hand, when the fractional order is approaching to integer value, the solutions tend to the exact solutions of the considered FODE. Therefore in each example, we have compared the exact and approximate solution and found that both the solutions were in closure contact with each other. Hence the established method can be very helpful in solving many classes and systems of FODEs under both initial and boundary conditions. In future the shifted Laguerre polynomials can be used to compute numerical solutions of partial differential equations of fractional order.

Author contribution

All authors equally contributed this paper and approved the final version.

Competing interests

We declare that no competing interests exist regarding this manuscript.

Author details

Adnan Khan[1], Kamal Shah[1,2]* and Danfeng Luo[3]

1 Department of Mathematics, University of Malakand, Dir(L), Khyber Pakhtunkhwa, Pakistan

2 Department of Mathematics and General Sciences, Prince Sultan University, Riyadh, Saudi Arabia

3 Key Laboratory of Computing and Stochastic Mathematics (Ministry of Education), School of Mathematics and Statistics, Hunan Normal University, Changsha, Hunan, P.R. China

*Address all correspondence to: kamalshah408@gmail.com

References

[1] Butzer PL, Westphal U. An Introduction to Fractional Calculus. Singapore: World Scientific; 2000

[2] Samko SG, Kilbas AA, Marichev OI. Fractional Integrals and Derivatives. Switzerland: Gordon and Breach; 1993

[3] Scalas E, Raberto M, Mainardi F. Fractional calculus and continous time finance. Physica A: Statistical Mechanics and its Applications. 2000; **284**:376-384

[4] Hilfer R. Applications of Fractional Calculus in Physics. Singapore: World Scientific; 2000

[5] Amairi M, Aoun M, Najar S, Abdelkrim MN. A constant enclosure method for validating existence and uniqueness of the solution of an initial value problem for a fractional differential equation. Applied Mathematics and Computation. 2010; **217**(5):2162-2168

[6] Deng J, Ma L. Existence and uniqueness of solutions of initial value problems for nonlinear fractional differential equations. Applied Mathematics Letters. 2000;**23**:676-680

[7] Girejko E, Mozyrska D, Wyrwas M. A sufficient condition of viability for fractional differential equations with the Caputo derivative. Journal of Mathematical Analysis and Applications. 2011;**38**:146-154

[8] Baleanu D, Diethelm K, Scalas E, Trujillo JJ. Fractional Calculus Models and Numerical Methods. Singapore: World Scientific; 2009

[9] Guy J. Modeling fractional stochastic systems as non-random fractional dynamics driven by Brownian motions. Applied Mathematical Modelling. 2008; **32**:836-859

[10] Sabatier JATMJ, Agrawal OP, Machado JT. Advances in Fractional Calculus. Dordrecht: Springer; 2007

[11] Kilbas AA, Srivastava HM, Trujillo JJ. Theory and Applications of Fractional Differential Equations. Amsterdam: Elsevier; 2006

[12] Lakshmikantham I, Leela S. Theory of Fractional Dynamical Systems. Cambridge, UK: Cambridge Scientific Publishing; 2009

[13] Li CP, Deng WH. Remarks on fractional derivatives. Applied Mathematics and Computation. 2007; **187**:777-784

[14] Li CP, Dao XH, Guo P. Fractional derivatives in complex planes. Nonlinear Analysis: Theory Methods & Applications. 2009;**71**:5-6

[15] Li C, Gong Z, Qian D, Chen Y. On the bound of the Lyapunov exponents for the fractional differential systems. Chaos: An Interdisciplinary Journal of Nonlinear Science. 2010;**20**(1):013127

[16] Oldham KB, Spanier J. The Fractional Calculus. New York: Acad. Press; 1974

[17] Ortigueira MD. Comments on modeling fractional stochastic systems as non-random fractional dynamics driven Brownian motions. Applied Mathematical Modelling. 2009;**33**: 2534-2537

[18] Qian DL, Li CP, Agarwal RP, Wong PJY. Stability analysis of fractional differential system with Riemann-Liouville derivative. Mathematical and Computer Modelling. 2010;**52**:862-874

[19] West BJ, Bologna M, Grigolini P. Physics of Fractional Operators. New York: Springer; 2003

[20] Yang S, Xiao A, Su H. Convergence of the variational iteration method for solving multi-order fractional differential equations. Computers & Mathematics with Applications. 2010; **60**:2871-2879

[21] Ray SS, Bera RK. Solution of an extraordinary differential equation by adomian decomposition method. Journal of Applied Mathematics. 2004; **4**:331338

[22] Hashim I, Abdulaziz O, Momani S. Homotopy analysis method for fractional IVPs. Communications in Nonlinear Science and Numerical Simulation. 2009;**14**:674-684

[23] Bengochea G. Operational solution of fractional differential equations. Applied Mathematics Letters. 2014;**32**: 48-52

[24] Khalil H, Khan RA. The use of Jacobi polynomials in the numerical solution of coupled system of fractional differential equations. International Journal of Computer Mathematics. 2015;**92**(7): 1452-1472

[25] Doha EH, Bhrawy AH, Ezz-Eldien SS. Efficient Chebyshev spectral methods for solving multi-term fractional orders differential equations. Applied Mathematical Modelling. 2011; **35**:5662-5672

[26] Esmaeili S, Shamsi M, Luchko Y. Numerical solution of fractional differential equations with a collocation method based on Muntz polynomials. Computers & Mathematics with Applications. 2011; **62**:918-929

[27] Odibat Z, Momani S, Erturk VS. Generalized differential transform method an application to differential equations of fractional order. Applied Mathematics and Computation. 2008; **197**:467-477

[28] Baleanu D, Mustafa OG, Agarwal RP. An existence result for a superlinear fractional differential equation. Applied Mathematics Letters. 2010;**23**:1129-1132

[29] Baleanu D, Mustafa OG, Agarwal RP. On the solution set for a class of sequential fractional differential equations. Journal of Physics A. 2010; **43**:385-209

[30] Doha EH, Abd-Elhameed WM. Efficient solutions of multidimensional sixth-order boundary F value problems using symmetric generalized Jacobi-Galerkin method. Abstract and Applied Analysis. 2012;**2012**:12

[31] Bhrawy AH, Al-Shomrani MM. A Jacobi dual-Petrov Galerkin-Jacobi collocation method for solving Korteweg-de Vries equations. Abstract and Applied Analysis. 2012;**2012**:14

[32] Singh AK, Singh VK, Singh VK. The Bernstein operational matrix of integration. Applied Mathematical Sciences. 2009;**3**:2427-2436

[33] Bhrawy AH, Alofi AS, Ezz-Eldien SS. A quadrature tau method for fractional differential equations with variable coefficients. Applied Mathematics Letters. 2011;**24**:2146-2152

[34] Bhrawy AH, Mohammed MA. A shifted Legendre spectral method for fractional-order multi-point boundary value problems. Advances in Difference Equations. 2012;**2012**:8

[35] Khalil H, Khan RA. New operational matrix of integration and coupled system of Fredholm integral equations. Chinese Journal of Mathematics. 2014; **16**:12

[36] Khan RA, Khalil H. A new method based on Legendre polynomials for solution of system of fractional order partial differential equations.

International Journal of Computer Mathematics. 2014;**91**(12):2554-2567

[37] Khalil H, Khan RA. A new method based on Legendre polynomials for solutions of the fractional two-dimensional heat conduction equation. Computers & Mathematics with Applications. 2014;**67**:1938-1953

[38] Guo BY, Wang LL. Modified Laguerre pseudospectral method refined by multidomain Legendre pseudospectral approximation. Journal of Computational and Applied Mathematics. 2006;**190**:304-324

[39] Gulsu M, Gurbuz B, Ozturk Y, Sezer M. Laguerre polynomial approach for solving linear delay difference equations. Applied Mathematics and Computation. 2011;**217**:6765-6776

[40] Bhrawy AH, Taha TM, Machado JAT. A review of operational matrices and spectral techniques for fractional calculus. Nonlinear Dynamics. 2015;**81**(3):1023-1052

[41] Diethelm K, Ford NJ. Numerical solution of the Bagley Torvik equation. BIT Numerical Mathematics. 2002; **42**(1):490-500

[42] Akyuz-Dascioglu A, Isler N. Bernstein collocation method for solving nonlinear differential equations. Mathematical and Computational Applications. 2013;**18**:293-300

[43] Shah K. Using a numerical method by omitting discretization of data to study numerical solutions for boundary value problems of fractional order differential equations. Mathematical Methods in the Applied Sciences. 2019; **42**:6944-6959. DOI: 10.1002/mma.5800

Chapter 3

A Shamanskii-Like Accelerated Scheme for Nonlinear Systems of Equations

Ibrahim Mohammed Sulaiman, Mustafa Mamat and Umar Audu Omesa

Abstract

Newton-type methods with diagonal update to the Jacobian matrix are regarded as one most efficient and low memory scheme for system of nonlinear equations. One of the main advantages of these methods is solving nonlinear system of equations having singular Fréchet derivative at the root. In this chapter, we present a Jacobian approximation to the Shamanskii method, to obtain a convergent and accelerated scheme for systems of nonlinear equations. Precisely, we will focus on the efficiency of our proposed method and compare the performance with other existing methods. Numerical examples illustrate the efficiency and the theoretical analysis of the proposed methods.

Keywords: Newton method, Shamanskii method, diagonal updating scheme, nonlinear equations, Jacobian matrix

1. Introduction

A large aspect of scientific and management problems is often formulated by obtaining the values of x of which the function evaluation of that variable is equal to zero [1]. The above description can be represented mathematically by the following system of nonlinear equations:

$$\begin{aligned} f_1(x_1, x_2, \ldots, x_n) &= 0 \\ f_2(x_1, x_2, \ldots, x_n) &= 0 \\ \vdots \quad \vdots \quad \vdots &= \vdots \\ f_n(x_1, x_2, \ldots, x_n) &= 0 \end{aligned} \tag{1}$$

where $x_1, x_2, \ldots, x_n \in \mathbb{R}^n$ are vectors and f_i is nonlinear functions for $i = 1, 2, \ldots, n$. The above system of equations (1) can be written as

$$F(x) = 0 \tag{2}$$

where $F : R^n \to R^n$ is continuously differentiable in an open neighborhood of the solution x^*. These systems are seen as natural description of observed phenomenon of numerous real-life problems whose solutions are seen as an important goal in

 IntechOpen

mathematical study. Recently, this area has been studied extensively [2, 3]. The most powerful techniques for handling nonlinear systems of equations are to linearize the equations and proceed to iterate on the linearized set of equations until an accurate solution is obtained [4]. This can be achieved by obtaining the derivative or gradient of the equations. Various scholars stress that the derivatives should be obtained analytically rather than using numerical approach. However, this is usually not always convenient and, in most cases, not even possible as equations may be generated simply by a computer algorithm [2]. For one variable problem, system of nonlinear equation defined in (2) represents a function $F: R \to R$ where f is continuous in the interval $f \in [a, b]$.

Definition 1: Consider a system of equations $f_1, f_2, \dots, f_n$; the solution of this system in one variable, two variables, and n variable is referred to as a point $(a_1, a_2, \dots, a_n) \in R^n$ such that $f_1(a_1, a_2, \dots, a_n) = f_2(a_1, a_2, \dots, a_n) = \dots = f_n(a_1, a_2, \dots, a_n) = 0$.

In general, the problem to be considered is that for some function $f(x)$, one wishes to evaluate the derivative at some points x, i.e.,

$$\text{Given } f(x), \text{ Evaluate; deriv} = \frac{df}{dx}$$

This often used to represent an instantaneous change of the function at some given points [5].

Definition 2: For a function $f(x)$ that is smooth, then there exists, at any point x, a vector of first-order partial derivative or gradient vector:

$$\nabla f(x) = \begin{bmatrix} \frac{\partial f}{\partial x_1} \\ \frac{\partial f}{\partial x_2} \\ \cdot \\ \cdot \\ \cdot \\ \frac{\partial f}{\partial x_n} \end{bmatrix} = g(x).$$

The Taylor's series expansion of the function $f(x)$ about point x_0 is an ideal starting point for this discussion [1].

Definition 3: Let f be a differentiable function; the Taylor's $f(x)$ around a point a is the infinite sum:

$$f(x) = f(a) + f'(a)(x-a) + \frac{f''(a)}{2}(x-a)^2 + \frac{f'''(a)}{3!}(x-a)^3 + \dots$$

However, continuous differentiable vector valued function does not satisfy the mean value theorem (MVT), an essential tool in calculus [6]. Hence, academics suggested the use of the following theorem to replace the mean valued theorem.

Theorem 1: Let $F: \mathbb{R}^n \to \mathbb{R}^m$ be continuously differentiable in an open convex set $D \subset \mathbb{R}^n$. For any $x, x+s \in D$

$$F(x+s) - F(x) = \int_0^1 J(x+ts)s\,dt \equiv \int_x^{x+s} F'(z)dz$$

Definition 4: Suppose $F : R^n \to R^n$ is continuously differentiable at the point $x \in R^n$ and each component function $f_1, f_2, \ldots, f_m$ is also continuously differentiable at x; then the derivative of $F\,x$ is defined as

$$J_x(F) = \begin{pmatrix} \frac{\partial f_1}{\partial x_1} & \frac{\partial f_1}{\partial x_2} & \cdots & \frac{\partial f_1}{\partial x_m} \\ \frac{\partial f_2}{\partial x_1} & \frac{\partial f_2}{\partial x_2} & \cdots & \frac{\partial f_2}{\partial x_m} \\ \vdots & \vdots & \ddots & \vdots \\ \frac{\partial f_n}{\partial x_1} & \frac{\partial f_n}{\partial x_2} & \cdots & \frac{\partial f_n}{\partial x_m} \end{pmatrix}$$

Most of the algorithms employ for obtaining the solution of Eq. (1) centered on approximating the Jacobian matrix which often provides a linear map $T(x) : R^n \to R^n$ defined by Eq. (3)

$$T(x) = J_x F(x)\, \forall x \in R^n \tag{3}$$

Also, if F is differentiable at point x_*, then the affine function $A(x) = f(x_*) + J_*(x - x_*)$ is a good approximation to $F(x)$ near $x = x_*$ in such a way that

$$\lim_{x \to x_*} \frac{\|F(x) - F(x_*) - J_*(x - x_*)\|}{\|x - x_*\|} \tag{4}$$

where $\|x - x_*\| = \sqrt{(x_1 - x_*)^2 + (x_2 - x_*)^2 + \ldots + (x_n - x_*)^2}$.

If all the given component functions $f_1, f_2, \ldots, f_m$ of $J_x(F)$ are continuous, then we say the function F is differentiable.

The most famous method for solving nonlinear systems of equations $F(x) = 0$ is the Newton method which generates a sequence $\{x_k\}$ from any given initial point x_0 via the following:

$$x_{k+1} = x_k - F'(x_k)^{-1} F(x_k) \tag{5}$$

where $F'(x_k)$ is the Jacobian for $F(x_k)$. The above sequence Eq. (5) is said to converge quadratically to the solution x^* if x_0 is sufficiently near the solution point and the Jacobian $F'(x_k)$ is nonsingular [7, 8]. This convergent rate makes the method outstanding among other numerical methods. However, Jacobian evaluation and solving the linear system for the step $s(x_n) = -F'(x_n)^{-1} F(x_n)$ are expensive and time-consuming [9]. This led to the study of different variants of Newton methods for systems of nonlinear equations. One of the simplest and low-cost variants of the Newton method that almost entirely evades derivate evaluation at every iteration is the chord method. This scheme computes the Jacobian matrix $F'(x_0)$ once throughout the iteration process for finite dimensional problem as presented in Eq. (6),

$$x_{k+1} = x_k - F'(x_0)^{-1} F(x_k) \tag{6}$$

The rate of convergence is linear and improves as the initial point gets better. Suppose x_0 is sufficiently chosen near solution point x_* and $F(x^*)$ is nonsingular; then, for some $K_c > 0$, we have

$$\|x_{n+1} - x^*\| \le K_c \|x_0 - x^*\| \|x_n - x^*\| \tag{7}$$

The convergence theorems and proof of Eq. (7) can be referred to [9, 10]. Motivated by the excellent convergence of Newton method and low cost of Jacobian evaluation of chord method, a method due originally to Shamanskii [11, 12] that lies between Newton method and chord method was proposed and has been analyzed in Kelly [9, 13–15]. Other variants of Newton methods with different Jacobian approximation schemes include [9, 14, 16–18]. However, most of these methods require the computation and storage of the full or approximate Jacobian, which become very difficult and time-consuming as the dimension of systems increases [10, 19].

It would be worthwhile to construct a derivative-free approach and analyze with existing techniques [20–22]. The aim of this work is to derive a diagonal matrix for the approximate Jacobian of Shamanskii method by means of variational techniques. The expectation would be to reduce computational cost, storage, and CPU time of evaluating any problem. The proposed method works efficiently by combining the good convergence rate of Shamanskii method and the derivate free approach employed, and the results are very encouraging. The next section presents the Shamanskii method for nonlinear systems of equations.

2. Shamanskii method

It is known that the Newton method defined in Eq. (2) converges quadratically to x^* when the initial guess is sufficiently close to the root [7, 10, 19]. The major concern about this method is the evaluation and storage of the Jacobian matrix at every iteration [23]. A scheme that almost completely overcomes this is the chord method. This method factored the Jacobian matrix only once in the case of a finite dimensional problem, thereby reducing the evaluation cost of each iteration as in Eq. (3) and thereby degrading the convergence rate to linear [10].

Motivated by this, a method due originally to Shamanskii [11] was developed and analyzed by [7, 13, 14, 16, 24]. Starting with an initial approximation x_0, this method uses the multiple pseudo-Newton approach as described below:

$$x_{k+\frac{1}{2}} = x_k - F'(x_k)^{-1}F(x_k) \tag{8}$$

$$x_{k+1} = x_{k+\frac{1}{2}} - F'(x_k)^{-1}F\left(x_{k+\frac{1}{2}}\right) \tag{9}$$

after little simplification, we have

$$x_{k+1} = x_k - F'(x_k)^{-1}\left[F(x_k) + F\left(x_k - F'(x_k)^{-1}F(x_k)\right)\right] \tag{10}$$

This method converges superlinearly with q-order of at least $t+1$ when the initial approximation x_0 is sufficiently chosen near the solution point x^* and $F'(x^*)$ is nonsingular. This implies that there exists $K_s > 0$, such that

$$\|x_{n+1} - x^*\| \le K_s \|x_n - x^*\|^{t+1} \tag{11}$$

Combining Eq. (7) and the quadratic convergence of Newton method produces the convergence rate of the Shamanskii method as in Eq. (8). Thus, the balance is between the reduced evaluation cost of Fréchet derivative and Jacobian computation for Shamanskii method and Newton method rapid convergence. This low-cost derivative evaluation as well as the rapid convergence rate of several methods including the Shamanskii method has been studied and analyzed in [13, 15]. From the analysis, the researchers conclude that that Shamanskii method has shown superior performance compared to Newton method in terms of efficiency whenever work is measured in terms of function evaluations [9]. Also, if the value of t is sufficiently chosen, then, as the dimension increases, the performance of the Shamanskii method improves and thus reduces the limit of complexity of factoring the approximate Jacobian for two pseudo-Newton iterations [14]. Please refer to [15] for the proof of the convergence theorem described below.

Theorem 2 [15]: Let $F : D \subset R^n \to R^n$ conform hypotheses H1(2), H2, and H3. Then the solution point x^* is a point of attraction of the Shamanskii iteration, i.e., Eq. (10), and this method possesses at least cubic order of convergence.

3. Diagonal updating scheme for solving nonlinear systems

Evaluation or inversion of the Jacobian matrix at every iteration or after few iterations does not seem relevant even though the computational cost has generally been reduced as in Shamanskii method [14, 25–28]. As a matter of fact, it can be easily shown that by adding a diagonal updating scheme to a method, we would have a new low memory iterative approach which would approximate the Jacobian $F'(x_k)$ into nonsingular diagonal matrix that can be updated in every iteration [29–31]. Indeed, using the Shamanskii procedure, the proposed method avoids the main complexity of the Newton-type methods by reusing the evaluated Jacobian during the iteration process. This is the basic idea of the Shamanskii-like method which is described as follows.

Given an initial approximation x_0, we compute Eq. (2) to obtain the Jacobian $F'(x_k)$ and present a diagonal approximation to the Jacobian say D_k as follows:

$$F'(x_{k+1}) \approx D_{k+1} \tag{12}$$

Suppose $s_k = x_{k+1} - x_k$ and $y_k = F(x_{k+1}) - F(x_k)$; by mean value theorem (MVT), we have

$$D_{k+1}s_k \approx y_k \tag{13}$$

Substituting Eq. (12) in Eq. (13), we have

$$F'(x_k)s_k \approx y_k \tag{14}$$

Since D_{k+1} is the update of diagonal matrix D_k, let us assume D_{k+1} satisfy the weak secant equation:

$$s_k^T D_{k+1} s_k = s_k^T y_k \tag{15}$$

which would be used to minimize the deviation between D_{k+1} and D_k under some norms. The updated formula for D_k follows after the theorem below:

Theorem 3: Suppose D_{k+1} is the update of the diagonal matrix D_k and $\Delta_k = D_{k+1} - D_k$, $s_k \neq 0$. Consider the problem

$$\min \frac{1}{2} \|\Delta_k\|_F^2 \tag{16}$$

such that Eq. (15) holds and $\|.\|_F$ denotes the Frobenius norm. From Eq. (16), we have the following solution also regarded as the optimal solution:

$$\Delta_k = \frac{s_k^T y_k - s_k^T D_{k+1} s_k}{tr(\Omega_k^2)} \Omega_k \tag{17}$$

where $\Omega_k = \text{diag}\left(\left(s_k^{(1)}\right)^2, , \left(s_k^{(2)}\right)^2, \ldots, , \left(s_k^{(n)}\right)^2\right)$, $\sum_{i=1}^n \left(s_k^{(i)}\right)^4 = tr\left(\Omega_k^2\right)$, and Tr is the trace operation.

Proof: It is known that the objective function and the constraint of Eq. (16) are convex; thus, we intend to use its Lagrangian function to obtain the unique solution as follows:

$$\mathrm{L}(\Delta_k, \mu) = \frac{1}{2}\|\Delta_k\|_F^2 + \mu\left(s_k^T \Delta_k s_k - s_k^T y_k - s_k^T D_{k+1} s_k\right) \tag{18}$$

where μ is the corresponding Lagrangian multiplier. Simplifying Eq. (18), we have

$$\mu = \frac{s_k^T y_k - s_k^T D_{k+1} s_k}{\sum_{i=1}^n \left(s_k^{(i)}\right)^4} \tag{19}$$

and

$$\Delta_k^{(i)} = \frac{s_k^T y_k - s_k^T D_{k+1} s_k}{\sum_{i=1}^n \left(s_k^{(i)}\right)^4}\left(s_k^{(i)}\right)^2 \quad \forall i = 1, 2, .., n \tag{20}$$

Also, for diagonal matrix D_k, the element of the diagonal component is given as $D_k^{(i)}$, and the ith component of the vector s_k is $s_k^{(i)}$. Then $\Omega_k = \text{diag}\left(\left(s_k^{(1)}\right)^2, , \left(s_k^{(2)}\right)^2, \ldots, , \left(s_k^{(n)}\right)^2\right)$, and $\sum_{i=1}^n \left(s_k^{(i)}\right)^4 = tr\left(\Omega_k^2\right)$. To complete the proof, we rewrite Eq. (20) as follows:

$$\Delta_k = \frac{\left(s_k^T y_k - s_k^T D_{k+1} s_k\right)}{tr\left(\Omega_k^2\right)}\Omega_k. \tag{21}$$

This completes the proof. ∎

Now, from the above description of the theorem, we deduce that the best possible diagonal update D_{k+1} is as follows:

$$\mathrm{D}_{k+1} = \mathrm{D}_k + \frac{\left(s_k^T y_k - s_k^T D_{k+1} s_k\right)}{tr\left(\Omega_k^2\right)}\Omega_k \tag{22}$$

However, for possibly small $\|s_k\|$ and $tr\Omega_k$, we need to define a condition that would be applied for such cases. To this end, we require that $\|s_k\| \geq s_1$ for some chosen small $s_1 > 0$. Otherwise, we set the updated diagonal $\mathrm{D}_{k+1} = \mathrm{D}_k$ where D_{k+1} is defined as

$$\mathrm{D}_{k+1} = \begin{cases} \mathrm{D}_k + \dfrac{\left(s_k^T y_k - s_k^T D_{k+1} s_k\right)}{tr\left(\Omega_k^2\right)}\Omega_k; \ \|s_k\| \geq \epsilon_1 \\ D_k; \ Otherwise \end{cases} \tag{23}$$

Thus, the proposed accelerated method is described as follows:

$$x_{k+1} = x_k - D_k^{-1}\left[F(x_k) + F\left(x_k - D_k^{-1}F(x_k)\right)\right] \tag{24}$$

The performance of this proposed method would be tested on well-known benchmark problems employed by researchers on existing methods. This would be

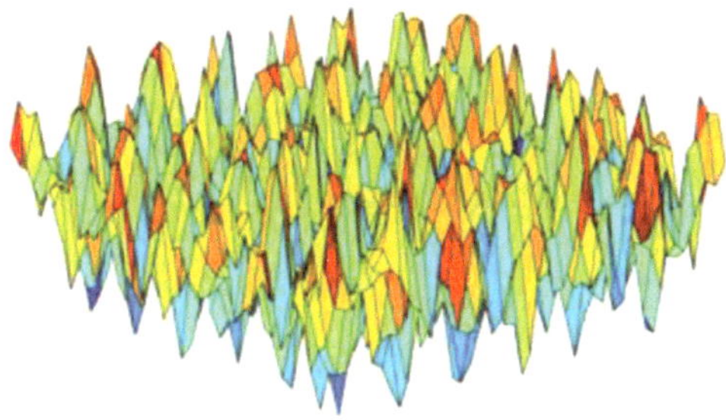

Figure 1.
Functions with a huge number of significant local optima.

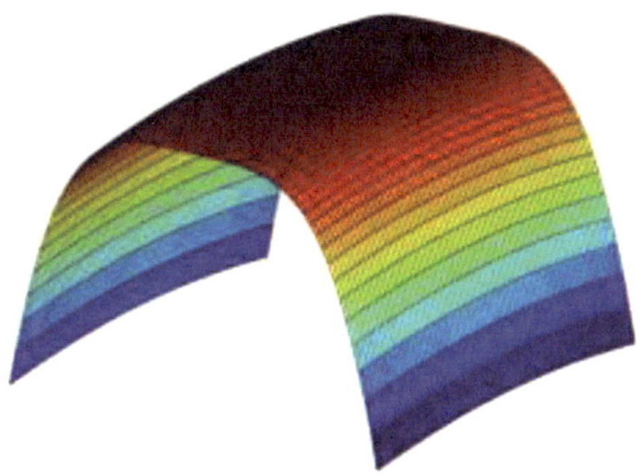

Figure 2.
Functions with significant null space.

Figure 3.
Essentially unimodal function.

carried out using a designed computer code for its algorithm. The problems could be artificial or real-life problems. The artificial problems check the performance of any algorithm in situation such as point of singularity, function with many solutions, and null space effect as presented in **Figures 1–3** [7, 32].

While the real-life problems emerge from fields such as chemistry, engineering, management, etc., the real-life problems often contain large data or complex algebraic expression which makes it difficult to solve.

4. Numerical results

This section demonstrates the proposed method and illustrates its advantages on some benchmark problems with dimensions ranging from 25 to 1,000 variables.

These include problems with restrictions such as singular Jacobian or problems with only one point of singularity. To evaluate the performance of the proposed diagonal updating Shamanskii method (DUSM), we employ some tools by Dolan and Moré [33] and compare the performance with two classical Newton-type methods based on the number of iterations and CPU time in seconds. The methods include:

1. The Newton method (NM)

2. The Shamanskii method (SM)

These tools are used to represent the efficiency, robustness, and numerical comparisons of different algorithms. Suppose there exist n_s solvers and n_p problems; for each problem p and solver s, they define:

$t_{p,s}$ = computing time needed to solve a problem by solver (the number of iteration or CPU time)

Requiring a baseline for comparisons, they compared the performance on problem p by solver s with the best performance by any solver for this problem using the performance ratio:

$$r_{p,s} = \frac{t_{p,s}}{\min\{t_{p,s} : s \in S\}}$$

We suppose that parameter $r_m \geq r_{p,s}$ for all p, s is chosen and $r_{p,s} = r_M$ if and only if solver s does not solve problem p. The performance of solvers s on any given problem might be of interest, but because we would prefer obtaining the overall assessment of the performance of the solver, then it was defined as

$$p_s(t) = \frac{1}{n_p} size\{p \in P : r_{p,s} \leq t\}.$$

Thus, $p_s(t)$ was the probability for solver $s \in S$ that a performance ratio $r_{p,s}$ was within a factor $t \in R$ of the best possible ratio. Then, function p_s was the cumulative distribution function for the performance ratio. The performance profile $p_s : R \rightarrow [0, 1]$ for a solver was nondecreasing, piecewise, and continuous from right. The value of $p_s(1)$ is the probability that the solver will win over the rest of the solvers. In general, a solver with high value of $p(\tau)$ or at the top right of the figure is preferable or represents the best solver.

All problems considered in this research are solved using MATLAB (R 2015a) subroutine programming [37]. This was run on an Intel® Core™ i5-2410M CPU @ 2.30 GHz processor, 4GB for RAM memory and Windows 7 Professional operating system. The termination condition is set as

$$\|s_k\| + \|F(x_k)\| \leq 10^{-6}$$

and the program has been designed to terminate whenever:

- The number of iterations exceeds 500, and no point of x_k satisfies the termination condition.

- The CPU time in seconds reaches 500.

- Insufficient memory to initiate the run.

At the point of failure due to any of the above conditions as in the tabulated results, it is assumed the number of iteration and CPU time is zero and thus that point has been denoted by " $*$." The following are the details of the standard test problems, the initial points used, and the exact solutions for systems of nonlinear equations.

Problem 1 [31]: System of n nonlinear equations

$$F_i(x) = \left(1 - x_i^2\right) + x_i(1 + x_i x_{n-2} x_{n-1} x_n) - 2$$
$$i = 1, 2, 3, \ldots, n, \qquad x_0 = (0.3, 0, 3, \ldots, 0.3)$$

Problem 2 [34]: Systems of n nonlinear equations

$$F_i(x) = x_i^2 - \cos(x_i - 1)$$
$$i = 1, 2, 3, \ldots, n, \qquad x_0 = (0.2, 0.2, \ldots, 0.2)$$

Problem 3 [31]: Structured exponential function

$$F_i(x) = e^{x_i} - 1$$
$$F_n(x) = x_n - 0.1x_n^2$$
$$i = 1, 2, 3, \ldots, n, \quad x_0 = (0.05, 0.05, \ldots, 0.05)$$

Problem	Dim	NM		DUSM		SM	
		NI	CPU	NI	CPU	NI	CPU
1	25	13	0.016102	8	0.034777	13	0.015999
2	25	6	0.009522	7	0.028231	7	0.010412
3	25	*	*	4	0.023766	*	*
4	25	16	0.019679	17	0.077072	22	0.022889
5	25	4	0.006605	16	0.061750	4	0.005761
1	50	13	0.032998	8	0.090310	13	0.032271
2	50	10	0.022134	7	0.089785	7	0.017036
3	50	4	0.010350	4	0.052899	4	0.010238
4	50	30	0.054640	17	0.228077	23	0.041569
5	50	4	0.012361	16	0.201262	4	0.010735
1	100	13	0.073565	8	0.339333	13	0.066363
2	100	10	0.054075	7	0.292001	7	0.044512
3	100	*	*	4	0.175300	*	*
4	100	15	0.075073	18	0.770165	25	0.118170
5	100	4	0.029221	17	0.755556	4	0.023154
1	1000	13	1.868606	8	27.171776	13	2.042222
2	1000	10	1.444533	7	24.295632	7	1.045329
3	1000	*	*	4	27.1250	*	*
4	1000	52	6.757533	19	63.981376	39	5.138997
5	1000	4	0.610145	18	62.364143	4	0.612590

Table 1.
Numerical comparison of NM, DUSM, and SM.

Problem 4 [35]: Extended trigonometric of Byeong-Chun

$$F_i(x) = \cos\left(x_i^2 - 1\right) - 1$$
$$i = 1, 2, 3, \dots, n, \qquad x_0 = (0.06, 0.06, \dots, 0.06)$$

Problem 5 [36]: Extended spare system of Byeong

$$F_i(x) = x_i^2 - x_i - 2$$
$$i = 1, 2, 3, \dots, n, \qquad x_0 = (1.1, 11.1, \dots, 1.1)$$

Table 1 shows the number of iterations (NI) and CPU time for Newton method (NM), Shamanskii method (SM), and the proposed diagonal updating method (DUSM), respectively. The performance of these methods was analyzed via storage locations and execution time. It can be observed that the proposed DUSM was able to solve the test problems perfectly, while NM and SM fail at some points due to the matrix being singular to working precision. This shows that the diagonal scheme employed has provided an option in the case of singularity, thereby reducing the computational cost of the classical Newton-type methods.

5. Conclusion

This chapter proposes a diagonal updating formula for systems of nonlinear equations which attributes to reduction in Jacobian evaluation cost. By computational experiments, we reach the conclusion that the proposed scheme is reliable and efficient and reduces Jacobian computational cost during the iteration process. Meanwhile, the proposed scheme is superior compared to the result of the classical and existing numerical methods for solving systems of equations.

Author details

Ibrahim Mohammed Sulaiman*, Mustafa Mamat and Umar Audu Omesa
Universiti Sultan Zainal Abidin, Kuala Terengganu, Malaysia

*Address all correspondence to: sulaimanib@unisza.edu.my

References

[1] Rainer K. Numerical Analysis. New York: Springer Science+Business Media, LLC; 1998

[2] Sulaiman IM. New iterative methods for solving fuzzy and dual fuzzy nonlinear equations [PhD thesis]. Malaysia: Faculty of Informatics and Computing, Universiti Sultan Zainal Abidin; 2018

[3] Wenyu S, Ya-Xiang Y. Optimization Theory and Methods, Springer Optimization and Its Applications. Boston, MA: Springer; 2006

[4] John RH. Numerical Methods for Nonlinear Engineering Models. Netherlands: Springer; 2009

[5] Burden RL, Faires JD. Numerical Analysis. 8th ed. USA: Thomson; 2005

[6] Wright SJ, Nocedal J. Numerical Optimization. 2nd ed. Berlin, Germany: Springer; 1999

[7] Dennis JE Jr, Schnabel RB. Numerical Method for Unconstrained Optimization and Nonlinear Equations. Houston, Texas: SIAM; 1996

[8] Griva I, Nash SG, Sofer A. Linear and Nonlinear Optimization. 2nd ed. Philadelphia: SIAM; 2009

[9] Kelley CT. A Shamanskii-like acceleration scheme for nonlinear equations at singular roots. Mathematics of Computation. 1986;**47**:609-623

[10] Kelley CT. Iterative Methods for Linear and Nonlinear Equations. Philadelphia, PA: SIAM; 1995

[11] Shamanskii VE. A modification of Newton's method. Ukrains'kyi Matematychnyi Zhurnal. 1967;**19**:133-138

[12] Shamanskii VE. On a realization of Newton's method on electronic computers. Ukrains'kyi Matematychnyi Zhurnal. 1966;**18**(6):135-140

[13] Traub JF. Iterative Methods for the Solution of Equations. Englewood Cliffs: Prentice-Hall; 1964

[14] Kchouk B, Dussault J. The Chebyshev–Shamanskii method for solving systems of nonlinear equations. Journal of Optimization Theory and Applications. 2013;**157**:148-167

[15] Ortega JM, Rheinboldt WC. Iterative Solution of Nonlinear Equations in Several Variables. New York: Academic Press; 1970

[16] Brent RP. Some efficient algorithms for solving systems of nonlinear equations. Journal of Nucleic Acids. 1973;**10**(2):327-344

[17] Waziri MY, Leong WJ, Mamat M, Moyi AU. Two-step derivative-free diagonally Newton's method for large-scale nonlinear equations. World Applied Sciences Journal. 2013;**21**:86-94. DOI: 10.5829/idosi.wasj.2013.21.am.2045

[18] Broyden CG. A class of methods for solving nonlinear simultaneous equations. Mathematics of Computation. 1965;**19**(92):577-593

[19] Chong EKP, Zak SH. An Introduction to Optimization, Wiley Series in Discrete Mathematics and Optimization. New York: John Wiley and Sons; 2013

[20] Jose LH, Eulalia M, Juan RM. Modified Newton's method for systems of nonlinear equations with singular Jacobian. Journal of Computational and Applied Mathematics. 2009;**224**:77-83

[21] Leong WJ, Hassan MA, Waziri MY. A matrix-free quasi-Newton method for solving large-scale nonlinear systems.

Computational and Applied Mathematics. 2011;**625**:2354-2363

[22] Natasa K, Zorana L. Newton-like methods with modification of the right-hand side vector. Mathematics of Computation. 2002;**71**:237-250

[23] Waziri MY, Leong WJ, Hassan MA, Monsi M. A new Newton's method with diagonal Jacobian approximation for systems of nonlinear equations. Journal of Mathematics and Statistics. 2010;**6** (3):246-252

[24] Lampariello F, Sciandrone M. Global convergence technique for the Newton method with periodic Hessian evaluation. Journal of Optimization Theory and Applications. 2001;**111**(2): 341-358

[25] Sulaiman IM, Mamat M, Nurnadiah Z, Puspa LG. Solving dual fuzzy nonlinear equations via Shamanskii method. International Journal of Engineering & Technology. 2018;**7** (3.28):89-91

[26] Ypma TJ. Historical development of the Newton-Raphson method. SIAM Review. 1995;**37**(4):531-551

[27] Hao L, Qin N. Incomplete Jacobian Newton method for nonlinear equation. Computers and Mathematics with Applications. 2008;**56**(1):218-227

[28] Chency E, Kincaid D. Numerical Mathematics and Computing. Asia: Nelson Education, Cengage Learning; 2012

[29] Sulaiman IM, Mamat M, Afendee MM, Waziri MY. Diagonal updating Shamanskii-like method for solving singular fuzzy nonlinear equation. Far East Journal of Mathematical Sciences. 2017;**103**(10):1619-1629

[30] Waziri MY, Leong WJ, Hassan MA, Monsi M. Jacobian-free Newton's method for systems of nonlinear equations. Journal of Numerical Mathematics and Stochastics. 2010;**2**(1): 54-63

[31] Waziri MY, Abdulmajid Z. An improved diagonal Jacobian approximation via a quasi-Cauchy condition for solving large-scale systems of nonlinear equations. Journal of Applied Mathematics. 2013;**3**:1-6

[32] Andrei N. An unconstrained optimization test functions collection. Advanced Modeling and Optimization. 2008;**10**:147-161

[33] Dolan E, Moré JJ. Benchmarking optimization software with performance profiles. Mathematical Programming. 2002;**91**(2):201-213

[34] Hafiz MA, Muhammad SMB. An efficient two-step iterative method for solving system of nonlinear equation. Journal of Mathematics Research. 2012; **4**(4):28-34

[35] Shin BC, Darvishi M, Kim CH. A comparison of the Newton-Krylov method with high order newton-like methods to solve nonlinear systems. Applied Mathematics and Computation. 2010;**217**(7):3190-3198

[36] Mamat M, Muhammad K, Waziri MY. Trapezoidal Broyden's method for systems of nonlinear equations. Applied Mathematical Sciences. 2014;**8**(6):54-63

[37] Sulaiman I, Mamat M, Waziri MY, Umar AO, et al. Journal of Advanced Research in Modelling and Simulations. 2018;**1**(1):13-18

Chapter 4

Modified Moving Least Squares Method for Two-Dimensional Linear and Nonlinear Systems of Integral Equations

Massoumeh Poura'bd Rokn Saraei and Mashaallah Matinfar

Abstract

This work aims at focusing on modifying the moving least squares (MMLS) methods for solving two-dimensional linear and nonlinear systems of integral equations and system of differential equations. The modified shape function is our main aim, so for computing the shape function based on the moving least squares method (MLS), an efficient algorithm is presented. In this modification, additional terms is proposed to impose based on the coefficients of the polynomial base functions on the quadratic base functions ($m = 2$). So, the MMLS method is developed for solving the systems of two-dimensional linear and nonlinear integral equations at irregularly distributed nodes. This approach prevents the singular moment matrix in the context of MLS based on meshfree methods. Also, determining the best radius of the support domain of a field node is an open problem for MLS-based methods. Therefore, the next important thing is that the MMLS algorithm can automatically find the best neighborhood radius for each node. Then, numerical examples are presented that determine the main motivators for doing this so. These examples enable us to make comparisons of two methods: MMLS and classical MLS.

Keywords: moving least squares, modified moving least squares, systems of integral equations, algorithm of shape function, numerical solutions

MSC 2010: 45G15, 45F05,45F35, 65D15

1. Introduction

In mathematics, there are many functional equations of the description of a real system in the natural sciences (such as physics, biology, Earth science, meteorology) and disciplines of engineering. For instance, we can point to some mathematical model from physics that describe heat as a partial differential equation and the inverse problem of it's as integro-differential equations. Also, another example in nature is Laplace's equation which corresponds to the construction of potential for a vector field whose effect is known at the boundary of Domain alone. Especially, the integral equations have wide applicability which has been cited in [1–4].

However, there are many significant analytical methods for solving integral equations but most of them especially in nonlinear cases, finding an analytical representation of the solution is so difficult, therefore, it is required to obtain approximate solutions. The interested reader can find several numerical methods for approximating the solution of these problems in [5–14] and the references therein.

Moreover, there are various numerical and analytical methods have been used to estimate the solution of integrodifferential equations or Abels integral equations [12, 15–18]. Recently the meshless based methods, particularly Moving Least Squares (MLS) method, for a solution of partial differential equations and ordinary differential equations have been paid attention. Using this approach some new methods such as meshless local boundary integral equation method [19], Boundary Node Method (BNM) [20], moving least square reproducing polynomial meshless method [21] and other relative methods are constructed. The new class of meshless methods has been developed which only relied on a set of nodes without the need for an additional mesh in the solution of a one-dimensional system of integral equations [22].

A local approximation of unknown function presented in the MLS method give us to possible choose the compact support domain for each data point as a sphere or a parallelogram box centered on a point [23, 24]. So each data point has an associated with the size of its compact support domain as dilatation parameter. Therefore the number of data point and dilatation parameter are direct effects on the MLS, Also by increasing the degree of the polynomial base function for complex data distributions give a more validated fashion. Nevertheless, in this case, it becomes more difficult to ensure the independence of the shape functions, and the least-squares minimization problem becomes ill-posed.

The common solution for increased the number of admissible node distribution is increasing the size of the support domains (a valid node distribution is referred to as an œadmissible node distribution [23]). There have been several proposed for choosing the radius of support domain [25], but one of the efficient suggestion was raised by Chen shen [26]. The author in [27] has introduced a new algorithm for selecting the suitable radius of the domain of influence. Also in [28], presented a modified MLS(MMLS) approximation on the shape function generation algorithm with additional terms based on the coefficients of the polynomial basis functions. It is an efficient method which has been proposed for handling a singular moment matrix in the MLS based methods. The advantage of this method compared to methods based on mesh such as a finite element or finite volume is this the domain of the problem is not important because this approximation method is based on a set of scattered points instead of domain elements for interpolation or approximation. So the geometry of the domain does not interfere in the MLS.

2. Methodology

2.1 Introduction of the MLS approximation

The Moving Least Square (MLS) method is a feasible numerical approximation method that is an extension of the least squares method, also it is the component of the class of meshless schemes that have a highly accurate approximation. The MlS approximation method is a popular method used in the many meshless methods [12, 19, 21, 22, 29, 30]. In many procedures used to construct the MLS shape function is used support-domain concept. The support domain of the shape function is a set of nodes in the problem domain that just those points directly contributes to the construction of the shape function, so the MLS shape function is locally supported. According to the classical least squares method, an optimization problem should be solved as follows

$$min\left(\sum_{j=1}^{m}\left(u^h(\mathbf{x}_j)-u(\mathbf{x}_j)\right)^2\right)$$

Where Ideally the approximation function $u^h(x)$ should match the function $u(\mathbf{x})$. Therefore, in the MLS approach, a weight optimization problem will be solved which is dependent on nodal points. We start, with the basic idea of taking a set of the nodal points in Ω so that $\Omega \subseteq \mathbb{R}^d$. Also $\Omega_{\mathbf{x}} \subseteq \Omega$ is neighboring nodes of point $\mathbf{x}$ and finding an approximation function with m basis functions, in a system with n equations as

$$T(U)=F$$

where T consists of linear and nonlinear operators and $U=(u_1,u_2,\dots,u_n)$ is the unknown vector of functions, also $F=\left(f_1,f_2,\dots,f_n\right)$ is the known vector of functions.

So for the approximation of any of the $u_i, i=1,2,\dots,n$ in $\Omega_{\mathbf{x}}$, $\forall x \in \Omega_{\mathbf{x}}$, $u_i^h(\mathbf{x})$ can be defined as

$$u_i^h(\mathbf{x})=\sum_{j=1}^{m}a_j(\mathbf{x})p_j(\mathbf{x})=P^T(\mathbf{x})a(\mathbf{x}). \tag{1}$$

Let $\mathbf{P}=\{p_1,p_2,\dots p_m\}$ a set of polynomial of degree at most $m, m\in\mathbb{N}$. Let $\mathbf{a}(\mathbf{x})$ is a vector containing unknown coefficients $a_j(\mathbf{x}), j=1,2,\dots m$ dependent on the intrest point position. Also m unknown functions $\mathbf{a}(\mathbf{x})=(a_1(\mathbf{x}),a_2(x),\dots a_m(\mathbf{x}))$ are chosen such that:

$$J(\mathbf{x})=\sum_{j=1}^{m}\left(\mathbf{P}^T(\mathbf{x}_j)\mathbf{a}(\mathbf{x})-u_i(\mathbf{x}_j)\right)^2 w_i(\mathbf{x})=[P.\mathbf{a}-\mathbf{u}_i]^T.W.[P.\mathbf{a}-\mathbf{u}_i], \tag{2}$$

is minimized. Note that the weight function $w_i(\mathbf{x})$ is associated with node j. As we know, each redial basis function that define in [31] can be used as weight function, we can define $w_j(r)=\phi\left(\frac{r}{\delta}\right)$ where $r=\|\mathbf{x}-\mathbf{x}_i\|_2$ (the Euclidean distance between $\mathbf{x}$ and $\mathbf{x}_j$) and $\phi:\mathbb{R}^d\to\mathbb{R}$ is a nonnegative function with compact support. In this chapter, we will use following weight functions and will compare them to each other, corresponding to the node j, in the numerical examples.

a. Guass weight function

$$w(r)=\begin{cases}\dfrac{\exp\left(\dfrac{-r^2}{c^2}\right)-\exp\left(\dfrac{-\delta^2}{c^2}\right)}{1-\exp\left(\dfrac{-\delta^2}{c^2}\right)} & 0\le r\le\delta \\ 0 & elsewhere.\end{cases} \tag{3}$$

b. RBF weight function

$$w(r) = \begin{cases} (1-r)^6(6+36r+82r^2+72r^3+30r^4+5r^5) & 0 \le r \le \delta \\ 0 & elsewhere. \end{cases} \tag{4}$$

c. Spline weight function

$$w(r) = \begin{cases} 1-6\left(\frac{r}{\delta}\right)^2+8\left(\frac{r}{\delta}\right)^3-3\left(\frac{r}{\delta}\right)^4 & 0 \le r \le \delta \\ 0 & elsewhere. \end{cases} \tag{5}$$

Where c is constant and is called shape parameter. Also δ is the size of support domain.

N is the number of nodes in $\Omega_{\mathbf{x}}$ with $w_i(x) > 0$, the matrices P and W are defined as

$$P = \left[\mathbf{p}^T(\mathbf{x}_1), \mathbf{p}^T(\mathbf{x}_2), \ldots \mathbf{p}^T(\mathbf{x}_N)\right]^T_{N\times(m+1)} \tag{6}$$

$$W = diag((w_i(\mathbf{x})), i = 1, 2, \ldots, N \tag{7}$$

and

$$\mathbf{u}^h = \left[u_1^h, u_2^h, \ldots u_n^h\right]. \tag{8}$$

It is important to note that $u_i^T, i = 1, 2, \ldots n$, in (2) and (8) are the artificial nodal values, and not the nodal values of the unknown trial function $u^h(\mathbf{x})$ in general. With respect to $\mathbf{a}(\mathbf{x})$ and u_i^T will be obtained,

$$A(\mathbf{x})a(\mathbf{x}) = B(\mathbf{x})\mathbf{u}_i, \tag{9}$$

where the matrices $A(\mathbf{x})$ and $B(\mathbf{x})$ are defined by:

$$B(\mathbf{x}) = [w_1\mathbf{p}(\mathbf{x}_1), w_2\mathbf{p}(\mathbf{x}_2), \ldots, w_N\mathbf{p}(\mathbf{x}_N)] \tag{10}$$

$$A(\mathbf{x}) = \sum_{i=1}^{N} w_i(\mathbf{x})\mathbf{p}^T(\mathbf{x}_i)\mathbf{p}(\mathbf{x}_i) = \mathbf{p}^T(\mathbf{x})w(\mathbf{x})\mathbf{p}(\mathbf{x}). \tag{11}$$

The matrix $A(\mathbf{x})$ in (11) is non-singular when the rank of matrix $P(\mathbf{x})$ equals to m and vice versa. In such a case, the MLS approximation is well-defined. With computing $\mathbf{a}(\mathbf{x}), u_i^h$ can be obtained as follows:

$$u_i^h(\mathbf{x}) = \sum_{j=1}^{N} \phi_j(\mathbf{x})u_i(\mathbf{x}_j) = \varphi^T.\mathbf{u}_i \tag{12}$$

$\phi_j(\mathbf{x})$ is called the shape function of the MLS approximation corresponding to the nodal point $\mathbf{x}_j$, where

$$\varphi(\mathbf{x}) = \mathbf{p}^T(\mathbf{x})A^{-1}(\mathbf{x})B(\mathbf{x}) \tag{13}$$

Also with use the weight function, matrix A, B are computed and then $\phi_i(x)$ is determined from (13), If, further, ϕ is sufficiently smooth, derivatives of U are usually approximated by derivatives of U^h,

$$D^{\alpha}u_i\mathbf{x} \approx D^{\alpha}u_i^h(\mathbf{x}) = \sum_{j=1}^{N} D^{\alpha}a_j(\mathbf{x})u_i(\mathbf{x}_j), x \in \Omega \tag{14}$$

2.2 Modify algorithm of MLS shape function

In the MLS approximation method, a local evaluation of the approximating unknown function is desired, and therefore for any nodal points the compact support domain is chosen as a sphere or a parallelogram box centered on the point [23, 29, 32]. This finding which the support domains contain what points. Each data point has a connected dilatation parameter λ which is given $\delta_i = \lambda h_i$. Also, δ_i is the size of compact support domain in a node point $\mathbf{x}_i$.

Also, the necessary condition for that the moment matrix A be nonsingular is that for any point $x_i \in \Omega, i = 1, 2, \dots, N$, [31].

$$\aleph\left(\{j | x_j \in \Omega_{\delta_i}\}\right) \geq m, \; j = 1, 2, \dots, N$$

So the dilatation parameters λ determine the number of points of support domain, Also these points participate in the production of the shape function Therefore, λ is very important and should be chosencorrectly so that the moment matrix A is nonsingular.

In the remainder of this section, we introduce the new algorithm, with the aim of avoiding the singularity of the matrix A by choosing the correct λ parameter by the algorithm.

Algorithm 1

Require: $X = \{x_i : i = 1, 2, \dots, N\}$- Coordinates of points whose MLS shape function to be evaluated.

```
procedure MATRIX A
  λ_new ← λ
  α ← 0.01 (This value selected experimentally.)
  δ = λ_new × h (h: the fill distance is defined to be h = sup_{x∈Ω} min_{1≤j≤N}
  ||x − x_i||_2)
  Loop
  set I(x) = {j ∈ {1, 2, ..., N}, ||x − x_i||_2 ≤ δ} (Using set of indices I, by
  localization at a fixed point x)
  for j ∈ I(x) do
     for i = 1 : N do
        Compute w_i for any x_j ∈ Ω_i
        A = A + w_i p_i^2
     end
  end
  if cond(A) ≥ 1/ε then
     {
     λ_new = λ_new + αλ
     δ = λ_new × h
  else
     goto end
     }
  if δ_i ≤ ||X_Ω||_2 then
     goto Loop
  end
```

In Algorithm 1.

X: is a set containing N scattered points which are called centers or data site and I (x) is the Index of points which MLS shape function is evaluated.

α: is a small positive number that is selected experimentally.

Then in every node points, matrix A is computed.

By running the algorithm the new value is assigned to λ, this value is related to the condition number of matrix A and its amount will increase. Therefore, the size of the support domain is increased and then the matrix A with new nodal points in the support domain is reproduced. This loop is repeated until $\frac{1}{cond(A)} \geq \varepsilon$.

The main idea of the moving least squares approximation is that for every point x can solve a locally weighted least squares problem [30], it is a local approximation, thus the additional condition to stop the loop is the size of the local support domain, the value of λ should be well enough to pave the local approximation, Line 20 is said to satisfy this condition.

2.3 Modified MLS approximation method

One of the common problems in Classic MLS method is the singularity of the moment matrix A in irregularity chosen nodal points. To avoid the nodal configurations which lead to a singular moment matrix, the usual solution is to enlarge the support domains of any nodal point. But it is not an appropriate solution, in [31] to tackle such problems is proposed a modified Moving least squares(MMLS)approximation method. This modifies allows, base functions in $m \geq 2$ to be used with the same size of the support domain as linear base functions $(m = 1)$. We should note that,impose additional terms based on the coefficients of the polynomial base functions is the main view of the modified technique. As we know, in the basis function $\mathbf{p}(\mathbf{x})$ is

$$\mathbf{p}(\mathbf{x}) = \left[1, x, x^2, \dots, x^m\right]^T \tag{15}$$

where $\mathbf{x} \in \mathbb{R}$, Also the correspond coefficients a_j, that should be determined are [24]:

$$\mathbf{a}(\mathbf{x}) = \left[a_1, a_x, a_{x^2}, \dots, a_{x^m}\right]^T \tag{16}$$

For obtaining these coefficients, the functional (2) rewrite as follows:

$$\bar{J}(\mathbf{x}) = \sum_{j=1}^{m} \left(\mathbf{P}^T(\mathbf{x}_j)\mathbf{a}(\mathbf{x}) - u_i(\mathbf{x}_j)\right)^2 w_i(\mathbf{x}) + \sum_{\nu=1}^{m-2} \overline{w}_\nu(x)\bar{\mathbf{a}}_\nu^2(\mathbf{x}), i = 1, 2, \dots, n \tag{17}$$

Where $\overline{w}$ is a vector of positive weights for the additional constraints, also $\bar{\mathbf{a}} = \left[a_{x^2}, a_{x^3}, \dots, a_{x^m}\right]^T$ is the modified matrix.

The matrix form of (17) is as follows:

$$\bar{J}(\mathbf{x}) = [P.\mathbf{a} - \mathbf{u}_i]^T.W.[P.\mathbf{a} - \mathbf{u}_i] + \mathbf{a}^T H\mathbf{a}, i = 1, 2, \dots, n \tag{18}$$

where H is as,

$$H = \begin{bmatrix} O_{2,2} & O_{m-2,m-2} \\ O_{2,2} & diag(\overline{w}) \end{bmatrix}, \tag{19}$$

where, $O_{i,j}$ is the null matrix. By minimizing the functional (18), the coefficients $a(\mathbf{x})$ will be obtained. So we have

$$\overline{A}(\mathbf{x})\mathbf{a}(x) = B(\mathbf{x})\mathbf{u}_i, \tag{20}$$

where

$$\overline{A} = P^T.W.P + H \tag{21}$$

And the matrics $B(\mathbf{x})$ is determined as the same before. So we have

$$\varphi_m(\mathbf{x}) = \mathbf{a}(\mathbf{x}) = p^T(\mathbf{x})\overline{A}^{-1}(\mathbf{x})B(\mathbf{x}) \tag{22}$$

where $\varphi_m(\mathbf{x})$ is the shape function of the MMLS approximation method.

3. Stiff systems of ordinary differential equations

In this section, we use MLS approximation method for numerical solution of the Stiff system of ordinary differential equations so consider the following differential equation

$$A(U) - F(\mathbf{x}) = 0, U(0) = U_0, \mathbf{x} \in \Omega \tag{23}$$

with boundary conditions,

$$B\left(U, \frac{\partial U}{\partial \mathbf{x}}\right) = 0, \mathbf{x} \in \partial\Omega.$$

where A is a general differential operator, U_0 is an initial approximation of (23), $F(\mathbf{x})$ is a vector of known analytical functions on the domain Ω and $\partial\Omega$ is the boundary of Ω. The operator can be divided by $A = L + N$, where L is the linear part, and N is the nonlinear part of its. So (23) can be, rewritten as follows;

$$L(U) + N(U) - F(\mathbf{x}) = 0 \tag{24}$$

We assume that $\mathbf{a} = \{a_1, a_2, \dots, a_m\}$ are the MLS shape functions so in order to solve system (24), N nodal points x_i are selected on the Ω, which $\{x_i | i = 1, 2, \dots, N\}$ is q-unisolvent. The distribution of nodes could be selected regularly or randomly. Then instead of u_j from U, we can replace u_j^h from (13). So we have

$$u_j^h(\mathbf{x}) = \sum_{i=1}^{N} a_i(\mathbf{x}) u_j(\mathbf{x}_i) \tag{25}$$

where $j = 1, 2, \dots, n$ is the number of unknown functions. we estimate the unknown functions u_i by (25), so the system (24) becomes the following form

$$L\left(\sum_{i=1}^{N} a_i(\mathbf{x})u_1(\mathbf{x}_i), \sum_{i=1}^{N} a_i(\mathbf{x})u_2(\mathbf{x}_i), \dots, \sum_{i=1}^{N} a_i(\mathbf{x})u_n(\mathbf{x}_i)\right) +$$
$$N\left(\sum_{i=1}^{N} a_i(\mathbf{x})u_1(\mathbf{x}_i), \sum_{i=1}^{N} a_i(\mathbf{x})u_2(\mathbf{x}_i), \dots, \sum_{i=1}^{N} a_i(\mathbf{x})u_n(\mathbf{x}_i)\right) = (f_1(\mathbf{x}), f_2(\mathbf{x}), \dots, f_n(\mathbf{x})) + \mathbf{r}(\mathbf{x}). \tag{26}$$

where $\mathbf{r}(\mathbf{x})$ is residual error function which vanishes to zero in collocation points thus by installing the collocation points $\mathbf{y}_r; r = 1, 2, \dots, N$, so

$$\begin{aligned}
&L\left(\sum_{i=1}^{N} a_i(\mathbf{y}_r)u_1(\mathbf{x}_i), \sum_{i=1}^{N} a_i(\mathbf{y}_r)u_2(\mathbf{x}_i), \dots, \sum_{i=1}^{N} a_i(\mathbf{y}_r)u_n(\mathbf{x}_i)\right) + \\
&N\left(\sum_{i=1}^{N} a_i(\mathbf{y}_r)u_1(\mathbf{x}_i), \sum_{i=1}^{N} a_i(\mathbf{y}_r)u_2(\mathbf{x}_i), \dots, \sum_{i=1}^{N} a_i(\mathbf{y}_r)u_n(\mathbf{x}_i)\right) = \\
&\sum_{i=1}^{N} L(a_i(\mathbf{y}_r))u_1(\mathbf{x}_i), \sum_{i=1}^{N} L(a_i(\mathbf{y}_r))u_2(\mathbf{x}_i), \dots, \sum_{i=1}^{N} L(a_i(\mathbf{y}_r))u_n(\mathbf{x}_i)) + \\
&N\left(\sum_{i=1}^{N} a_i(\mathbf{y}_r)u_1(x_i), \sum_{i=1}^{N} a_i(\mathbf{y}_r)u_2(\mathbf{x}_i), \dots, \sum_{i=1}^{N} a_i(\mathbf{y}_r)u_n(\mathbf{x}_i)\right) = \\
&(f_1(\mathbf{y}_r), f_2(y_r), \dots, f_n(\mathbf{y}_r))
\end{aligned} \tag{27}$$

therefore

$$CU = \begin{bmatrix} L(a_1(y_1)) & L(a_2(y_1)) & \dots & L(a_N(y_1)) \\ L(a_1(y_2)) & L(a_2(y_2)) & \dots & L(a_N(y_2)) \\ \vdots & & & \\ L(a_1(y_N)) & L(a_2(y_N)) & \dots & L(a_N(y_N)) \end{bmatrix} \begin{bmatrix} u_1(x_1) & u_2(x_1) & \dots & u_n(x_1) \\ u_1(x_2) & u_2(x_2) & \dots & u_n(x_2) \\ \vdots & & & \\ u_1(x_N) & u_2(x_N) & \dots & u_n(x_N) \end{bmatrix} \tag{28}$$

And the matrix form of (27) as follows

$$C_{N\times N}U_{N\times n} + N_{N\times n}(\mathbf{a}, U) = F_{N\times n}(y_r) \tag{29}$$

where

$$\begin{aligned}
C_i &= [L(a_1(\mathbf{y}_r)), \dots, L(a_N(\mathbf{y}_r))]_{i=1}^{n} \\
U_i &= \left[(u_i(x_1), u_i(x_2), \dots, u_i(x_N))^T\right]_{i=1}^{n} \\
F(\mathbf{y}_r) &= \left(\left[(f_1(\mathbf{y}_r))_{r=1}^{N}\right]^T, \left[(f_2(\mathbf{y}_r))_{r=1}^{N}\right]^T, \dots \left[(f_n(\mathbf{y}_r))_{r=1}^{N}\right]\right)^T.
\end{aligned} \tag{30}$$

by imposing the initial conditions at $t = 0$, we have

$$\left(\sum_{i=1}^{N} a_i(0)u_1(t_i), \sum_{i=1}^{N} a_i(0)u_2(t_i), \dots, \sum_{i=1}^{N} a_i(0)u_n(t_i)\right) = U_0 \tag{31}$$

and Solving the non-linear system (29) and (31), lead to finding quantities $u_j(x_i)$. Then the values of $u_j(x)$ at any point $x \in \Omega$, can be approximated by Eq. (25) as:

$$u_j(\mathbf{x}) \simeq \sum_{i=1}^{N} a_i(\mathbf{x})u_j(\mathbf{x}_i)$$

Remark

Note that, for simplicity, the modification derivation is made for bivariate data, but can be easily extended to higher dimensions. Also, for implementation, the modified moving least squares approximation method it is sufficient to use φ_m instead of φ.

3.1 Error analysis

The convergence analysis of the method in matrix norm has been investigated for the systems of one and two-dimensional Fredholm integral equations by authors of [22]. It should be noted that The convergence analysis of the method for implementation classic moving least squares approximation method on a system of integral equations has been discussed and it should be investigated for modified Mls method. we can use the results for this type of equations.

So in continuation of this section, the error estimations for the system of differential equations is developed. In [26], has obtained error estimates for moving least square approximations in the one-dimensional case. Also in [33], is developed for functional in n-dimensional and was used the error estimates to prove an error estimate in Galerkin coercive problems. In this work, have improved error estimate for the systems of stiff ordinary differential equations.

Given $\delta > 0$ let $W_\delta \geq 0$ be a function such that $supp(w_\delta) \subset \overline{B_\delta(0)} = \{z \| z| \leq \delta\}$ and $X_\delta = \{x_1, x_2, \dots, x_n\}$, $n = n(\delta)$, a set of points in $\Omega \subset \mathbb{R}$ an open interval and let $U = (u_1, u_2, \dots, u_N)$ be the unknown function such that $u_{i1}, u_{i2}, \dots, u_{in}$ be the values of the function u_i in those points, i.e., $u_{i,j} = u_i(x_j), i = 1, \dots, N, j = 1, \dots, n$. A class of functions $W = \{\omega_j\}_{j=1}^N$ is called a partition of unity subordinated to the open covering I_N if it possesses the following properties:

- $W_j \in C_s^0, s > 0\, or\, s = \infty$,
- $\sup(\omega_j) \subseteq \overline{\Lambda}_j$,
- $\omega_j(x) > 0, x \in \Lambda_j$,
- $\sum_{i=1}^{N} \omega_j = 1 for\ \ every\, x \in \overline{\Omega}$

There is no unique way to build a partition of unity as defined above [34].

As we know, the MLS approximation is well defined if we have a unique solution at every point $x \in \overline{\Omega}$. for minimization problem. And non-singularity of matrix $A(x)$, ensuring it is. In [33] the error estimate was obtained with the following assumption about the system of nodes and weight functions $\{\Theta_N, W_N\}$:

We define

$$\langle u, v \rangle = \sum_{j=1}^{n} w(x - x_j) u(x_j) v(x_j)$$

then

$$\|u\|_x^2 = \sum_{j=1}^{n} w(x - x_j) u(x_j)^2$$

Also for vector of unknown functions, we define

$$\|U\|_\infty = max\left\{ |u_i|_x, i = 1, 2, \dots, N \right\}$$

are the discrete norm on the polynomial space $\mathbb{P}_m^1$ if the weight function w satisfy the following properties.

a. For each $x \in \Omega$, $w(x - x_j) > 0$ at least for $(m+1)$ indices j.

b. For any $x \in \Omega$, the moment matrix $A(x) = w(x)P^T$ is nonsingular.

Definition 3.1. *Given $\mathbf{x} \in \overline{\Omega}$, the set $ST(\mathbf{x}) = \{j : \omega_j \neq 0\}$ will be called the star of x.*

Theorem 3.1. *[34, 35] A necessary condition for the satisfaction of Property **b** is that for any $\mathbf{x} \in \overline{\Omega}$*

$$n = card(ST(\mathbf{x})) \geq card(\mathbb{P}_m) = m + 1$$

For a sample point $\mathbf{c} \in \overline{\Omega}$, if $ST(\mathbf{c}) = \{j_1, \ldots j_k\}$, the mesh-size of the star $ST(\mathbf{c})$ defined by the number is $h(ST(\mathbf{c})) = \max\{h_{j1}, \ldots h_{jk}\}$.

Assumptions. Consider the following global assumptions about parameters. There exist

(a_1) An over bound of the overlap of clouds:

$$E = \sup_{c \in \overline{\Omega}}\{card(ST(c))\}.$$

(a_2) Upper bounds of the condition number:

$$CB_q = \sup_{c \in \overline{\Omega}}\{CN_q(ST(c)), q = 1, 2\}.$$

where the numbers $CN_q(ST(\mathbf{c}))$ are computable measures of the quality of the star $ST(c)$ which defined in Theorem 7 of [19].

(a_3) An upper bound of the mesh-size of stars:

$$R = \sup_{c \in \overline{\Omega}}(hST(c)).$$

(a_4) An uniform bound of the derivatives of $\{w_j\}$. That is the constant $G_q > 0, q = 1, 2$, such that

$$\|D^\mu W_j\|_{L_\infty} \leq \frac{G_q}{R^{|\mu|}} \quad 1 < \mu < q,$$

(a_5) There exist the number $\gamma > 0$ such that any two points $\mathbf{x}, \mathbf{y} \in \Omega$ can be joined by a rectifiable curve Γ in Ω with length $|\Gamma| \leq \gamma\|\mathbf{x}\text{-}\mathbf{y}\|$.

Assuming all these conditions, Zuppa [34] proved.

Lemma 3.1. *$U = (u_1, u_2, \ldots u_n)$ such that $u_i \in C^{m+1}(\overline{\Omega})$ and $\|U\|_\infty = u_k, \; 1 < k < n$, There exist constants $C_q, q = 1$ or 2,*

$$C_1 = C_1(\gamma, d, E, G_1, CB_1),$$

$$C_2 = C_1(\gamma, d, E, G_2, CB_1, CB_2),$$

then

$$\|D^\mu U - D^\mu U^h\|_\infty < C_q R^{q+1-|\mu|}\|u_k^{(m+1)}\|_{L^\infty(\Omega)} \quad 0 < \mu < q \tag{32}$$

Proof: see [36].

3.2 System of ODE

If in (24) the non-linear operator N be zero, we have

$$L(U) = (f_1, f_2, \dots, f_n) \tag{33}$$

where U is the vector of unknown function and L is a matrix of derivative operators,

$$L(U(.)) = \sum_{\varsigma=1}^{n} \lambda_\varsigma \frac{\partial^\varsigma}{(.)^\varsigma} U(.). \tag{34}$$

And from (25), we define

$$U^h(t) = \sum_{i=1}^{N} a_i(t) U(t_i)$$

where $(a_i)_{i=1}^N$ are the MLs shape functions defined on the interval $[0, 1]$ satisfying the homogeneous counterparts of the boundary conditions in (23). Also if the weight function w possesses k continuous derivatives then the shape functions a_j is also in C^k [33]. By the collocation method, is obtained an approximate solution $U^h(t)$. And demand that

$$L^h(U(.)) = \sum_{\varsigma=0}^{n} \lambda_\varsigma \frac{\partial^\varsigma}{(.)^\varsigma} U^h(.) \tag{35}$$

where ($\lambda = 0$ *or* 1). It is assumed that in the system of ODE derivative of order at most $n = 2$. Each of the basis functions a_i has compact support contained in $(0, 1)$ then the matrix C in (30) is a bounded matrix. If δ be fixed, independent of N, then the resulting system of linear equations can be solved in $O(N)$ arithmetic operations.

Lemma 3.2. *Let* $U = (u_1, u_2, \dots u_n)$ *and* $F = (f_1, f_2, \dots f_n)$ *so that* $u_i \in C^{m+1}(\overline{\Omega})$ $m \geq 1$ *and* $\|u_i\|_\infty = u_k, k \in \{1, 2, \dots, n\}$ *where* Ω *be a closed, bounded set in R. Assume the quadrature scheme is convergent for all continuous functions on* Ω*. Further, assume that the stiff system of ODE (23) with the fixd initial condition is uniquely solvable for given* $f_i \in C(\Omega)$*. Moreover take a suitable approximation* U^h *of* U *Then for all sufficiently large n, the approximate matrix L for linearly case exist and are uniformly bounded,* $|L| \leq M$ *with a suitable constant* $M < \infty$*. For the equations* $L(U) = F$ *and* $L^h(U) = F$ *we have*

$$E_t = \|L(U(t)) - L^h(U(t))\|_\infty$$

so that

$$\|E_t\|_\infty \leq C_q K(\lambda, \varsigma) R^{m+1-\mu} \|u_k^{(m+1)}\|_{L_\infty}.$$

Proof. we have

$$\|L(U(t)) - L^h(U(t))\|_\infty = \|\sum_{\varsigma=0}^{n} \lambda_\varsigma \frac{\partial^\varsigma}{t^\varsigma} U(t) - \sum_{\varsigma=0}^{n} \lambda_\varsigma \frac{\partial^\varsigma}{t^\varsigma} U^h(t)\|_\infty$$

so due to the lemma (36),

$$\|L(U(t)) - L^h(U(t))\|_\infty \leq \sum_{\varsigma=0}^{n} |\lambda_\varsigma| \| \frac{\partial^\varsigma}{t^\varsigma} U(t) - \frac{\partial^\varsigma}{t^\varsigma} U^h(t) \|_\infty$$
$$\leq \max_i \sum_{\varsigma=0}^{n} |\lambda_\varsigma| \| \frac{\partial^\varsigma}{t^\varsigma} u_i(t) - \frac{\partial^\varsigma}{t^\varsigma} u_i^h(t) |$$
$$\leq \sum_{\varsigma=0}^{n} C_q |\lambda_\varsigma| \| u_k^{(m+1)} \|_{L_\infty} R^{m+1-\varsigma}$$

where should be $m \geq \varsigma$ so,

$$\sum_{\varsigma=0}^{n} |\lambda_\varsigma| R^{m+1-\varsigma} \leq K(\lambda, \varsigma) R^{m+1-\mu}$$

where μ is the highest order derivative And $K(\lambda, \varsigma) = \sum_{\varsigma=0}^{n} |\lambda_\varsigma|$, so demanded that

$$\|E_t\|_\infty \leq C_q K(\lambda, \varsigma) R^{m+1-\mu} \| u_k^{(m+1)} \|_{L_\infty}.$$

It should be noted that in the nonlinear system the upper bound of error depends on the nonlinear operator.

4. Two-dimensional linear systems of integral equations

4.1 Fredholm type

In this section, we will provide an approximation solution of the 2-D linear system of Fredholm integral equations by the MLS method. The matrix form of this system could be considered as

$$\mathbf{U}(x,y) = \mathbf{F}(x,y) + \int_\Omega K(x,y,\theta,s)\mathbf{U}(\theta,s)d\theta ds, \quad (x,y) \in \Omega, \tag{36}$$

where $\Omega = [a,b] \times [c,d]$ as $\Omega \subset \mathbb{R}^2$, Also $K(x,y,\theta,s) = [\kappa_{ij}(x,y,\theta,s)]$, $i,j = 1,2, \dots, n$ is the matrix of kernels, $\mathbf{U}(x,y) = (u_1(x,y), u_2(x,y), \dots u_n(x,y))^T$ is the vector of unknown function and $\mathbf{F}(x,y) = (f_1(x,y), f_2(x,y), \dots f_n(x,y))^T$ is the vector of known functions.

In addition, is took that two cases for the domain, the rectangular shape, and nonrectangular one and three cases relative to the geometry of the nonrectangular domain are considered where can be transformed into the rectangular shape [35].

The first one is $\Omega = \{(\theta,s) \in \mathbb{R}^2 : a \leq s \leq b, g_1(s) \leq \theta \leq g_2(s)\}$ where $g_1(s)$ and $g_2(s)$ are continues functions of s, the second one can be consider as $\Omega = \{(\theta,s) \in \mathbb{R}^2 : c \leq \theta \leq d, g_1(\theta) \leq s \leq g_2(\theta)\}$ where $g_1(\theta)$ and $g_2(\theta)$ are continues functions of θ, Also the last one is a domain which is neither of the first nor the second kinds but could be separated to finite numbers of first or second sub-domains, it is labeled as a domain of third kind. Without loss of generality, the first kind domain can be assumed as

$$\Omega = \{(\theta, s) \in \mathbb{R}^2 : -1 \leq s \leq 1, g_1(s) \leq \theta \leq g_2(s)\} \tag{37}$$

by the following linear transformation

$$\theta(t, s) = \frac{g_2(\theta) - g_1(\theta)}{2} t + \frac{g_2(\theta) - g_1(\theta)}{2}, \tag{38}$$

the interval $[g_1(\theta), g_2(\theta)]$ is converted to the fixed interval $[-1, 1]$, so we have

$$\mathbf{U}(x, y) = \mathbf{F}(x, y) + \int_{-1}^{1}\int_{-1}^{1} \mathrm{K}(x, y, t, s)\mathbf{U}(t, s)dtds, \ \textit{such that}(x, y) \in [-1, 1] \times [-1, 1] \tag{39}$$

where

$$\mathrm{K}(x, y, t, s) = \frac{g_2(\theta) - g_1(\theta)}{2} \mathrm{K}(x, y, \theta, s) \tag{40}$$

Also, the second kind is straight similarly by commuting the variables and the third kind can be separated to finite numbers of sub-domains of the first or second kinds, so the method can be applied in each sub-domain as described earlier.

So, for the numerical integration $\Omega = \bigcup_{l=1}^{L} \Omega_l$ and $\Omega_l \bigcap \Omega_k \neq \varnothing$, $1 \leq k, l \leq L$

$$\int_{\Omega} g(s)ds = \sum_{l=1}^{L} \int_{\Omega_l} g(s)ds \tag{41}$$

Here, the MLS method is applied for the general case where the domain is $[a, b] \times [c, d]$.

To apply the method, as described in section 2.1, instead of u_i from U, we can replace u_i^h from (12). So we have

$$U^h(\mathbf{x}) = \left(u_1^h(\mathbf{x}), u_2^h(\mathbf{x}), \ldots, u_n^h(\mathbf{x})\right)^T \tag{42}$$

Also, obviously from (12)

$$u_j^h(x, y) = \sum_{i=1}^{N} \phi_i(x, y) u_j(x_i, y_i) \tag{43}$$

in this section, is assumed that the domain has rectangular shape, so system (36) becomes as follows

$$\left(u_1^h(x, y), u_2^h(x, y), \ldots u_n^h(x, y)\right)^T = f(x, y) + \int_c^d\int_a^b [k_{ij}(x, y, t, s)] \left(u_1^h(t, s), u_2^h(t, s), \ldots u_n^h(t, s)\right)^T dtds. \tag{44}$$

By substituting (42) in (44), and it holds at points $(x_r, y_r), r = 1, 2, \dots, N$ we have

$$\mathbf{f}(x_r, y_r) = \left(\sum_{i=1}^{N} \phi_i(x_r, y_r) u_1(x_i, y_i), \sum_{i=1}^{N} \phi_i(x_r, y_r) u_2(x_i, y_i), \dots \sum_{i=1}^{N} \phi_i(x_r, y_r) u_n(x_i, y_i)\right)^T$$
$$- \int_c^d \int_a^b \left[k_{ij}(x_r, y_r, t, s)\right] \left(\sum_{i=1}^{N} \phi_i(t, s) u_1(x_i, y_i), \dots, \sum_{i=1}^{N} \phi_i(t, s) u_n(x_i, y_i)\right)^T dtds. \tag{45}$$

We consider the m_1-point numerical integration scheme over Ω relative to the coefficients (τ_k, ς_p) and weights ω_k and ω_p for solving integrals in (45), i.e.,

$$(F_N)_j u_j(x, y) = \sum_{p=1}^{m_1} \sum_{k=1}^{m_1} \left(\sum_{i=1}^{N} k_{ji}(x_r, y_r, \tau_k, \varsigma_k) \phi_i(\tau_k, \varsigma_k) \omega_k \omega_p\right), (x, y) \in \Omega, u_i \in (-\infty, \infty) \tag{46}$$

Applying the numerical integration rule (46) yields

$$\mathbf{f}(x_r, y_r) = \left(\sum_{i=1}^{N} \left(\phi_i(x_r, y_r) - \sum_{p=1}^{m_1} \sum_{k=1}^{m_1} \left(\sum_{j=1}^{N} k_{j1}(x_r, y_r, \tau_k, \varsigma_k) \phi_i(\tau_k, \varsigma_k) \omega_k \omega p\right)\right) u_{1i},\right.$$
$$\sum_{i=1}^{N} \left(\phi_i(\tau_k, \varsigma_k) - \sum_{p=1}^{m_1} \sum_{k=1}^{m_1} \left(\sum_{j=1}^{N} k_{j2}(x_r, y_r, \tau_k, \varsigma_k) \phi_i(\tau_k, \varsigma_k) \omega_k \omega p\right)\right) u_{2i},$$
$$\left.\dots \sum_{i=1}^{N} \left(\phi_i(\tau_k, \varsigma_k) - \sum_{p=1}^{m_1} \sum_{k=1}^{m_1} \left(\sum_{j=1}^{N} k_{jn}(x_r, y_r, \tau_k, \varsigma_k) \phi_i(\tau_k, \varsigma_k) \omega_k \omega p\right)\right) u_{ni}\right)^T, r = 1, 2, \dots N$$

where $(u_j)_i$ are the approximate quantities of u_j when we use a quadrature rule instead of the exact integral. Now if we set $F_l, l = 1, 2, \dots n$ as a N by N matrices defined by:

$$(F_l)_{i,j} = \phi_i(x_r, y_r) - \sum_{p=1}^{m_1} \sum_{k=1}^{m_1} \left(\sum_{j=1}^{N} k_{jl}(x_r, y_r, \tau_k, \varsigma_p) \phi_i(\tau_k, \varsigma_p) \omega_k\right) \omega_p \tag{47}$$

So, the moment matrix F is defined by (47) as follows

$$\mathrm{F} = [F_1, F_2, \dots F_n]_{nN \times nN} \tag{48}$$

And

$$U = \left[(u_{11}, u_{12}, \dots u_{1N})^T, (u_{21}, u_{22}, \dots u_{2N})^T, \dots (u_{n1}, u_{n2}, \dots u_{nN})^T\right]^T$$
$$\mathbf{f}(x_r, y_r) = \left(\left[(f_1(x_r, y_r))_{r=1}^N\right]^T, \left[(f_2(x_r, y_r))_{r=1}^N\right]^T, \dots \left[(f_n(x_r, y_r))_{r=1}^N\right]\right)^T. \tag{49}$$

So by solving the following linear system of equations with a proper numerical method such as Gauss elimination method or etc. leads to quantities, u_{ji}.

$$\mathrm{F}U = \mathbf{f} \tag{50}$$

Therefore any $u_j(x,y)$ at any arbitrary point (x,y) from the domain of the problem, can be approximated by Eq. (43) as

$$u_j(x,y) \approx \sum_{i=1}^{N} \phi_i(x,y) u_{ji}(x_i,y_i) \tag{51}$$

4.2 Volterra type

Implementation of the proposed method on the Volterra integral equations is very simple and effective. In this case, the domain under study is as $\Omega = [a,x] \times [c,y]$ such that $0 \le x \le 1, 0 \le y \le 1$ and a, c are constant, so a Volterra system type of integral equations can be consider as

$$\mathbf{U}(x,y) = \mathbf{F}(x,y) + \int_{\Omega} \mathrm{K}(x,y,t,s)\mathbf{U}(t,s)dtds, \, (x,y) \in \Omega, \tag{52}$$

like the Fredholm type, it is the matrix form of a system, so we have

$$\begin{aligned} \mathbf{U}(x,y) &= (u_1(x,y), u_2(x,y), \dots u_n(x,y))^T, \textit{ the vector of unknown functions} \\ \mathbf{F}(x,y) &= (f_1(x,y), f_2(x,y), \dots f_n(x,y))^T, \textit{ the vector of known functions} \\ \mathrm{K}(x,y,t,s) &= [\kappa_{ij}(x,y,t,s)] i,j = 1,2, \dots, n \textit{ the matrix of kernels.} \end{aligned} \tag{53}$$

By the following transformation the interval $[a,x]$ and $[c,y]$ can be transferred to a fixed interval $[a,b]$ and $[c,d]$,

$$t(x,\theta) = \frac{x-a}{b-a}\theta + \frac{b-x}{b-a}a. \tag{54}$$

$$s(y,\xi) = \frac{y-c}{d-c}\xi + \frac{d-y}{d-c}c. \tag{55}$$

Then instead of u_i from U, we can replace u_i^h from (12). So we have

$$U^h(\mathbf{x}) = (u_1^h(\mathbf{x}), u_2^h(\mathbf{x}), \dots u_n^h(\mathbf{x}))^T \tag{56}$$

where

$$u_i^h(\mathbf{x}) = \sum_{j=1}^{N} \phi_j(\mathbf{x}) u_i(\mathbf{x}_j) \tag{57}$$

where $\mathbf{x} = (x,y) \in [a,b] \times [c,d]$, thus, system (52) becomes

$$(u_1^h(\mathbf{x}), u_2^h(\mathbf{x}), \dots u_n^h(\mathbf{x}))^T = F(\mathbf{x}) + \int_a^{\mathbf{x}}\int_c^{\mathbf{y}} [\kappa_{ij}(x,y,t,s)].(u_1^h(x,y), u_2^h(x,y), \dots u_n^h(x,y))^T dtds. \tag{58}$$

Therefore from (54) and (55), the system (58) takes the following form

$$(u_1^h(\mathbf{x}), u_2^h(\mathbf{x}), \dots u_n^h(\mathbf{x}))^T = F(\mathbf{x}) + \int_a^{\mathbf{b}}\int_c^{\mathbf{d}} [\kappa_{ij}(x,y,t,s)].(u_1^h(x,y), u_2^h(x,y), \dots u_n^h(x,y))^T d\theta d\xi. \tag{59}$$

Where

$$K(.,.,.,.) = \frac{x-a}{b-a}\frac{y-c}{d-c}\mathrm{K}(.,.,.,.), \tag{60}$$

Using techniques employed in the Fredholm case yields the same final linear system where

$$(F_l)_{i,j} = \phi_i(x_r, y_r) - \sum_{p=1}^{m_1}\sum_{k=1}^{m_1}\left(\sum_{j=1}^{N} k_{jl}\left(x_r, y_r, \tau_k, \varsigma_p\right)\phi_i\left(\tau_k, \varsigma_p\right)\omega_k\right)\omega_p \tag{61}$$

where $l = 1, 2, \dots, n$.

5. Nonlinear systems of two-dimensional integral equation

5.1 Fredholm type

In the nonlinear system, the unknown function cannot be written as a linear combination of the unknown variables or functions that appear in them, so the matrix form of Fredholm integral equations defined as the following form [27].

$$\mathbf{U}(x,y) = \mathbf{F}(x,y) + \int_{\Omega}\mathbf{K}(x,y,\theta,s,\mathbf{U}(\theta,s))d\theta ds, (x,y)\in\Omega, \tag{62}$$

Where $\mathbf{U}(x,y)$, K and $\mathbf{F}$ are defined as,

$$\mathbf{U}(x,y) = (u_1(x,y), u_2(x,y), \dots, u_n(x,y))^T$$

$$\mathbf{K}(x,y,\theta,s,\mathbf{U}(\theta,s)) = \left[k_{ij}(x,y,\theta,s,\mathbf{U}(\theta,s))\right], i,j = 1,2,\dots,n$$

$$\mathbf{F} = \left(f_1, f_2, \dots, f_n\right)^T$$

As mentioned above, we assume that $\Omega = [a,b]\times[c,d]$.

To apply the aproximation MLS method, we estimate the unknown functions u_i by (12), so the system (62) becomes the following form

$$\left(u_1^h(x,y), u_2^h(x,y), \dots u_n^h(x,y)\right)^T = f(x,y) + \int_c^d\int_a^b\left[k_{ij}\left(x,y,t,s,\mathbf{U}^h(t,s)\right)\right]dtds \tag{63}$$

We consider the m_1-point numerical integration scheme over the domain under study relative to the coefficients $\left(\tau_k, \varsigma_p\right)$ and weights ω_k and ω_p for solving tow-dimentional integrals in (63), i.e.,

$$(\mathrm{F}_N)_i u_i(x,y) = \sum_{p=1}^{m_1}\sum_{k=1}^{m_1} k_{ji}\left(x, y, \tau_k, \varsigma_p, \sum_{j=1}^{N}\phi_j\left(\tau_k, \varsigma_p\right)u_i(x,y)\omega_k\omega_p\right), (x,y)\in\Omega, u_i\in(-\infty,\infty) \tag{64}$$

Applying the numerical integration rule (64) in (63) yields

$$\left(u_1^h(x_r,y_r),u_2^h(x_r,y_r),\ldots,u_n^h(x_r,y_r)\right)^T=f(x_r,y_r)+\sum_{p=1}^{m_1}\sum_{k=1}^{m_1}\left[k_{ij}\left(x_r,y_r,\tau_k,\varsigma_p,\mathbf{U}^h\left(\tau_k,\varsigma_p\right)\right)\right]dtds \tag{65}$$

Finding unknowns $\mathbf{U}^h$ by solving the nonlinear system of algebraic Eq. (65) yields the following approximate solution at any point $(x,t)\in\Omega$.

$$u_j(x,y)\approx\sum_{i=1}^{N}\phi_i(x,y)u_{ji}(x_i,y_i) \tag{66}$$

5.2 Volterra type

Two-dimensional nonlinear system of Volterra integral equations can be considered as the following form

$$\mathbf{U}(x,y)=\mathbf{F}(x,y)+\int_a^x\int_c^y \mathrm{K}(x,y,\theta,s,\mathbf{U}(\theta,s))d\theta ds,\ (x,y)\in\Omega, \tag{67}$$

where K, **F** are known function and **U** the vector of unknown functions are defined in (63) [27]. In order to apply the MLS approximation method, as same as the linear type, the interval $[a,x]$ and $[c,y]$ transferred to a fixed interval $[a,b]$ and $[c,d]$. Then $u_i^h, i=1,2,\ldots,n$ instead of u_i in $U=(u_1,u_2,\ldots,u_n$ from (12) is replaced. So the nonlinear system (67) is converted to

$$\left(u_1^h(x,y),u_2^h(x,y),\ldots,u_n^h(x,y)\right)^T=f(x,y)+\int_c^d\int_a^b \overline{K}(x,y,t,s,\mathbf{U}^h(t,s))dtds \tag{68}$$

where

$$\overline{K}(x,y,t,s,\mathbf{U}^h(t,s))=\frac{\mathbf{x}-a}{b-a}\frac{\mathbf{y}-c}{d-c}K(x,y,t,s,\mathbf{U}^h(t,s)), \tag{69}$$

Using the numerical integration technique (64) which applied in the Fredholm case yields the same final nonlinear system (65), so the approximation solution of **U** would be found by solving this system of equations.

6. Examples

In this section, the proposed method can be applied to the system of 2-dimensional linear and nonlinear integral equations [37] and the system of differential equations. Also, the results of the examples illustrate the effectiveness of the proposed method Also the relative errors for the collocation nodal points is used.

$$\|e_i\|_\infty=\frac{\|u_{iex}(x,y)-u_i^h(x,y)\|}{\|u_{iex}(x,y)\|}$$

where u_i^h is the approximate solution of the exact solution u_{iex}. Linear and quadratic basis functions are utilized in computations.

6.1 Example 1

As the first example, we consider the following system of nonlinear Fredholm integral equations [27].

$$u_1(x,y) = f_1(x,y) + \int_\Omega u_1(s,t)u_2(s,t)dsdt$$

$$u_2(x,y) = f_2(x,y) + \int_\Omega u_1(s,t)u_2(s,t) + u_2^2(s,t)dsdt$$

where $\Omega = [0,1] \times [0,1]$. The exact solutions are $U(x,y) = (x+y, x)$ and the $F(x,y) = \left(x + y - \frac{7}{12}, x - \frac{11}{12}\right)$. The distribution of randomly nodes is shown in **Figure 1**. By attention to the irregular nodal points distribution, unsuitable δ can lead to a singular matrix A. So in this example, the adapted algorithm can tackle such problems. The MLS and MMLS shape functions are computed by using Algorithm 1, so the exact value of the radius of the domain of influence is not important; in fact, it is chosen as an initial value.

The condition numbers of the matrix A is shown in **Figure 2** and the determinant of A at sample points p is shown in **Figure 3**, where the radius of support domains for any nodal points is started from $\delta = 0.05$. Note that, there is a different radius of support domain for any node point, it might be increased due to the inappropriate distribution of scattered points by the algorithm. These variations are shown in **Table 1** for sample points (x,y), where $Cond(A)$ is the conditions number A, its initial case ($\delta = 0.005$) and final case ($New\delta,$) and $N.O.iteration$ is the number of iteration of the algorithm for determining a suitable radius of support domain.

In computing, $\delta = 2r$ where $r = 0.05$ and $c = \frac{2}{\sqrt{3}}r$. Also In MMLS, $\overline{w}_\nu = 0.1$, $\nu = 1, 2, 3$. It should be noted that, these values were also selected experimentally. Relative errors of the MLS method for different Gauss-Legendre number points at $m = 1, 2$ and 3 compared in **Table 2**, also investigating the proposed methods shown that increasing the number of numerical integration points does not guarantee the error decreases. jkjk.

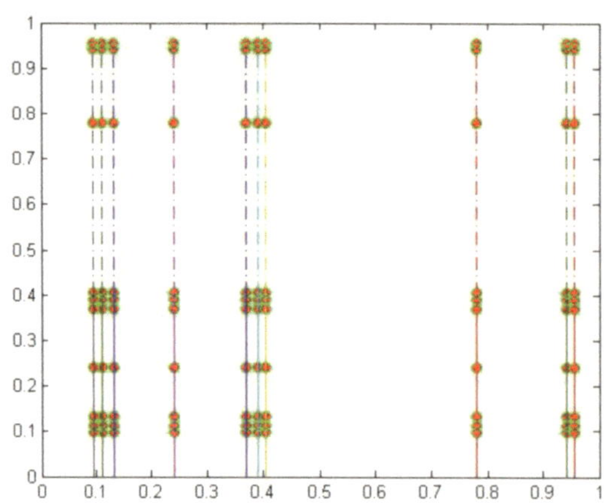

Figure 1.
The scatter data of example 1.

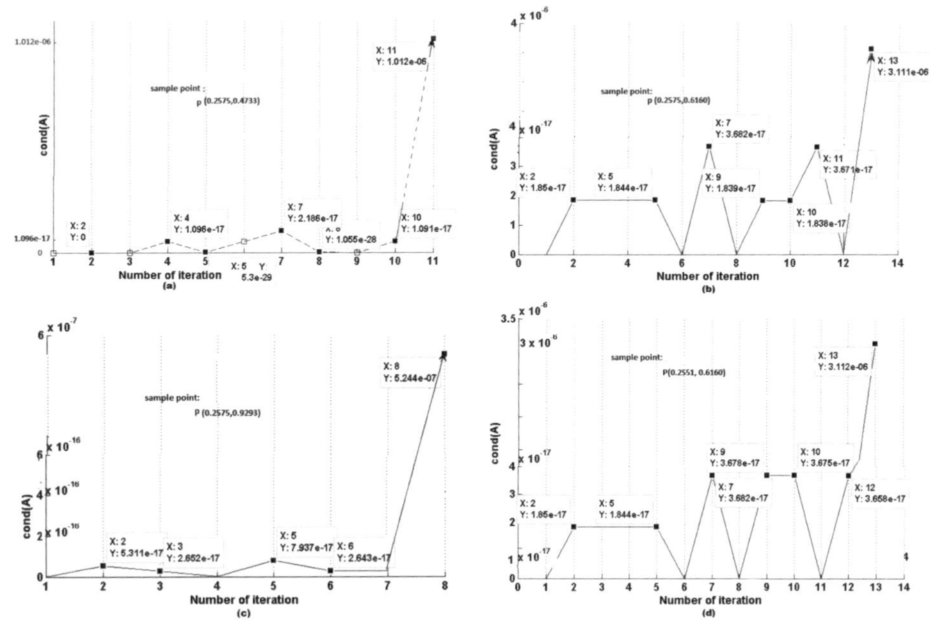

Figure 2.
The condition numbers of a at a sample point p and $\delta = 0.05$ for example 1. Using algorithm 1.

	Sample points		Cond(A)		Result of algorithm	
n	x	y	*initial*	*final*	*New*δ	*N.O.iteration*
1	0.2575	0.4733	$1.1005e-17$	$1.0117e-06$	0.1297	11
2	0.2575	0.6160	0	$3.1111e-06$	0.1569	13
3	0.2575	0.9293	$5.3204e-17$	$5.2445e-07$	0.0974	8
4	0.2551	0.6160	0	$3.1115e-06$	0.1569	13
5	0.6991	0.2435	$2.7166e-17$	$1.4652e-07$	0.0550	2

Table 1.
Change of the radius and the condition number A at sample points (x,y) using algorithm 1, for example 1.

	m = 1 $\|\|e\|\|_\infty$		m = 2 $\|\|e\|\|_\infty$		m = 3 $\|\|e\|\|_\infty$	
N	u_1	u_2	u_1	u_2	u_1	u_2
5	6.28×10^{-4}	2.7×10^{-3}	3.45×10^{-7}	1.15×10^{-6}	3.1×10^{-6}	2.8×10^{-6}
10	1.2×10^{-4}	5.9×10^{-4}	2.46×10^{-7}	8.41×10^{-7}	2.1×10^{-7}	5.41×10^{-7}
15	2.3×10^{-4}	5.9×10^{-4}	4.47×10^{-8}	1.34×10^{-7}	4.47×10^{-8}	2.15×10^{-7}
20	2.3×10^{-4}	5.14×10^{-4}	6.12×10^{-7}	2.46×10^{-6}	3.2×10^{-7}	1.9×10^{-6}
30	3.2×10^{-4}	5.9×10^{-4}	1.84×10^{-6}	6.64×10^{-6}	2.34×10^{-6}	7.14×10^{-6}

Table 2.
Maximum relative errors for different points Gauss-Legendre quadrature rule $\delta = 2r$, for example 1.

	MMLS $\|e\|_\infty$		MLS $\|e\|_\infty$		CPU times	
N	u_1	u_2	u_1	u_2	MLS	MMLS
10	3.78×10^{-10}	8.13×10^{-10}	2.46×10^{-7}	8.41×10^{-7}	389.9574	2.9776×10^3
15	1.35×10^{-10}	2.00×10^{-11}	4.47×10^{-8}	1.34×10^{-7}	410.9083	2.1109×10^3
20	8.63×10^{-11}	2.51×10^{-9}	6.12×10^{-7}	2.46×10^{-6}	634.8373	3.2115×10^3
30	3.99×10^{-10}	1.58×10^{-9}	1.84×10^{-6}	6.64×10^{-6}	1.0331×10^3	2.5844×10^3

Table 3.
Compare relative errors and CPU times of MLS and MMLS for different points Gauss-Legendre quadrature rule, $(m = 2)$, *for example 1.*

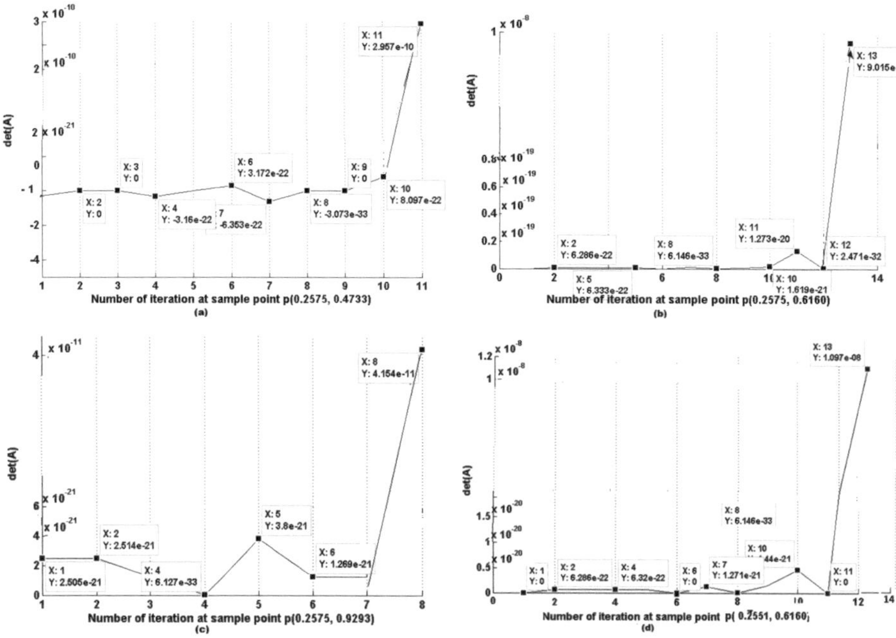

Figure 3.
The determinant of a at a sample point p and $\delta = 0.05$ *for example 1. Using algorithm 1.*

In **Table 3**, we can see that the CPU times for solving the nonlinear system (65) are much larger in MMLS method; but, the errors are very smaller (**Figure 3**).

6.2 Example 2

Consider the system of linear Fredholm integral equations with [27].

$$K(x,y,t,s) = \begin{pmatrix} x(t+s) & -t \\ ts & (y+x)t \end{pmatrix}, \tag{70}$$

Such that $U(x,y) = (x+y, x)$ is the vector of The exact solutions and the vector of unknown function is $F(x,y) = \left(-\frac{1}{6}(x+y) + \frac{1}{3}, \frac{4}{3}x - \frac{1}{3} + \frac{1}{3}y\right)$. Also the domain of the problem determine by $\Omega = [0,1] \times [0,1]$. In this example, initial value of r as radius of support domain set by 0.05. Also Algorithm 1 is used for producing shape function at $m = 1, 2, 3$. In computing, we put $\overline{w}_\nu = 0.1, \nu = 1, 2, 3$.

	m = 1			m = 3		
N	u_1	u_2	*CPU.T.*	u_1	u_2	*CPU.T.*
5	2.94×10^{-4}	6.02×10^{-4}	108.957	2.08×10^{-4}	2.91×10^{-4}	152.1474
10	1.79×10^{-4}	3.89×10^{-4}	159.258	1.7×10^{-4}	2.37×10^{-4}	170.7879
15	1.86×10^{-4}	3.989×10^{-4}	198.135	2.47×10^{-4}	3.30×10^{-4}	252.75424
20	1.86×10^{-4}	3.986×10^{-4}	221.321	2.41×10^{-4}	3.29×10^{-4}	247.0093
30	1.85×10^{-4}	3.987×10^{-4}	308.987	2.25×10^{-4}	3.37×10^{-4}	314.8173
40	1.85×10^{-4}	3.980×10^{-4}	395.125	1.87×10^{-4}	2.86×10^{-4}	402.9594

Table 4.
Relative errors and CPU times of MLS for different points Gauss-Legendre quadrature rule at $m = 1, 3$, for example 2.

		MLS			MMLS		
m	N	u_1	u_2	*CPU.T.*	u_1	u_2	*CPU.T.*
2	5	1.24×10^{-7}	3.89×10^{-6}	289.75	4.47×10^{-7}	6.12×10^{-6}	297.7
	10	3.16×10^{-7}	5.23×10^{-7}	189.99	2.16×10^{-8}	6.26×10^{-8}	210.09
	15	1.68×10^{-7}	4.13×10^{-7}	389.95	2.02×10^{-8}	8.12×10^{-8}	390.15
	20	1.61×10^{-7}	3.88×10^{-7}	410.90	2.04×10^{-8}	5.24×10^{-8}	425.87
	30	1.45×10^{-7}	3.80×10^{-7}	604.80	1.98×10^{-8}	3.25×10^{-8}	618.44
	40	1.44×10^{-7}	3.84×10^{-7}	689.9574	1.04×10^{-8}	6.87×10^{-8}	697.04

Table 5.
Compare relative errors and CPU times of MLS and MMLS for different points Gauss-Legendre quadrature rule at $m = 2$, for example 2.

Table 4 shows relative errors and CPU times of MLS for different Gauss-Legendre number points at $m = 1, 3$. As shown in **Table 5**, comparing the errors of MMLS and MLS method determines the capability and accuracy of the proposed technique to solve systems of linear Fredholm integral equations. This indicates the advantage of the proposed method over these systems of equations.

Comparing the errors of MMLS and MLS method determines the capability and accuracy of the proposed method to solve systems of linear Fredholm integral equations.

6.3 Example 3

The third example that we want to approximate is the system of linear Volterra-Fredholm integral equations with [27].

$$K(x,y,t,s) = \begin{pmatrix} (x+y)\exp^{(t+s)} & (x+y)\exp^{(t+s)} \\ 1 & -1 \end{pmatrix}, \tag{71}$$

The domain is considered as $\Omega = \{(x,y) \in \mathbb{R}^2 : 0 \le x \le 1, 0 \le y \le (1-y)\}$ so that $y \in [0,1]$. Also the exact solutions are $(\exp^{x+y}, \exp^{x-y})$. It is important to note that the linear transformation used in the experiment is only (55) and from (40) the kernel becomes

$$K(.,.,.,.) = \frac{y-c}{d-c} \mathrm{K}(.,.,.,.). \tag{72}$$

In this example, the effect of increasing radius of the domain of influence δ_i in MLS method on error has been investigated. Therefore the δ_i was considered as follows

$$\delta_i = \lambda r \; i = 1, 2, \ldots, N \tag{73}$$

In this way, useful information is obtained about the performance of the proposed method. By investigating the results in **Tables 6** and **7** we found that the relative error in MLS was also related to the radius of the domain of influence (i.e. $\delta = \lambda r$ so that $\lambda = 3, 5, 7$); however, it cannot be greater than 7. For example, the relative errors by choosing $\lambda = 10$ (i.e. $\delta = 0.05\lambda$) and 10−point GaussLegendre quadrature rule are $\|e\|_{\infty u_1} = 3.3 \times 10^{-2}$ and $\|e\|_{\infty u_2} = 1.8 \times 10^{-2}$ at $m = 1$. Also, **Table 8** depicts, the number of points in the numerical integration rule cannot be effective to increase the accuracy of the method.

	$\lambda = 3\|e\|_\infty$		$\lambda = 5\|e\|_\infty$		$\lambda = 7\|e\|_\infty$	
r	u_1	u_2	u_1	u_2	u_1	u_2
0.2	9×10^{-3}	5×10^{-3}	1.9×10^{-3}	1×10^{-3}	6.02×10^{-4}	3.36×10^{-4}
0.1	1.2×10^{-3}	6.87×10^{-4}	1.34×10^{-4}	7.46×10^{-5}	4.57×10^{-6}	2.91×10^{-6}
0.05	1.44×10^{-4}	8.14×10^{-5}	2.23×10^{-5}	1.26×10^{-5}	6.27×10^{-6}	3.6×10^{-6}

Table 6.
Maximum relative errors of MLS for 10 Gauss-Legendre points and different values of $\delta = \lambda r$ at $m = 2$, for example 3.

	$\lambda = 3\|e\|_\infty$		$\lambda = 5\|e\|_\infty$		$\lambda = 7\|e\|_\infty$	
r	u_1	u_2	u_1	u_2	u_1	u_2
0.2	1.5×10^{-3}	8.77×10^{-3}	6.56×10^{-5}	3.01×10^{-5}	1.09×10^{-4}	6.19×10^{-5}
0.1	1.08×10^{-4}	6.51×10^{-5}	3.47×10^{-5}	2.08×10^{-5}	1.24×10^{-5}	7.22×10^{-6}
0.05	6.63×10^{-6}	3.93×10^{-6}	3.9×10^{-6}	2.31×10^{-6}	2.41×10^{-6}	1.38×10^{-6}

Table 7.
Maximum relative errors of MLS for 10 Gauss-Legendre points and different values of $\delta = \lambda r$ at $m = 3$ for example 3.

	$\lambda = 3\|e\|_\infty$		$\lambda = 5\|e\|_\infty$		$\lambda = 7\|e\|_\infty$	
N	u_1	u_2	u_1	u_2	u_1	u_2
5	2.2×10^{-3}	1.5×10^{-3}	6.33×10^{-5}	4.52×10^{-5}	2.21×10^{-5}	1.59×10^{-5}
10	1.1×10^{-3}	7.4×10^{-4}	4.17×10^{-5}	2.89×10^{-5}	2.96×10^{-6}	2.72×10^{-6}
15	2×10^{-3}	1.3×10^{-3}	7.9×10^{-5}	5.07×10^{-5}	6.31×10^{-6}	5.11×10^{-6}
20	2.1×10^{-3}	1.3×10^{-3}	8.06×10^{-5}	5.17×10^{-5}	3.96×10^{-6}	3.69×10^{-6}
30	2.1×10^{-3}	1.3×10^{-3}	8.16×10^{-5}	5.17×10^{-5}	4.11×10^{-6}	3.74×10^{-6}

Table 8.
Maximum relative errors of MLS for different values of $\delta = \lambda r, r = 0.05$ and Gauss-Legendre points at $m = 1$, using algorithm 1 for example 3.

	MLS		MMLS		CPU time	
r	u_1	u_2	u_1	u_2	MLS	MMLS
0.2	1.09×10^{-4}	6.19×10^{-5}	5.52×10^{-4}	3.08×10^{-4}	4.0556	3.3893
0.1	1.24×10^{-5}	1.08×10^{-6}	1.25×10^{-5}	7.36×10^{-6}	38.0616	31.7945
0.05	1.91×10^{-6}	1.085×10^{-6}	1.54×10^{-5}	8.77×10^{-6}	496.1746	409.0342

Table 9.
Compare relative errors and CPU times of MLS and MMLS for $\overline{w}_\nu = 10$ and 10 Gauss-Legendre points and different values of $\delta = 7r$ at $m = 2$ for example 3.

Then the relative error by MMLS shape function described in section (2.3) were obtained, using $\delta = 7r$ and $\overline{w}_\nu = 10$, such that $\nu = 1, 2, 3$ as weights of additional coefficients for MMLS. We can see that in **Table 9** the errors of the system of linear Volterra-Fredholm integral equations are similar in both methods (i.e. MLS and MMLS methods).

6.4 Example 4

Consider the following nonlinear stiff systems of ODEs [38].

$$\begin{cases} u_1'(t) = -1002u_1(t) + 1000u_2^2(t) \\ u_2'(t) = u_1(t) - u_2(t) - u_2^2(t) \end{cases}$$

With the initial condition $u_1(0) = 1$ and $u_2(0) = 1$. The exact solution is

$$u_1(t) = exp\,(-2t)$$
$$u_2(t) = exp\,(-t).$$

In this numerical example, two scheme are compared and as explained the main task of the modified method tackle the singularity of the moment matrix. **Table 10** presents the maximum relative error by MLS on a set of evaluation points (with $h = 0.1$ *and* 0.02) and $\delta = 4h$ *and* $3h$. Also in **Table 11** MLS and MMLS at different number of nodes for $h = 0.004$ and $\delta = 5h$ *and* $8h$, were compared (**Tables 10–12**).

	m = 2, δ = 4 r		m = 2, δ = 3 r	
r	u_1	u_2	u_1	u_2
0.1	5×10^{-3}	4.1×10^{-4}	8.85×10^{-4}	2.2×10^{-3}
0.02	5.8×10^{-2}	6.5×10^{-5}	5.42×10^{-4}	6.52×10^{-5}

Table 10.
Maximum relative errors by MLS, example 4.

	m = 3, δ = 5 r		m = 3, δ = 8 r	
Type	u_1	u_2	u_1	u_2
MLS	1.03×10^{0}	0.98×10^{1}	1.01×10^{0}	9.2×10^{0}
MMLS	9.23×10^{-4}	9.22×10^{-4}	6.89×10^{-4}	6.96×10^{-4}

Table 11.
Maximum relative errors for h = 0.004 by MMLS and MLS, example 4.

	MLS			MMLS		
weight type	u_1	u_2	**Cputime**	u_1	u_2	**Cputime**
Guass	3.06×10^{-3}	9.92×10^{-5}	61.1706	8.5×10^{-4}	6.5×10^{-4}	0.5598
Spline	5.06×10^{-4}	5.3×10^{-4}	64.5897	1.93×10^{-2}	4.23×10^{-4}	0.6714
RBF	6.407×10^{-4}	3.02×10^{-4}	59.1790	6.9×10^{-3}	6.9×10^{-3}	0.5768

Table 12.
Maximum relative errors by MLS $t \in [0,5], h = 0.004$,, *example 2.*

6.5 Example 5

In this example, we consider $U(t) = \left(\frac{1}{47}\left(95\exp^{(-2t)} - 48\,exp\,(-96t)\right), \frac{1}{47}\right.$ $(48\exp(-96t) - \exp(-2t)))$ as the exact solution and $U(0,0) = (1,1)$ as the initial conditions for the following system of ODE,

$$\begin{cases} x'(t) = -x(t) + 95y(t) \\ y'(t) = -x(t) - 97y(t) \end{cases}$$

Table 12 presents the maximum relative norm of the errors on a fine set of evaluation points (with $h = 0.004$) and $\delta = 5h$ for MLS and MMLS at different type of weight functions. As seen in this table, one major advantage of MMLS is that the computational time used by MMLS is less than MLS.

7. Conclusion

In this paper, two meshless techniques called moving least squares and modified Moving least-squares approximation are applied for solving the system of functional equations. Comparing the results obtained by these methods with the results obtained by the exact solution shows that the moving least squares methods are the reliable and accurate methods for solving a system of functional equations. Meshless methods are free from choosing the domain and this makes it suitable to study real-world problems. Also, the modified algorithm has changed the ability to select the support range radius In fact, the user can begin to solve any problem with an arbitrary radius from the domain and the proposed algorithm can correct it during execution.

Author details

Massoumeh Poura'bd Rokn Saraei* and Mashaallah Matinfar
Department of Mathematics, Science of Mathematics Faculty, University of Mazandaran, Iran

*Address all correspondence to: m.pourabd@gmail.com and m.matinfar@umz.ac.ir

References

[1] Scudo FM. Vito Volterra and theoretical ecology. Theoretical Population Biology. 1971;**2**:1-23

[2] Small RD. Population growth in a closed model. In: Mathematical Modelling: Classroom Notes in Applied Mathematics. Philadelphia: SIAM; 1989

[3] TeBeest KG. Numerical and analytical solutions of Volterra's population model. SIAM Review. 1997; **39**:484-493

[4] Wazwaz AM. Partial Differential Equations and Solitary Waves Theory. Beijing and Berlin: HEP and Springer; 2009

[5] Mckee S, Tang T, Diogo T. An Euler-type method for two-dimensional Volterra integral equations of the first kind. IMA Journal of Numerical Analysis. 2000;**20**:423-440

[6] Hanson R, Phillips J. Numerical solution of two-dimensional integral equations using linear elements. SIAM Journal on Numerical Analysis. 1978;**15**: 113-121

[7] Babolian E, Masouri Z. Direct method to solve Volterra integral equation of the first kind using operational matrix with block-pulse functions. Journal of Computational and Applied Mathematics. 2008;**220**:51-57

[8] Beltyukov BA, Kuznechichina LN. A RungKutta method for the solution of two-dimensional nonlinear Volterra integral equations. Differential Equations. 1976;**12**:1169-1173

[9] Masouri Z, Hatamzadeh-Varmazyar S, Babolian E. Numerical method for solving system of Fredholm integral equations using Chebyshev cardinal functions. Advanced Computational Techniques in Electromagnetics. 2014:1-13

[10] Jafarian A, Nia SAM, Golmankhandh AK, Baleanu D. Numerical solution of linear integral equations system using the Bernstein collocation method. Advances in Difference Equations. 2013:1-23

[11] Singh P. A note on the solution of two-dimensional Volterra integral equations by spline. Indian Journal of Mathematics. 1979;**18**:61-64

[12] Assari P, Adibi H, Dehghan M. A meshless method based on the moving least squares (MLS) approximation for the numerical solution of two-dimensional nonlinear integral equations of the second kind on non-rectangular domains. Numerical Algorithm. 2014;**67**:423-455

[13] Brunner H, Kauthen JP. The numerical solution of two-dimensional Volterra integral equations by collocation and iterated collocation. IMA Journal of Numerical Analysis. 1989;**9**:47-59

[14] Wazwaz A. Linear and Nonlinear Integral Equations Methods and Applications. 2011

[15] Dehghan M, Abbaszadeha M, Mohebbib A. The numerical solution of the two-dimensional sinh-Gordon equation via three meshless methods. Engineering Analysis with Boundary Elements. 2015;**51**:220-235

[16] Mirzaei D, Schaback R, Dehghan M. On generalized moving least squares and diffuse derivatives. IMA Journal of Numerical Analysis. 2012;**32**:983-1000

[17] Salehi R, Dehghan M. A generalized moving least square reproducing kernel method. Journal of Computational and Applied Mathematics. 2013;**249**:120-132

[18] Saadatmandi A, Dehghan M. A collocation method for solving Abels

integral equations of first and second kinds. Zeitschrift für Naturforschung. 2008;**63a**(10):752-756

[19] Dehghan M, Mirzaei D. Numerical solution to the unsteady two-dimensional Schrodinger equation using meshless local boundary integral equation method. International Journal for Numerical Methods in Engineering. 2008;**76**:501-520

[20] Mukherjee YX, Mukherjee S. The boundary node method for potential problems. International Journal for Numerical Methods in Engineering. 1997;**40**:797-815

[21] Salehi R, Dehghan M. A moving least square reproducing polynomial meshless method. Applied Numerical Mathematics. 2013;**69**:34-58

[22] Matin far M, Pourabd M. Moving least square for systems of integral equations. Applied Mathematics and Computation. 2015;**270**:879-889

[23] Li S, Liu WK. Meshfree Particle Methods. Berlin: Springer-Verlag; 2004

[24] Liu GR, Gu YT. A matrix triangularization algorithm for the polynomial point interpolation method. Computer Methods in Applied Mechanics and Engineering. 2003;**192**: 2269-2295

[25] Belinha J. Meshless Methods in Biomechanics. Springer; 2014

[26] Chen S. Building interpolating and approximating implicit surfaces using moving least squares [Phd thesis] EECS-2007-14. Berkeley: EECS Department, University of California. 2007

[27] Matin far M, Pourabd M. Modified moving least squares method for two-dimensional linear and nonlinear systems of integral equations. Computational and Applied Mathematics. 2018;**37**:5857-5875

[28] Joldesa GR, Chowdhurya HA, Witteka A, Doylea B, Miller K. Modified moving least squares with polynomial bases for scattered data approximation. Applied Mathematics and Computation. 2015;**266**:893-902

[29] Lancaster P, Salkauskas K. Surfaces generated by moving least squares methods. Mathematics of Computation. 1981;**37**:141-158

[30] Wendland H. Local polynomial reproduction, and moving least squares approximation. IMA Journal of Numerical Analysis. 2001;**21**:285-300

[31] Zuppa C. Error estimates for moving least square approximations. Bulletin of the Brazilian Mathematical Society, New Series. 2001;**34**:231249

[32] Liu GR. Mesh Free Methods: Moving beyond the Finite Element Method. Boca Raton: CRC Press; 2003

[33] Zuppa C. Error estimates for moving least squareapproximations. Bulletin of the Brazilian Mathematical Society, New Series. 2003;**34**(2):231-249

[34] Zuppa C. Good quality point sets error estimates for moving least square approximations. Applied Numerical Mathematics. 2003;**47**:575-585

[35] Mirzaei D, Dehghan M. A meshless based method for solution of integral equations. Applied Numerical Mathematics. 2010;**60**:245-262

[36] McLain D. Two dimensional interpolation from random data. Computer Journal. 1976;**19**:178181

[37] Pourabd M. Meshless method based moving least squares for solving systems of integral equations with investigating the computational complexity of algorithms [Phd thesis], IRANDOC-2408208. IRAN: Department of mathematic, University of Mazandaran. 2017

[38] Biazar J, Asadi MA, Salehi F. Rational Homotopy perturbation method for solving stiff systems of ordinary differential equations. Applied Mathematical Modelling. 2015;**39**: 12911299

Chapter 5

Informational Time Causal Planes: A Tool for Chaotic Map Dynamic Visualization

Felipe Olivares, Lindiane Souza, Walter Legnani and Osvaldo A. Rosso

Abstract

In the present chapter, we made a detailed analysis of the different regimes of certain chaotic systems and their correspondence with the change in the normalized Shannon entropy, Statistical Complexity, and Fisher information measure.
We construct a bidimensional plane composed of the selection of a pair of the informational tools mentioned above (a casual plane is defined), in which the different dynamical regimes appeared very clear and give more information of the underlying process. In such a way, a plane composed of the normalized Shannon entropy, statistical complexity, normalized Shannon entropy, and Fisher information measure can be applied to follow the changes in the behavior variations of the nonlinear systems.

Keywords: chaotic dynamics, statistical complexity, information theory quantifiers, Shannon entropy, Fisher information measure, Bandt-Pompe probability distribution function

1. Introduction

In the space of few decades, chaos theory has jumped from the scientific literature into the popular realm, being regarded as a new way of looking at complex systems like brains or ecosystems. It is believed that the theory manages to capture the disorganized order that pervades our world. Chaos theory is a facet of the complex system paradigm having to do with determinism randomness. As many other people before, we wish to approach it from the information theory viewpoint.

In 1959 Kolmogorov had pointed out that the probabilistic theory of information developed by Shannon could be applied to symbolic encodings of the phase space descriptions of physical non-linear dynamical systems and with line of rezoning it more or less direct characterize a process in terms of *its Kolmogorov-Sinai entropy* [1, 2]. It has been a cornerstone in the updated theory of dynamical systems that could be complimented with Pesin's theorem in 1977 [3]. With this theorem, Pesin has proven that for certain deterministic nonlinear dynamical systems exhibiting chaotic behavior, an estimation of the *Kolmogorov-Sinai entropy* can be computed as the sum of the positive Lyapunov exponents for the process.

As is well known, chaotic systems have sensitivity to initial conditions which means instability everywhere in the phase space and lead to nonperiodic motion

IntechOpen

(chaotic time series) [4]. One of the main characteristics of this kind of systems is its capability of long-term unpredictability despite the deterministic character of the temporal trajectory. In a system undergoing chaotic motion, two closeup neighboring points in the phase space after a short time elapsed show an exponential divergence of their respective trajectories. For example, let $\boldsymbol{X}_1(t)$ and $\boldsymbol{X}_2(t)$ be such two points, located within a ball of radius R at time t. Further, assume that these two points cannot be resolved within the ball due to poor instrumental resolution. At some later time t', the distance between the points will typically grow to $|\boldsymbol{X}_1(t') - \boldsymbol{X}_2(t')| \approx |\boldsymbol{X}_1(t) - \boldsymbol{X}_2(t)| \exp(\Lambda|t' - t|)$, in the case of chaotic dynamics, with $\Lambda > 0$, the average of Lyapunov exponents of the system. Clearly, if $|\boldsymbol{X}_1(t') - \boldsymbol{X}_2(t')| > R$, the points will be apart from each other, determining a non-zero distance between them. This fact could be interpreted by a certain kind of instability which reveals some information about the phase space population that was not available at earlier times [4]. This fact contributes to think that the chaotic behavior plays a role of *information source*.

As has been shown in the literature for a many of simple nonlinear processes, the Lyapunov exponents may be computed very precisely with different algorithms. In such a way, a nonlinear dynamical system may be considered as an information source from which information-related quantifiers may help visualize relevant details of the chaotic process. The existence of simple "calibrated" sources such as the logistic map provides a means for a precise evaluation of the performance of these information quantifiers. In this communication we take advantage such fact in order to show that planar representations constructed with two information theory-based quantifiers offer one possibility of easily visualizing many interesting details of chaos characteristics, including the fine structure of chaotic attractors. We exemplified their use showing the result on two chaotic maps: the logistic map and the delayed logistic map.

2. Information theory quantifier prescription

Many systems during its functioning generate a sequence of values that can be measured constituting what is called in science as time series (TS). The analysis concerns to extract the major quantity of information of them to accomplish the understanding of the meaning of the changes characterizing different dynamical regimes. It usually computes the experimental, or when the case permits the theoretical, probability distribution function (PDF) of the regimes exhibited by the TS, from here noted as $\mathcal{X}(t)$.

The mathematical tools applied once the PDF is available receive the name of informational tools; more precisely information theory quantifiers [5], the main feature of the quantifiers is exactly quantifying the amount of information coming from the TS, originating in the dynamical system.

2.1 Shannon entropy, Fisher information measure, and statistical complexity

The concept of entropy has many interpretations arising from a wide diversity of scientific and technological fields. Among them is associated with disorder, with the volume of state space, and with a lack of information too. There are various definitions according to ways of computing this important magnitude to study the dynamics of the systems, and one of the most frequent that could be considered of foundational definition is the denominated *Shannon entropy* [6], which can be interpreted as a measure of uncertainty. The *Shannon entropy* can be considered as one of the most representative examples of information quantifiers.

Let a continuous PDF be noted by $\rho(x)$ with $x \in \Omega \subset \mathbb{R}$ and $\int_{\Omega} \rho(x)dx = 1$; its associated *Shannon Entropy* $S[\rho]$ is defined by [7]:

$$S[\rho] = -\int_{\Omega} \rho(x) \ln(\rho(x))dx. \quad (1)$$

This concept means a global measure of the information contained in the TS; it has a low degree of sensitivity to strong changes in the distribution originating from a small-sized region of the set Ω.

For a time series $\mathcal{X}(t) \equiv \{x_t; t = 1, \ldots, M\}$, a set of M measures of the observable $\mathcal{X}$ and the associated PDF, given by $P = \{p_i; i = 1, \ldots, N\}$, with $\sum_{i=1}^{N} p_i = 1$ and N as the number of possible states of the system under study, the *Shannon entropy* (formally *Shannon's logarithmic information*) [7] is defined by

$$S[P] = -\sum_{i=1}^{N} p_i \ln(p_i). \quad (2)$$

Eq. (2) constitutes a function of the probability $P = \{p_i; i = 1, \ldots, N\}$, which is equal to zero when the outcomes of a certain experiment denoted by the index k associated with probabilities $p_k \approx 1$ will occur. Therefore, the known dynamics developed by the dynamical system under study is complete. If the knowledge of the system dynamics is minimal, all the states of the system can occur with equal probability; thus, this probability can be modeled by a uniform distribution $P_e = \{p_i = 1/N; \forall i = 1, \ldots, N\}$.

It is useful to define the so-called normalized Shannon entropy, denoted as $H[P]$ in which its expression is

$$H[P] = S[P]/S_{max}, \quad (3)$$

$(0 \leq H[P] \leq 1)$ with $S_{max} = S[P_e] = \ln N$.

In order to analyze the local aspects of variations in the content of information given by a TS is extended the use of the Fisher's Information Measure, which uses the gradient content of the PDF, and a difference that means the Shannon Entropy, the FIM as can be seen from its definition given in the expression (4) reflect tiny localized perturbations. It reads [8, 9]

$$F[\rho] = \int_{\Omega} \left| \frac{\partial}{\partial x} [\rho(x)] \right|^2 / \rho(x) dx = 4 \int_{\Omega} \left| \frac{\partial}{\partial x} [\psi(x)] \right|^2 dx, \quad (4)$$

where $\psi(x) = \sqrt{\rho(x)}$.

In this sense, the Fisher information is a local information quantifier. It has various interpretations, and, among others, it can be thought of as a measure of the ability to estimate a parameter. In other cases, it is applied to calculate the amount of information that can be extracted from a TS and also as a measure of the state of disorder of a system or phenomenon [8]. The so-called Cramer-Rao bound can be considered as the most important property in which the FIM participates [9]. The local sensitivity of FIM can contribute in such cases in which the analysis necessitates an appeal to a notion of *order*. When there are certain points in the set Ω at which the PDF $\rho(x) \rightarrow 0$ is convenient to redefine the FIM avoiding the division by $\rho(x)$, in such cases an alternative expression of can be found in [9].

The signal discretization carries a problem of loss of information. It was extended studies by several authors, for example, see [10, 11] and references therein. In particular, it entails the loss of Fisher's shift invariance, which has not been relevant in the present chapter. Taking in mind the considerations made above, the discrete normalized FIM runs over the interval [0,1] and [12] is given by

$$F[P] = F_0 \sum_{i=1}^{N-1} \left[(p_{i+1})^{1/2} - (p_i)^{1/2} \right]^2, \tag{5}$$

where the normalization constant F_0 is given by

$$F_0 = \begin{cases} 1 & \text{if } p_{i^*} = 1 \text{ for } i^* = 1 \text{ or } i^* = N \text{ and } p_i = 0 \forall i \neq i^* \\ 1/2 & \text{otherwise} \end{cases}. \tag{6}$$

The local sensitivity of FIM for discrete PDFs is reflected by the fact that the specific i-ordering of the discrete values in $P = \{p_i; i = 1, \ldots, N\}$ must be seriously taken into account in evaluating the sum in Eq. (5) [13]. Each term in Eq. (5) can be regarded as a kind of "distance" between two contiguous probabilities. Thus, a different ordering of the pertinent summands would lead to a different FIM value, thereby its local nature.

In a system with N different states which reach a very ordered state, we can think it generates a signal with a PDF given by $P_0 = \{p_k \cong 1, \text{and } p_i \cong 0; \forall k \neq i = 1, \ldots, N\}$, as it has a Shannon entropy $S[P_0] \cong 0$ and a normalized FIM $F[P_0] \cong F_0 = 1$. In the other extreme, if the system under analysis develops a very disordered state, it is natural to assume that this particular state is described by a PDF approximated by a uniform distribution $P_e = \{p_i = 1/N; \forall i = 1, \ldots, N\}$, and the corresponding Shannon entropy $S[P_e] \cong S_{max} = \ln N$ while $F[P_0] \cong 0$. In certain way it is easy to understand that the general behavior of the FIM is opposite to that of the Shannon entropy.

The third information quantifier applied in this chapter is the *statistical complexity measure* (SCM) which is a global informational quantifier. All the computations made in the present work were done with the definitions introduced by López-Ruiz et al., in their seminal paper [14] with improvements advanced by Lamberti et al. [15]. For a discrete probability distribution function (PDF), $P = \{p_i; i = 1, \ldots, N\}$, associated with a time series (TS), this functional $C[P]$ is given by

$$C[P] = Q_J[P, P_e].H[P], \tag{7}$$

where H denotes the amount of "disorder" given by the normalized Shannon entropy (Eq. (3)) and Q_J is called "disequilibrium," defined in terms of the Jensen-Shannon divergence, given by

$$Q_J[P, P_e] = Q_0 J[P, P_e] = Q_0 \{S[(P + P_e)/2] - S[P]/2 - S[P_e]/2\}. \tag{8}$$

The normalization condition Q_0 for the disequilibrium corresponds to the inverse of the maximum possible value of Jensen-Shannon divergence, that is, $Q_0 = J[P_0, P_e]$:

$$Q_0 = -2 \left\{ \left(\frac{N+1}{N} \right) \ln(N+1) - \ln(2N) + \ln N \right\}^{-1}. \tag{9}$$

In this way, we have $0 \leq H[P] \leq 1$ and $0 \leq Q_J[P, P_e] \leq 1$.

The $C[P]$ quantifies the existence of correlational structures giving a measure of the complexity of a TS. In the case of perfect order or total randomness of a signal coming of a dynamical system, the value of the $C[P]$ is identically null that means the signal possesses no structure. In between these two extreme instances, a large range of possible stages of physical structure may be realized by a dynamical system. These stages should be reflected in the features of the obtained PDF and quantified by a no-null $C[P]$.

The global character of the SCM arising in that its value does not change with different orderings of the PDF. So the $C[P]$ quantifies the disorder but also the degree of correlational structures. It is evident that the SCM adopted in this work is a not a trivial function of the entropy. It has consequences in the ranges that this information quantifier can take. For a given H value, the complexity C runs on a precise range limited by a minimum C_{min} and a maximum C_{max} [16]. These extreme values depend only on the probability space dimension and, of course, on the functional form adopted by H and Q_J.

2.2 The Bandt and Pompe approach to building up a PDF

In the beginning of this section, it was mentioned that during the analysis of a TS, one of the first steps is the computation of the PDF associated. Immediately a question emerges: What is the appropriate PDF that can be computed from the TS? The regrettable answer is not unique. There is no universal nonparametric algorithm given by the statistics in the literature to do with this task.

To give light in this subject, Bandt and Pompe (BP) [17] introduce a simple and robust symbolic method that takes into account the time causality connected with dynamics of the system. They proposed to use a symbol sequence from the TS that can be constructed in a natural way. So the PDF introduced by Bandt and Pompe (BP-PDF) did not use any kind of assumption about the model, in general unknown, in which of the underlying dynamics exists. To compute the BP-PDF, the "partitions" are constructed by comparing the order of neighboring relative values in the TS rather than by apportioning amplitudes according to different levels like in the usual amplitude statistic methodology.

One problem remains linked with the lack of information associated with the temporal causality in which origins are in the computed methodologies to calculate the amplitude of the histograms. To give an answer to this problem, Kowalski and co-workers [18] using the Cressie-Read family of divergence measure showed in quantitative assessment the advantages of the BP-PDF in relation to any scheme based upon the construction of the corresponding amplitude histogram of the PDF, and also the BP-PDF brought some insight information about the dynamics of the physical problem.

Two parameters are necessary to define at the time of computing the BP-PDF, namely, the embedding dimension and the embedding delay. To clarify these crucial concepts, we will give the following details. Let TS $\mathcal{X}(t) = \{x_t; t = 1, \dots, M\}$, with an embedding dimension $D > 1$ ($D \in \mathbb{N}$) and an embedding delay $\tau > 1$ ($\tau \in \mathbb{N}$); the BP pattern of order D generated by this selection of parameters shall be considered of the form

$$s \mapsto \left(x_{s-(D-1)\tau}, x_{s-(D-2)\tau}, x_{s-(D-3)\tau}, \dots, x_{s-\tau}, x_s\right). \tag{10}$$

So the methodology proposed by Bandt and Pompe has as a starting point for every time s, assigned with a D-dimensional vector that results from the evaluation of $\mathcal{X}(t)$ at times $s-(D-1)\tau, s-(D-2)\tau, ..., s-\tau, s$. It is easy to note that higher values of D imply more information about "the past" to contribute in the PDF.

Once time settled the ordinal pattern of order D related to the time sequence s, the next step is to compute the permutation pattern denoted by $\pi=(r_0, r_1, ..., r_{D-1})$ of $(0, 1, ..., D-1)$ that could be formalized by

$$x_{s-r_{(D-1)}\tau} \le x_{s-r_{(D-2)}\tau} \le ... \le x_{s-r_1\tau} \le x_{s-r_0\tau}. \tag{11}$$

At this stage of the BP-PDF procedure, the vector defined by Eq. (10) is converted into a definite symbol π. Then to get a unique result, BP considers that $r_k < r_{k-1}$ if $x_{s-r_k\tau} = x_{s-r_{k-1}\tau}$. This is justified if the values of $\{x_t\}$ have a continuous distribution so that equal values are very unusual.

Considering all the $D!$ possible orderings (permutations) π_i when embedding dimension D, their associated relative frequencies can be naturally computed according to the number of times; this sequence order is found in the TS, divided by the total number of sequences

$$p(\pi_i) = \frac{\#\{s | s \le M-(D-1)\tau; (s) \; has \; type \; \pi_i\}}{M-(D-1)\tau}. \tag{12}$$

In Eq. (12) the symbol # (usually applied to designate the set cardinality) means "number." In such a way, an ordinal pattern probability distribution $\Pi = \{p(\pi_i); i = 1, ..., D!\}$ is constructed from the TS.

Time series amplitude information is not considered, and it is a clear disadvantage of the methodology proposed by BP, but it is compensated by the valuable information given by the intrinsic structure of the process under analysis. The scheme proposed by BP can be understood as a symbolic representation of time series by recourse to a comparison of consecutive points ($\tau=1$) or nonconsecutive ($\tau>1$) points allowing for an accurate empirical reconstruction of the underlying phase space, even in the presence of weak (observational and dynamical) noise [17]. It is noticeable that the ordinal-pattern's associated PDF results invariant with respect to nonlinear monotonous transformations. Accordingly, nonlinear drifts or scaling artificially introduced by a measurement device will not modify the quantifier estimation, a nice property if one deals with experimental data (see [19]). Summing up all these advantages makes the BP methodology a better choice than conventional methods based on range partitioning.

Among other properties, we can mention the following characteristics to give reasons in the selection of the BP-PDF: (i) the reduced number of parameters needed contributes to its simplicity of implementation (D and τ the embedding length and delay, respectively), and (ii) the time required in the calculation process is in fact very short. The BP methodology has an extra advantage; it can be used to compute the PDF in TS arising in low-dimensional dynamical systems, and signals originated in a wide diversity of systems as well as chaotic, noisy, and regular reality-based ones, with a light analysis in the stationarity because there no mandatory condition to accomplish with a strong stationary assumption (for details see [17]).

Parameter D, required by the BP-PDF methodology, determines the number of accessible states which is given by $D!$. Moreover, the minimum length of the TS must satisfy the condition $M >> D!$ in order to achieve a reliable statistic and proper distinction between stochastic and deterministic dynamics [20]. The seminal work of BP [17] includes an advice on the choice of range of the parameters to compute the BP-PDF, when the selection of time lag is $\tau=1$, and recommends the other parameter (D) to pick up on the interval $3 \le D \le 6$.

2.3 Ordinal patterns for deterministic processes

There is a demonstrated fact, done by Amigó et al. [21, 22], that in the case of deterministic one-dimensional maps, independently of the TS length M, *not all possible ordinal patterns*, applying BP methodology [17], can effectively give orbits in the phase space. This is a kind of a new dynamical property that means the existence of *forbidden ordinal patterns.* The proximity of patterns as well as correlation is not linked with the abovementioned property [21, 22]. So the informational quantifiers give a new characteristic in the analysis of chaotic or deterministic TS.

2.4 Causal informational planes

To characterize a given dynamical system described by a TS, we are able to use two representation spaces: (a) one with global-global characteristics called causal entropy-complexity plane ($H \times C$) and (b) one with global-local characteristics called causality Shannon-Fisher plane ($H \times F$), respectively.

The time causal nature of the Bandt and Pompe PDF gives a criterion to separate and differentiate chaotic and stochastic systems in different regions in both informational planes ($H[\Pi] \times C[\Pi]$) [23] and ($H[\Pi] \times F[\Pi]$) [24, 25]. While the global plane gives information of the complexity of a system, the local one becomes able to separate different dynamical behaviors in function of a control parameter.

3. Description of the chaotic maps

We focus our attention on two chaotic maps, namely, the logistic map and logistic map with delay.

3.1 The logistic map

One of the most used examples of deterministic chaotic systems is the logistic map. Its simplicity and easy computational implementation had been one of the most useful tools to explain chaotic behavior. It is a quadratic map $\mathcal{F} : x_n \rightarrow x_{n+1}$ [26], described by the ecologically motivated, dissipative system given by the first-order difference equation:

$$x_{n+1} = r\, x_n (1 - x_n), \tag{13}$$

where $0 \leq x_n \leq 1$ and $0 \leq r \leq 4$ can be associated with a kind of growth rate in the population dynamics. The corresponding Lyapunov exponent can be evaluated numerically [26] via

$$\Lambda(r) = \lim_{N \to \infty} \frac{1}{N} \sum_{n=0}^{N-1} \ln |r(1 - 2x_n)|, \tag{14}$$

where N is the number of iterations. **Figure 1a** and **1b** displays the well-known bifurcation diagram and the corresponding Lyapunov exponent $\Lambda(r)$, respectively, as a function of the parameter $3.4 \leq r \leq 4.0$ with $\Delta r = 0.0005$. We evaluated numerically the logistic map starting from a random initial condition in the interval $0 < x_0 < 0.5$. The first $N_0 = 10^5$ iterations are disregarded (transitory states), and the next $N = 10^6$ ones are used for Lyapunov evaluation (Eq. (14)) and information theory quantifiers (Eqs. (3), (5), and (7)).

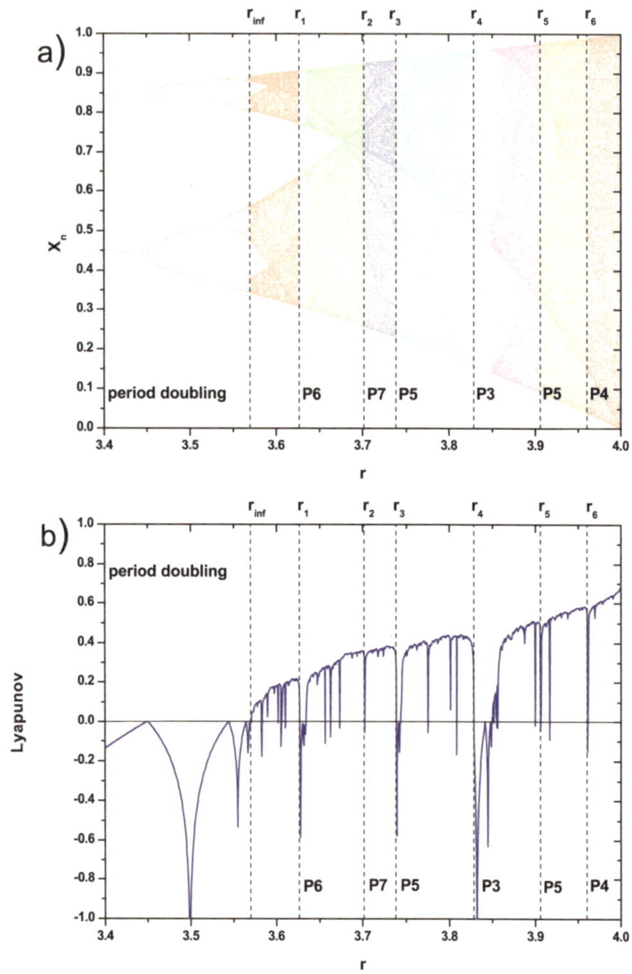

Figure 1.
(a) Bifurcation diagram and (b) Lyapunov exponent Λ for the logistic map as function of parameter r ($\Delta r = 0.0005$). The vertical segment lines delimited the different dynamical windows described in the text.

In the bifurcation diagram (**Figure 1a**), for fixed r, one appreciates that a periodic orbit consists of a countable set of points, while a chaotic attractor fills out dense bands within the unit interval. For $r \in [0, 1)$, one detects stable behavior $x_n = 0$. For $r \in [1, 3)$, there exist only a single steady-state solution given by $x_n = 1 - 1/r$. Increasing the control parameter, for $r \in [3, r_\infty)$, forces the system to undergo period-doubling bifurcations. Cycles of periods 2, 4, 8, 16, 32, etc. occur, and, if r_n denotes the values of r for which a 2^n cycle first appears, succesive r_ns converge to the limiting value $r_\infty \approx 3.5699456$ [26]. The value r_∞ splits the final-state diagram into two distinct parts: (a) the period-doubling zone on the left and (b) an area governed mainly by increasing chaotic behavior on the right. From **Figure 1b**, we see that period-doubling zone $r \in [3, r_\infty)$ Lyapunov are $\Lambda(r) \leq 0$, approximating to zero at each period-doubling bifurcation. The onset of chaos is apparent at r_∞ where Λ becomes positive for the first time. For $r = 4$ the iterates of the logistic map are represented by a random-looking distribution of dots which vertically span the range $x_n \in [0, 1]$, that is, complete developed chaos. For $r > r_\infty$ the Lyapunov exponent increases globally (see **Figure 1b**), except for dips one sees in the windows of periodic behavior. In the chaotic regime $r \in [r_\infty, 4]$, the period is infinitely long, and finite regions of the interval are visited by the orbits. Many periodic windows are observed, and all possible periods are represented, but the width of the window decreases as the period increases. Periodic windows suddenly appear as r increases, and they contain their own periodic-doubling route toward chaos. These facts exhibit the self-similar nature of the logistic map.

In **Figure 1a** and **1b**, we marked eight zones in order to analyze the logistic map behavior. They are *Zone 1*, $r \in [3.4, r_\infty)$ which corresponds to the period-doubling zone; *Zone 2*, $r \in [r_\infty, r_1)$, with $r_1 = 3.626557$, which corresponds to the start of periodic window of period 6; *Zone 3*, $r \in [r_1, r_2)$, with $r_2 = 3.701645$, which corresponds to the start of periodic window of period 7; *Zone 4*, $r \in [r_2, r_3)$, with $r_3 = 3.738177$, which corresponds to the start of periodic window of period 5; *Zone 5*, $r \in [r_3, r_4)$, with $r_4 = 3.828427$, which corresponds to the start of periodic window of period 3; *Zone 6*, $r \in [r_4, r_5)$, the largest periodic window, with $r_5 = 3.905573$, which corresponds to the start of periodic window of period 5; *Zone 7*, $r \in [r_5, r_6)$, with $r_6 = 3.960108$, which corresponds to the start of periodic window of period 4; and *Zone 8*, $r \in [r_6, 4]$, with $r = 4$ for fully developed chaos.

Periodic windows "interrupt" chaotic behavior in noticeable fashion. At the beginning of a window, there is a sudden and dramatic change in the long-term behavior of the logistic map. Consider, for example, the behavior for $r \geq r_4$ corresponding to the beginning of a period 3 window. We see three miniature copies of the whole final-state diagram (**Figure 1a**), and, indeed, we can reproduce the entire scenario of *period-doubling* $\rightarrow$ *chaos* (*band splitting*) $\rightarrow$ *chaos* (*band merging*) again, albeit at a much smaller scale. Same findings are encountered at all the other periodic windows, including miniature windows within the larger windows, as evidence of self-similarity.

3.2 The logistic map with delay

In 1948 Hutchinson [27] introduces a delay in the logistic equation to improve its applications in the study of population dynamics. The proposed model by Hutchinson has been applied in population dynamics [28], deterministic chaotic systems [29], the analysis of random discrete delay equations [30–32], etc. We face a discrete logistic equation with delay [33] given by the difference equation:

$$X_{n+1} = r\, X_n (1 - X_{n-1}), \tag{15}$$

with $0 \leq X_n \leq 1$ and $0 \leq r \leq 2.3$ (r the intrinsic growth). The equation resembles the logistic map (Eq. (13)) saved for the fact that the factor regulating population growth contains a one-generation time delay.

It is convenient to convert the second-order difference equation into an equivalent pair of first-order difference equations. The logistic map with delay is thus recasted as a two-dimensional map:

$$\begin{cases} x_{n+1} = r\, x_n (1 - y_n) \\ y_{n+1} = x_n \end{cases}, \tag{16}$$

and the corresponding Lyapunov exponents can be evaluated numerically [26] via

$$\Lambda_1(r) = \lim_{N \to \infty} \frac{1}{2N} \sum_{n=0}^{N-1} \ln \left| \frac{r^2 (1 - y_n - x_n z_n)^2 + 1}{1 + z_n^2} \right|, \tag{17}$$

and

$$\Lambda_1(r) + \Lambda_2(r) = \lim_{N\to\infty} \frac{1}{N} \sum_{n=0}^{N-1} \ln |\, r\, x_n|, \tag{18}$$

with $z_{n+1} = 1/\left[r\left(1 - y_n - x_n z_n\right]\right.$. In the previous equations, N is the number of iterations.

The pertinent bifurcation diagram and the corresponding Lyapunov exponents Λ_1 and Λ_2 are displayed in **Figure 2a** and **2b**, as a function of the parameter $0 \le r \le 2.3$ with $\Delta r = 0.0005$, respectively. We evaluated numerically the delayed logistic map starting from a random initial condition. The first $N_0 = 10^5$ iterations are disregarded (transitory states), and the next $N = 10^6$ ones are used for Lyapunov evaluation (Eqs. (15) and (16)) and information theory quantifiers (Eqs. (3), (5) and (7)).

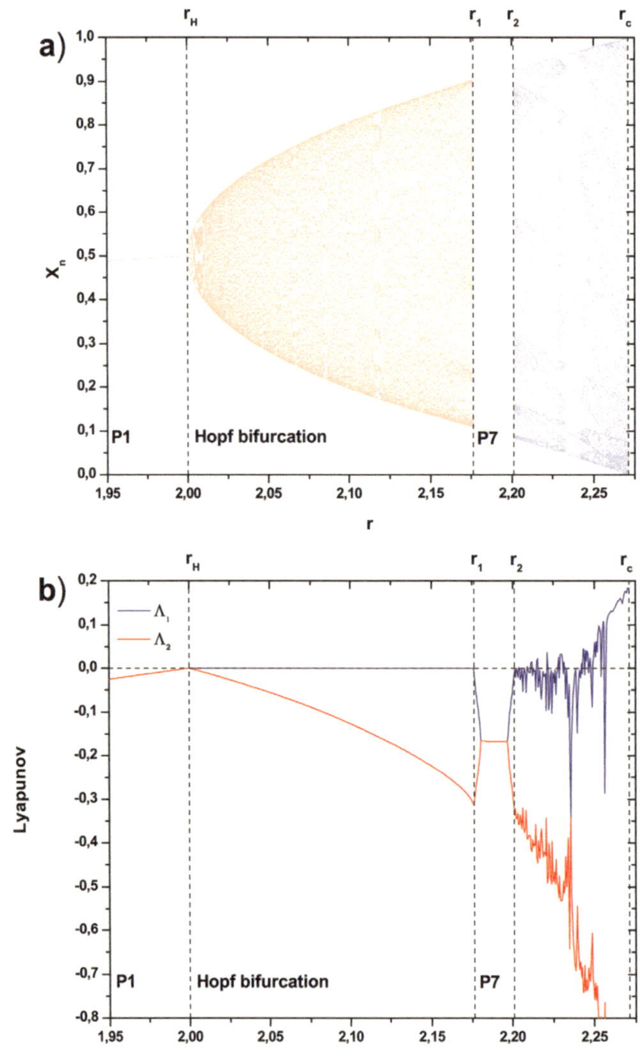

Figure 2.
(a) Bifurcation diagram and (b) Lyapunov exponents Λ_1 and Λ_2 for the delayed logistic map as function of parameter r ($\Delta r = 0.0005$). The vertical segment lines delimited the different dynamical windows described in the text.

This map has common characteristics with the usual logistic map. In particular, $X = 0$ is a fixed point for $r \in [0, 1)$, and $X_n = 1 - 1/r$ is a stationary state for $\in [1, r_H)$. In the delayed logistic map case, when the parameter value is $r_H = 2$, the system shows a Poincare-Andronov-Hopf bifurcation (see **Figure 2**). The quasiperiodic behavior persists over most of the range $r \in [r_H, r_1)$. A seven-cycle periodicity is observed for $r \in [r_1, r_2)$ (with $r_1 = 2.17640$ and $r_2 = 2.20071$). For the parameter $r \in [r_2, r_c)$, one mainly detects chaotic dynamics interspersed with regions of relative simplicity. For $r > r_c = 2.271$, the finite solutions are destabilized, and the system experiences a transition to $-\infty$. In the bifurcation diagram (see **Figure 2a**), it is difficult to distinguish quasiperiodicity from chaos, but the plot displaying Lyapunov exponents (see **Figure 2b**) indicates quasiperiodicity in the region where $\Lambda_1 = 0$.

4. Results and discussion

For each chaotic map previously described, the same time series of length $N = 10^6$ data, used for evaluating the corresponding Lyapunov exponents at each parameter value, are used now to build a Bandt-Pompe PDF (Π), taking an embedding dimension $D = 6$ and time lag $\tau = 1$. Then corresponding time causal information theory quantifiers, normalized Shannon entropy ($H[\Pi]$), statistical complexity ($C[\Pi]$), and Fisher information measure ($F[\Pi]$), were evaluated.

For ordinal entropic quantifiers of Shannon kind (global quantifiers), the BP-PDF provides univocal prescription. However, some ambiguities arise in the case in which one wishes to employ the BP-PDF to construct local quantifiers. The local sensitivity of the Fisher information measure for discrete PDFs is reflected in the fact that the specific "i-ordering" of the discrete values $p(\pi_i)$ must be taken into account in evaluating Eq. (5). If we are working with BP-PDF and consider patterns of length D, we will have $D!!$ possibilities for the i-ordering. We follow the Lehmer lexicographic order [34] in the generation of BP-PDF, because it provides the best graphic separation of different dynamics in the causal Shannon-Fisher plane [13, 24]. We display in **Figures 3** and **4** the causal information quantifiers (entropy, complexity and Fisher) as a function of the parameter r for the logistic map (see **Figure 1**) and delayed logistic map (see **Figure 2**), respectively. In these figures, the different dynamical zones and corresponding colors are both used in the original bifurcation diagrams (**Figures 1a** and **2a**).

For the logistic map, the period-doubling zone is detected for all the quantifiers. In particular for $r < r_\infty$, low entropy and complexity values and maximum Fisher value are found with the different periodic behaviors. A jump in the entropy and complexity value and a drop in Fisher value are observed when period doubling happens. This quantifier behavior is due to for periodic sequences the BP-PDF consisting of a very few $p(\pi_i) \neq 0$ values.

After r_∞ the dynamic becomes chaotic (positive Lyapunov exponent). An abrupt entropy and complexity growth and Fisher decreasing values are observed for $r > r_\infty$ reaching their maximum value at $r = 4$, where we face a totally developed chaotic dynamics. The several "drops" in the entropy and complexity, with the "jumps" in the Fisher values in the parameter interval $r_\infty < r \leq 4$, correspond to the periodic windows as can be easily confirmed compared with the bifurcation and Lyapunov exponent (see **Figure 1**).

For the delayed logistic map, a regular dynamic (steady state) is observed for parameter in the range $r < r_H$ and then has entropy and complexity null values and Fisher maximum value. For $r > r_H$ an oscillatory behavior appears, which is Hopf

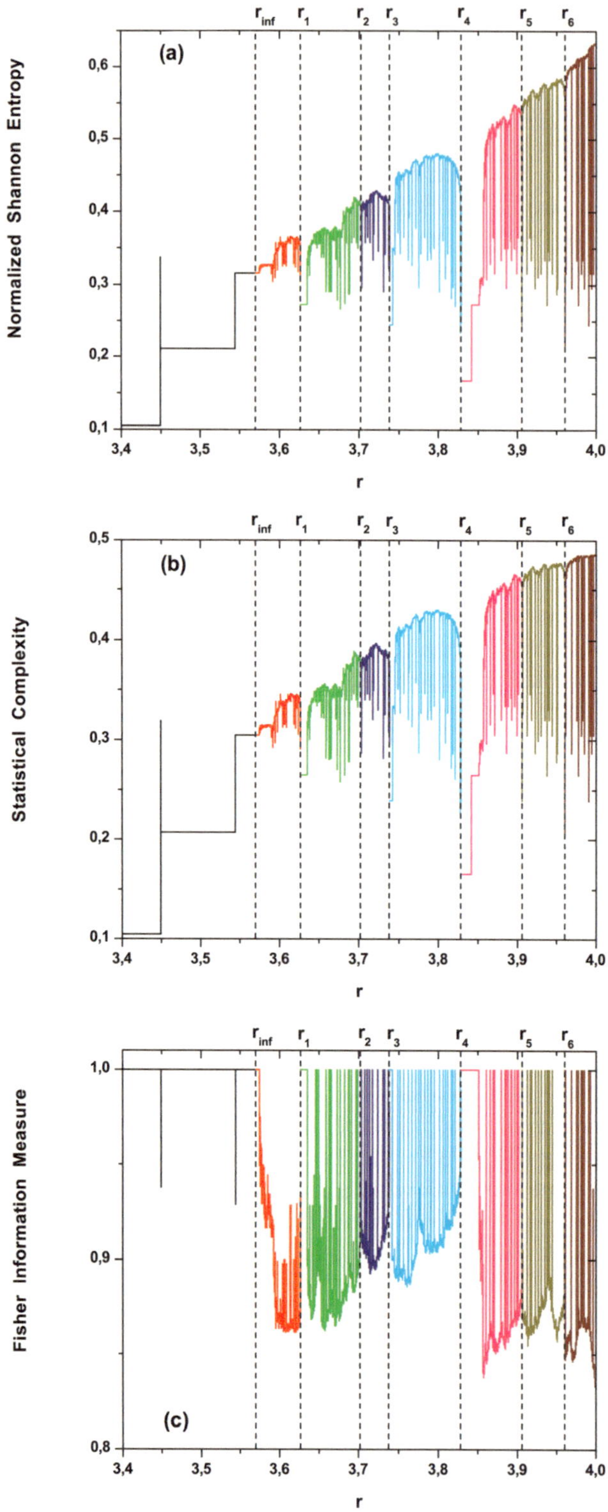

Figure 3.
Time causal information quantifiers for logistic map time series ($M = 10^6$ data) as a function of the parameter r ($\Delta r = 0.0005$): (a) italicized Shannon entropy; (b) statistical complexity; and (c) Fisher information measure, evaluated with Bandt-Pompe PDF, with $D = 6$, $\tau = 1$. The vertical segment lines delimited the different dynamical windows described in the text. The color code for the different zones is the same as in ***Figure 1a.***

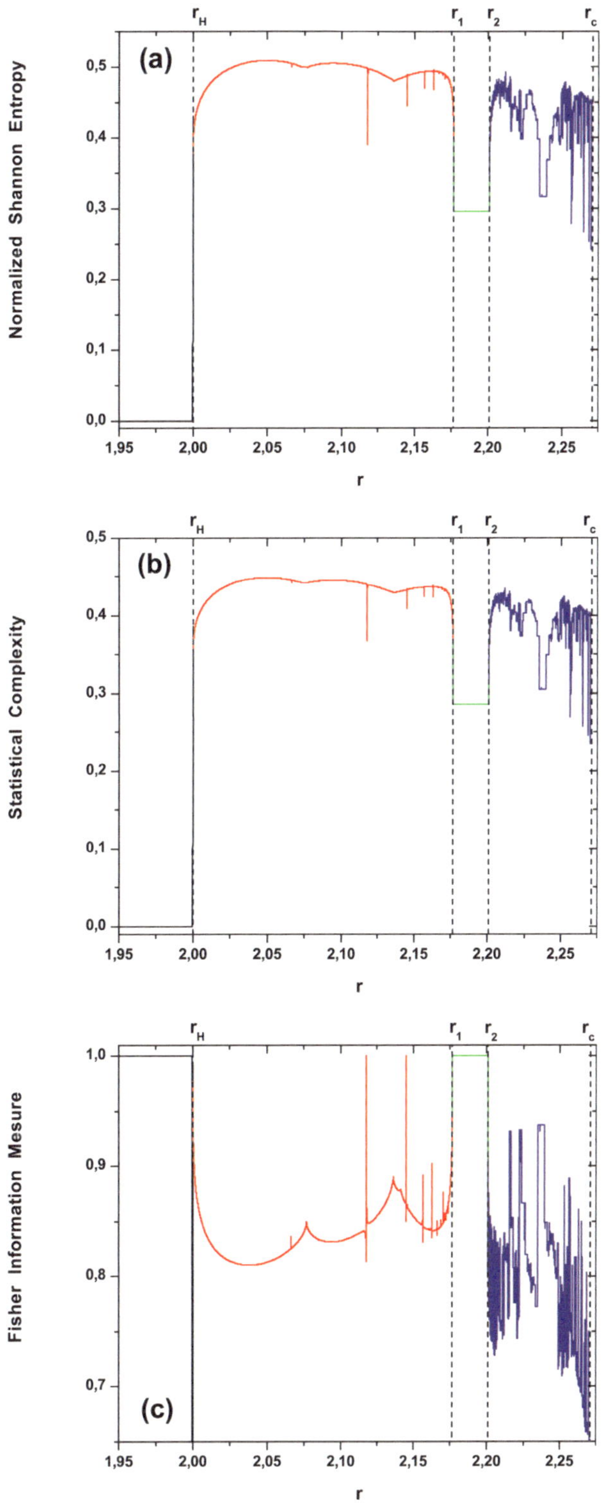

Figure 4.
Time causal information quantifiers for delayed logistic map time series ($M = 10^6$ data) as function of the parameter r ($\Delta r = 0.0005$): (a) italicized Shannon entropy; (b) statistical complexity; and (c) Fisher information measure, evaluated with Bandt-Pompe PDF, with $D = 6$ and $\tau = 1$. The vertical segment lines delimited the different dynamical windows described in the text. The color code for the different zones is the same as in ***Figure 2a***.

bifurcation with $\Lambda_1 = 0$. This quasiperiodic orbit can be thought as a mixture of periodic orbits of several different fundamental frequencies. The three quantifiers, entropy, complexity, and Fisher, are able to detect changes in the quasiperiodicity oscillations as a function of r, being Fisher the most sensitive (see **Figure 3b**). The growth in the amplitude of this oscillatory behavior as a function r is not detected by the quantifiers because of the independence of the BP-PDF on the amplitude values. Note that the same number of ordinal patterns (~30 patterns) is materialized for the whole quasiperiodic behavior, indicating its deterministic nature but not giving

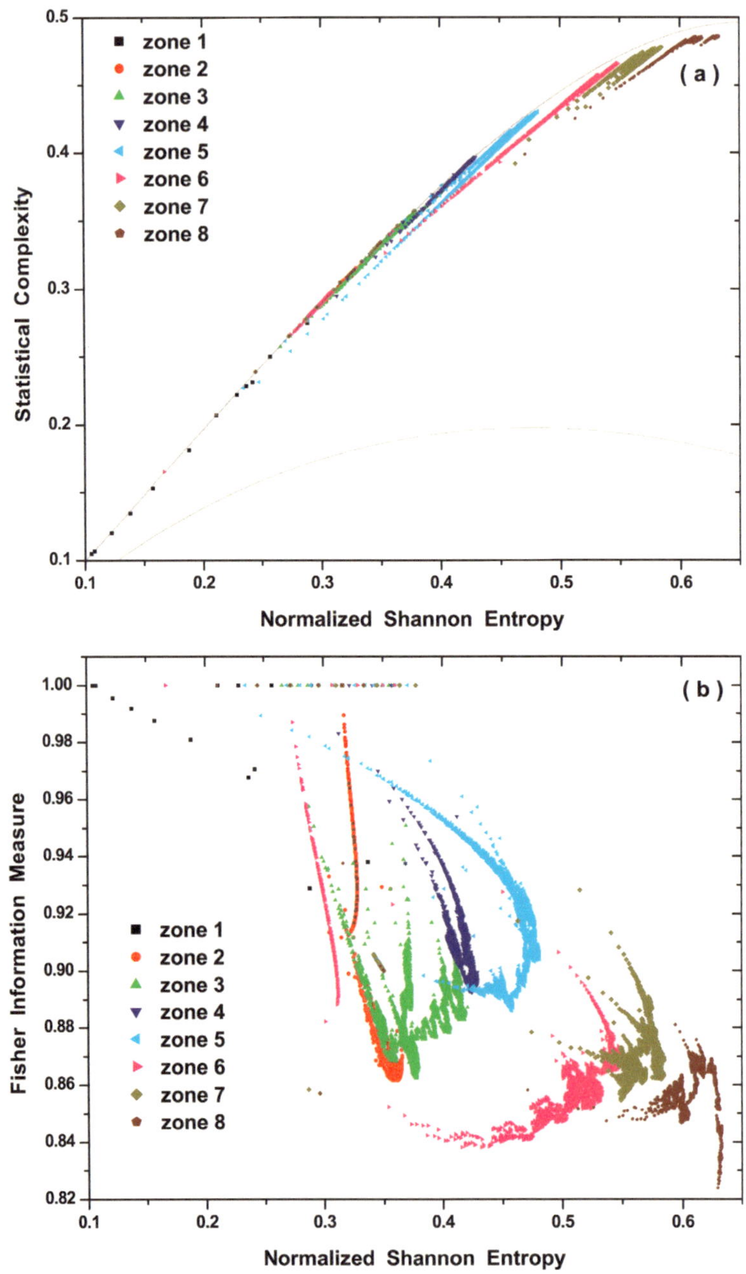

Figure 5.
Time causal information planes for logistic map time series ($M = 10^6$ data) for parameter r ($\Delta r = 0.0005$): (a) causal entropy-complexity plane and (b) causal Shannon-Fisher plane. The color code for the different zones is the same as in ***Figure 1a****.*

indications about the type of dynamics. For the parameter values $r_1 \leq r \leq r_2$, a period 7 window with $H = 0.299576$, $C = 0.288624$, and $F = 1$ is observed. In the parameter range $r_2 \leq r < r_c$, chaotic dynamics with some periodic windows is observed, characterized by higher values of entropy and complexity and lower values of Fisher, in relation to those previously obtained for the period 7 window.

The causal planes $H \times C$ and $H \times F$ for the logistic map in the parameter range $3.4 \leq r \leq 4.0$ are shown in **Figure 5a** and **5b**, respectively. Both planes provide a

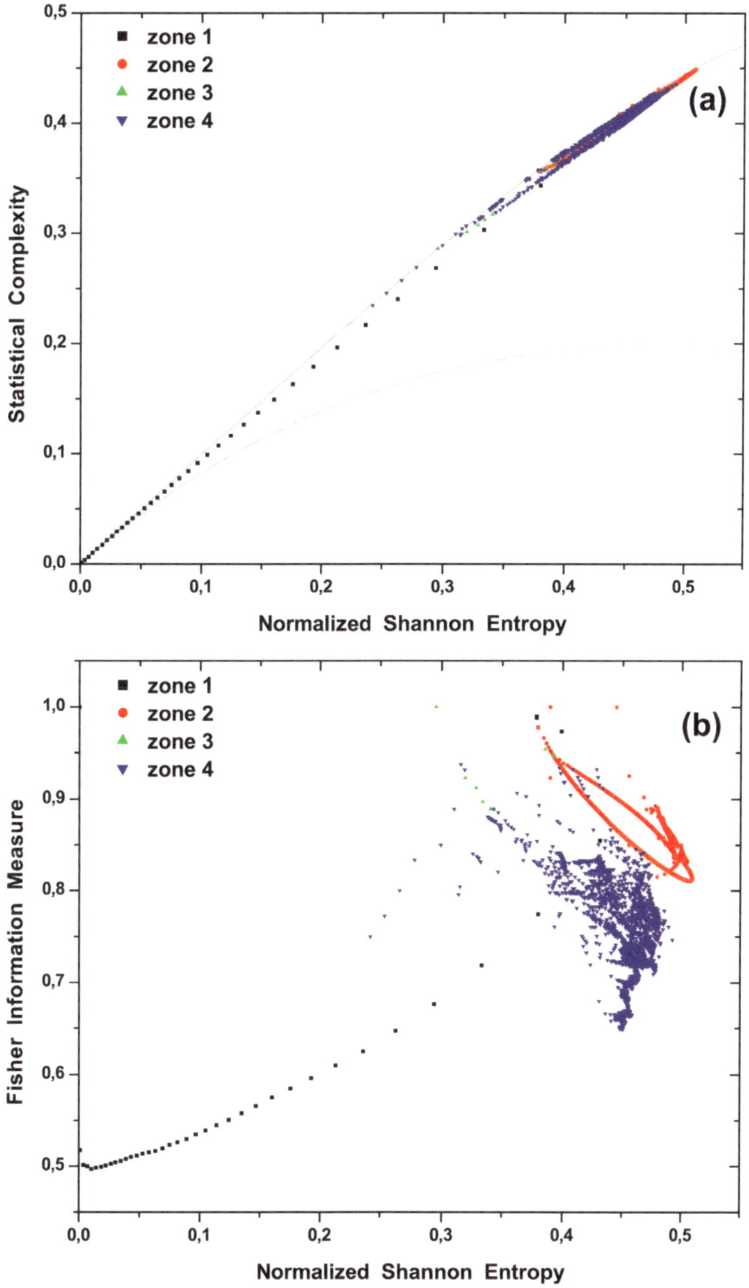

Figure 6.
*Time causal information planes for delayed logistic map time series ($M = 10^6$ data) for parameter r ($\Delta r = 0.0005$): (a) causal entropy-complexity plane and (b) causal Shannon-Fisher plane. The color code for the different zones is the same as in **Figure 2a**.*

characterization of the intrinsic information of the system, independently of the control parameter. From the causal plane $H \times C$ (**Figure 5a**), we can observed that the variation in the whole range of the parameter r locates the system very close to the maximum complexity curve C_{max}, reaching at $r = 4$ (totally developed chaos) and its maximum value $C = 0.48425$. Note also that low entropy values $H < 0.3$ correspond to periodic behavior values; however, these values have also associated high values of complexity with curve C_{max}, making difficult in this way the clear separation of the different dynamic behaviors, but a quantification of the global complexity of the logistic map is obtained. The causal plane $H \times F$ (see **Figure 5b**) shows a clear characterization of the various associated dynamics to different values of the control parameter r, locating them in different zones of the plane. It is in this instance, in which the Fisher permutation reveals its local character and, simultaneous with the global information delivered by Shannon entropy, gives us a sort of "topographical" plane of the dynamics.

In **Figure 6**, we show the behavior of the delayed logistic map for the whole range of the parameter r in the two causality planes. It is clearly that the $H \times C$ plane (**Figure 6a**) gives us just the information of the complexity of the map, which reaches the maximum curve (C_{max}), but does not differentiate between a Hopf bifurcation and the chaotic dynamics developed for $r > r_2$. On the other hand in the $H \times F$ plane (**Figure 6b**), one obtains a good defined structure for the quasiperiodic orbits, due to the oscillations in the quantifiers. The shapeless blue points for $r_2 < r < r_c$ is due to neither the H and the F are not detecting any kind of intermittency or bifurcation that can be present into the chaotic dynamic, at the present parameter resolution Δr.

5. Conclusions

We have shown that taken as starting point a probabilistic description of dynamical system considering the inherent temporal causality in the generated time series throughout Bandt-Pompe methodology, it is possible to evaluate information quantifiers of global or local character and a complete and detailed characterization of the dynamical system can be successfully archived with reference to an information causal plane, in which the two coordinate axes are different information quantifiers. The causal information planes defined are the global-global $H \times C$ plane and the global-local $H \times F$ plane, in which (i) the permutation normalized Shannon entropy ($H[\Pi]$) and the permutation statistical complexity ($C[\Pi]$) are responsible for the global features and (ii) the permutation Fisher information measure ($F[\Pi]$) accounts for the local attributes (all the information quantifiers are evaluated using BP-PDF denoted by Π).

For the discrete systems considered here, the logistic map and the delayed logistic map, we find that both H and C show a correspondence with one of the classic measures of chaoticity, the maximum exponent of Lyapunov, while the local sensitivity of F reveals details of the dynamics, invisible to the other quantifiers. The visualization of the location of the dynamics of the system under analysis, in the information planes, allows us to account for (a) the complexity of the system and (b) characterization of different dynamics in different locations of the plane, enabling the identification of different routes to chaos.

Author details

Felipe Olivares[1], Lindiane Souza[2], Walter Legnani[3] and Osvaldo A. Rosso[2,4]*

1 Instituto Física, Pontificia Universidad Católica de Valparaíso (PUCV), Valparaíso, Chile

2 Instituto Física, Universidad Federal de Alagoas (UFAL), Maceió, Brazil

3 Signals and Images Processsing Center (CSPSI), Facultad Regional Buenos Aires, Universidad Tecnológica Naciona (UTN), Ciudad Autónoma de Buenos Aires, Argentina

4 Instituto de Medicina Traslacional e Ingeniería Biomedica (IMTIB), Hospital Italiano de Buenos Aires (HIBA), CONICET, Ciudad Autónoma de Buenos Aires, Argentina

*Address all correspondence to: oarosso@gmail.com

References

[1] Kolmogorov AN. A new metric invariant for transitive dynamical systems and automorphisms in Lebesgue spaces. Doklady Akademii Nauk SSSR. 1959;**119**:861-864

[2] Sinai YG. On the concept of entropy for a dynamical system. Doklady Akademii Nauk SSSR. 1959;**124**:768-771

[3] Pesin YB. Characteristic Lyapunov exponents and smooth ergodic theory. Russian Mathematical Surveys. 1977;**32**: 55-114

[4] Abarbanel HDI. Analysis of Observed Chaotic Data. New York, USA: Springer-Verlag; 1996

[5] Gray RM. Entropy and Information Theory. Berlin-Heidelberg, Germany: Springer; 1990

[6] Shannon C, Weaver W. The Mathematical Theory of Communication. Champaign, IL: University of Illinois Press; 1949

[7] Brissaud JB. The meaning of entropy. Entropy. 2005;7:68-96

[8] Fisher RA. On the mathematical foundations of theoretical statistics. Philosophical Transactions of the Royal Society of London. Series A. 1922;**222**: 309-368

[9] Frieden BR. Science from Fisher information: A Unification. Cambridge: Cambridge University Press; 2004

[10] Zografos K, Ferentinos K, Papaioannou T. Discrete approximations to the Csiszár, Renyi, and Fisher measures of information. Canadian Journal of Statistics. 1986;**14**: 355-366

[11] Pardo L, Morales D, Ferentinos K, Zografos K. Discretization problems on generalized entropies and R-divergences. Kybernetika. 1994;**30**: 445-460

[12] Sánchez-Moreno P, Yáñez R, Dehesa J. Discrete densities and Fisher information. In: Proceedings of the 14th International Conference on Difference Equations and Applications. Istanbul, Turkey: Ugur-Bahçesehir University Press; 2009. pp. 291-298

[13] Olivares F, Plastino A, Rosso OA. Contrasting chaos with noise via local versus global information quantifiers. Physics Letters A. 2012;**376**:1577-1583

[14] López-Ruiz R, Mancini HL, Calbet X. A statistical measure of complexity. Physics Letters A. 1995;**209**:321-326

[15] Lamberti PW, Martín MT, Plastino A, Rosso OA. Intensive entropic non-triviality measure. Physica A: Statistical Mechanics and Its Applications. 2004; **334**:119-131

[16] Martín MT, Plastino A, Rosso OA. Generalized statistical complexity measures: Geometrical and analytical properties. Physica A: Statistical Mechanics and Its Applications. 2006; **369**:439-462

[17] Bandt C, Pompe B. Permutation entropy: A natural complexity measure for time series. Physical Review Letters. 2002;**88**:174102

[18] Kowalski AM, Martín MT, Plastino A, George Judge G. On extracting probability distribution Information from time series. Entropy. 2012;**14**: 1829-1841

[19] Saco PM, Carpi LC, Figliola A, Serrano E, Rosso AO. Entropy analysis of the dynamics of EL Niño/Southern Oscillation during the Holocene. Physica A: Statistical Mechanics and Its Applications. 2010;**389**:5022-5027

[20] Kowalski A, Martín MT, Plastino A, Rosso AO. Bandt-Pompe approach to the classical-quantum transition. Physica D: Nonlinear Phenomena. 2007; **233**:21-31

[21] Amigó JM, Zambrano S, Sanjuán MAF. True and false forbidden patterns in deterministic and random dynamics. Europhysics Letters. 2007;**79**:50001

[22] Amigó JM. Permutation Complexity in Dynamical Systems. Berlin, Germany: Springer-Verlag; 2010

[23] Rosso OA, Larrondo HA, Martín MT, Plastino A, Fuentes MA. Distinguishing noise from chaos. Physical Review Letters. 2007;**99**: 154102

[24] Olivares F, Plastino A, Rosso OA. Ambiguities in the Bandt and Pompe's methodology for local entropic quantifiers. Physica A: Statistical Mechanics and Its Applications. 2012; **391**:2518-2526

[25] Rosso OA, Olivares F, Plastino A. Noise versus chaos in a causal Fisher-Shannon plane. Papers in Physics. 2015; 7:070006

[26] Sprott JC. Chaos and Time Series Analysis. Oxford: Oxford University Press; 2004

[27] Hutchinson GE. Circular casual systems in ecology. Annals of New York Academy of Sciences. 1948;**50**:221-246

[28] Pounder JR, Rogers TD. The geometry of chaos: Dynamics of a nonlinear second order difference equation. Bulletin of Mathematical Biology. 1980;**42**:551-597

[29] Aronson DG, Chory MA, Hall GR, McGehee RP. Bifurcations from an invariant circle for two-parameter families of maps of the plane: A computer-assisted study. Communications in Mathematical Physics. 1982;**83**:303-354

[30] Cabrera JL, De La Rubia FJ. Numerical analysis of transient behavior in discrete random logistic equation with delay. Physics Letters A. 1995;**197**: 19-24

[31] Cabrera JL, De La Rubia FJ. Analysis of the behavior of a random nonlinear delay discrete equation. International Journal of Bifurcation and Chaos. 1996; **6**:1683-1690

[32] Cabrera JL, De La Rubia FJ. Resonance-like phenomena induced by exponentially correlated parametric noise. Europhysics Letters. 1997;**39**: 123-128

[33] Morimoto Y. Hopf bifurcation in the nonlinear recurrence equation xt+1= a xt (1-xt-1). Physics Letters A. 1988;**13**: 179-182

[34] Schwarz K. The Archive of Interesting Code. 2011. Available from: http://www.keithschwarz.com/interesting/code/?dir=factoradic-permutation

Chapter 6

On the Stabilization of Infinite Dimensional Semilinear Systems

El Hassan Zerrik and Abderrahman Ait Aadi

Abstract

This chapter considers the question of the output stabilization for a class of infinite dimensional semilinear system evolving on a spatial domain Ω by controls depending on the output operator. First we study the case of bilinear systems, so we give sufficient conditions for exponential, strong and weak stabilization of the output of such systems. Then, we extend the obtained results for bilinear systems to the semilinear ones. Under sufficient conditions, we obtain controls that exponentially, strongly, and weakly stabilize the output of such systems. The method is based essentially on the decay of the energy and the semigroup approach. Illustrations by examples and simulations are also given.

Keywords: semilinear systems, output stabilization, feedback controls, decay estimate, semigroups

1. Introduction

We consider the following semilinear system

$$\begin{cases} \dot{z}(t) = Az(t) + v(t)Bz(t), & t \geq 0, \\ z(0) = z_0, \end{cases} \tag{1}$$

where $A : D(A) \subset H \to H$ generates a strongly continuous semigroup of contractions $(S(t))_{t \geq 0}$ on a Hilbert space H, endowed with norm and inner product denoted, respectively, by $\|.\|$ and $\langle .,. \rangle$, $v(.) \in V_{ad}$ (the admissible controls set) is a scalar valued control and B is a nonlinear operator from H to H with $B(0) = 0$ so that the origin be an equilibrium state of system (1). The problem of feedback stabilization of distributed system (1) was studied in many works that lead to various results. In [1], it was shown that the control

$$v(t) = -\langle z(t), Bz(t) \rangle, \tag{2}$$

weakly stabilizes system (1) provided that B be a weakly sequentially continuous operator such that, for all $\psi \in H$, we have

$$\langle BS(t)\psi, S(t)\psi \rangle = 0, \quad \forall t \geq 0 \Rightarrow \psi = 0, \tag{3}$$

and if (3) is replaced by the following assumption

IntechOpen

$$\int_0^T |\langle BS(s)\psi, S(s)\psi\rangle| ds \geq \gamma \|\psi\|^2, \quad \forall \psi \in H \text{ (for some } \gamma, T > 0), \tag{4}$$

then control (2) strongly stabilizes system (1) [2].

In [3], the authors show that when the resolvent of A is compact, B self-adjoint and monotone, then strong stabilization of system (1) is proved using bounded controls.

Now, let the output state space Y be a Hilbert space with inner product $\langle ., .\rangle_Y$ and the corresponding norm $\|.\|_Y$, and let $C \in \mathcal{L}(H, Y)$ be an output operator.

System (1) is augmented with the output

$$w(t) := Cz(t). \tag{5}$$

The output stabilization means that $w(t) \to 0$ as $t \to +\infty$ using suitable controls. In the case when $Y = H$ and $C = I$, one obtains the classical stabilization of the state. If Ω be the system evolution domain and $\omega \subset \Omega$, when $C = \chi_\omega$, the restriction operator to a subregion ω of Ω, one is concerned with the behaviour of the state only in a subregion of the system evolution domain. This is what we call regional stabilization.

The notion of regional stabilization has been largely developed since its closeness to real applications, and the existence of systems which are not stabilizable on the whole domain but stabilizable on some subregion ω. Moreover, stabilizing a system on a subregion is cheaper than stabilizing it on the whole domain [4–8]. In [9], the author establishes weak and strong stabilization of (5) for a class of semilinear systems using controls that do not take into account the output operator.

In this paper, we study the output stabilization of semilinear systems by controls that depend on the output operator. Firstly we consider the case of bilinear systems, then we give sufficient conditions to obtain exponential, strong and weak stabilization of the output. Secondly, we consider the case of semilinear systems, and then under sufficient conditions, we obtain controls that exponentially, strongly, and weakly stabilize the output of such systems. The method is based essentially on the decay of the energy and the semigroup approach. Illustrations by examples and simulations are also given.

This paper is organized as follows: In Section 2, we discuss sufficient conditions to achieve exponential, strong and weak stabilization of the output (5) for bilinear systems. In Section 3, we study the output stabilization for a class of semilinear systems. Section 4 is devoted to simulations.

2. Stabilization for bilinear systems

In this section, we develop sufficient conditions that allow exponential, strong and weak stabilization of the output of bilinear systems. Consider system (1) with $B : H \to H$ is a bounded linear operator and augmented with the output (5).

Definition 1.1 The output (5) is said to be:

1. weakly stabilizable, if there exists a control $v(.) \in V_{ad}$ such that for any initial condition $z_0 \in H$, the corresponding solution $z(t)$ of system (1) is global and satisfies

$$\langle Cz(t), \psi\rangle_Y \to 0, \quad \forall \psi \in Y, \quad \text{as } t \to \infty,$$

2. strongly stabilizable, if there exists a control $v(.) \in V_{ad}$ such that for any initial condition $z_0 \in H$, the corresponding solution $z(t)$ of system (1) is global and verifies

$$\|Cz(t)\|_Y \to 0, \quad \text{as } t \to \infty,$$

and

3. exponentially stabilizable, if there exists a control $v(.) \in V_{ad}$ such that for any initial condition $z_0 \in H$, the corresponding solution $z(t)$ of system (1) is global and there exist $\alpha, \beta > 0$ such that

$$\|Cz(t)\|_Y \leq \alpha e^{-\beta t}\|z_0\|, \quad \forall t > 0.$$

Remark 1. It is clear that exponential stability of (5) $\Rightarrow$ strong stability of (5) $\Rightarrow$ weak stability of (5).

2.1 Exponential stabilization

The following result provides sufficient conditions for exponential stabilization of the output (5).

Theorem 1.2 Let A generate a semigroup $(S(t))_{t\geq 0}$ of contractions on H and if the condition:

1. $\mathcal{R}e(\langle C^* CAy, y\rangle) \leq 0, \ \forall y \in D(A)$, where C^* is the adjoint operator of C,
2. $\|CS(t)y\|_Y \leq \alpha\|Cy\|_Y$ and $\|CBy\|_Y \leq \beta\|Cy\|_Y$, for some $\alpha, \beta > 0$,
3. there exist $T, \gamma > 0$ such that

$$\int_0^T |\langle C^* CBS(t)y, S(t)y\rangle| dt \geq \gamma\|Cy\|_Y^2, \forall y \in H, \tag{6}$$

hold, then there exists $\rho > 0$ for which the control

$$v(t) = -\rho \operatorname{sign}(\langle C^* CBz(t), z(t)\rangle)$$

exponentially stabilizes the output (5).

Proof: System (1) has a unique mild solution $z(t)$ [10] defined on a maximal interval $[0, t_{\max}]$ by the variation of constants formula

$$z(t) = S(t)z_0 + \int_0^t v(s)S(t-s)Bz(s)ds. \tag{7}$$

From hypothesis 1, we deduce

$$\frac{d}{dt}\|Cz(t)\|_Y^2 \leq -2\rho \ |\langle C^* CBz(t), z(t)\rangle|.$$

Integrating this inequality, we get

$$\|Cz(t)\|_Y^2 - \|Cz(0)\|_Y^2 \leq -2\rho \int_0^t |\langle C^* CBz(\tau), z(\tau)\rangle| d\tau. \tag{8}$$

It follows that

$$\|Cz(t)\|_Y \leq \|Cz_0\|_Y. \tag{9}$$

For all $z_0 \in H$ and $t \geq 0$, we have

$$\langle C^*CBS(t)z_0, S(t)z_0\rangle = \langle C^*CBz(t), z(t)\rangle - \langle C^*CBz(t), z(t) - S(t)z_0\rangle \\ + \langle C^*CB(S(t)z_0 - z(t)), S(t)z_0\rangle.$$

Using hypothesis 2 and (9), we have

$$|\langle C^*CBS(t)z_0, S(t)z_0\rangle| \leq |\langle C^*CBz(t), z(t)\rangle| + 2\rho\alpha\beta\|C(z(t) - S(t)z_0)\|_Y\|Cz_0\|_Y.$$

It follows that from (7) and condition 2 that

$$|\langle C^*CBS(t)z_0, S(t)z_0\rangle| \leq |\langle C^*CBz(t), z(t)\rangle| + 2\rho\alpha^2\beta^2T\|Cz_0\|_Y^2. \tag{10}$$

Integrating (10) over the interval $[0, T]$ and replacing z_0 by $z(t)$ and using (6), we deduce that

$$\left(\gamma - 2\rho\alpha^2\beta^2T^2\right)\|Cz(t)\|_Y^2 \leq \int_t^{t+T} |\langle C^*CBz(s), z(s)\rangle| ds. \tag{11}$$

It follows from the inequality (8) that the sequence $\|Cz(n)\|_Y$ decreases and that for all $n \in \mathbb{N}$, we have

$$\|Cz(nT)\|_Y^2 - \|Cz((n+1)T)\|_Y^2 \geq 2\rho \int_{nT}^{(n+1)T} |\langle C^*CBz(s), z(s)\rangle| ds.$$

Using (11), we deduce

$$\|Cz(nT)\|_Y^2 - \|Cz((n+1)T)\|_Y^2 \geq 2\rho\left(\gamma - 2\rho\alpha^2\beta^2T^2\right)\|Cz(nT)\|_Y^2.$$

Taking $0 < \rho < \frac{\gamma}{2\alpha^2\beta^2T^2}$, we get

$$\|Cz(nT)\|_Y^2 \geq 2\rho\left(1 + 2\rho\left(\gamma - 2\rho\alpha^2\beta^2T^2\right)\right)\|Cz((n+1)T)\|_Y^2.$$

Then

$$\|Cz(nT)\|_Y^2 \leq \frac{1}{M^n}\|Cz_0\|_Y^2.$$

where $M = \left(1 + 2\rho\left(\gamma - 2\rho\alpha^2\beta^2T^2\right)\right) > 1$.
Since $\|Cz(t)\|_Y$ decreases, we deduce that

$$\|Cz(t)\|_Y \leq \sqrt{M}e^{\frac{-\ln(M)}{2T}t}\|z_0\|, \forall t \geq 0,$$

which gives the exponential stability of the output (5).
Example 1 On $\Omega =]0, 1[$, we consider the following system

$$\begin{cases} \dfrac{\partial z(x,t)}{\partial t} = Az(x,t) + v(t)z(x,t) & \Omega \times]0, +\infty[\\ z(x,0) = z_0(x) & \Omega, \end{cases} \tag{12}$$

where $H = L^2(\Omega)$ and $Az = -z$. The operator A generates a semigroup of contractions on $L^2(\Omega)$ given by $S(t)z_0 = e^{-t}z_0$. Let ω be a subregion of Ω. System (12) is augmented with the output

$$w(t) := \chi_\omega z(t), \tag{13}$$

where $\chi_\omega : L^2(\Omega) \to L^2(\omega)$, the restriction operator to ω and χ_ω^* is the adjoint operator of χ_ω. Conditions 1 and 3 of Theorem 1.2 hold, indeed: we have

$$\langle \chi_\omega^* \chi_\omega A y, y \rangle = -\|\chi_\omega y\|^2_{L^2(\omega)} \leq 0, \quad \forall y \in L^2(\Omega),$$

and for $T = 2$, we have

$$\int_0^2 \langle \chi_\omega^* \chi_\omega B e^{-t} y, e^{-t} y \rangle dt = \int_0^2 e^{-2t} dt \int_\omega |y|^2 dx = \left(\frac{1}{2} - \frac{1}{2e^4}\right) \|\chi_\omega y\|^2_{L^2(\omega)}.$$

We conclude that for all $0 < \rho < \frac{e^4 - 1}{16 e^4}$, the control

$$v(t) = \begin{pmatrix} -\rho & \text{if } \|\chi_\omega z(t)\|^2_{L^2(\omega)} \neq 0, \\ 0 & \text{if } \|\chi_\omega z(t)\|^2_{L^2(\omega)} = 0, \end{pmatrix}$$

exponentially stabilizes the output (13).

2.2 Strong stabilization

The following result will be used to prove strong stabilization of the output (5).

Theorem 1.3 Let A generate a semigroup $(S(t))_{t \geq 0}$ of contractions on H and $B : H \to H$ is a bounded linear operator. If the conditions:

1. $\mathcal{R}e(\langle C^* C A \psi, \psi \rangle) \leq 0, \quad \forall \psi \in D(A),$
2. $\mathcal{R}e(\langle C^* C B \psi, \psi \rangle \langle B \psi, \psi \rangle) \geq 0, \quad \forall \psi \in H$, hold, then control

$$v(t) = -\frac{\langle C^* C B z(t), z(t) \rangle}{1 + |\langle C^* C B z(t), z(t) \rangle|}, \tag{14}$$

allows the estimate

$$\left(\int_0^T |\langle C^* C B S(s) z(t), S(s) z(t) \rangle| ds \right)^2 = O\left(\int_t^{t+T} \frac{|\langle C^* C B z(s), z(s) \rangle|^2}{1 + |\langle C^* C B z(s), z(s) \rangle|} ds \right), \text{ as } t \to +\infty. \tag{15}$$

Proof: From hypothesis 1 of Theorem 1.3, we have

$$\frac{1}{2} \frac{d}{dt} \|C z(t)\|^2_Y \leq \mathcal{R}e(v(t) \langle C^* C B z(t), z(t) \rangle).$$

In order to make the energy nonincreasing, we consider the control

$$v(t) = -\frac{\langle C^* C B z(t), z(t) \rangle}{1 + |\langle C^* C B z(t), z(t) \rangle|},$$

so that the resulting closed-loop system is

$$\dot{z}(t) = A z(t) + f(z(t)), \quad z(0) = z_0, \tag{16}$$

where

$$f(y) = -\frac{\langle C^* CBy, y\rangle}{1 + |\langle C^* CBy, y\rangle|} By, \text{ for all } y \in H$$

Since f is locally Lipschitz, then system (16) has a unique mild solution $z(t)$ [10] defined on a maximal interval $[0, t_{max}]$ by

$$z(t) = S(t)z_0 + \int_0^t S(t-s) f(z(s))ds. \tag{17}$$

Because of the contractions of the semigroup, we have

$$\frac{d}{dt}\|z(t)\|^2 \leq -2\frac{\langle C^* CBz(t), z(t)\rangle \langle Bz(t), z(t)\rangle}{1 + |\langle C^* CBz(t), z(t)\rangle|}.$$

Integrating this inequality, we get

$$\|z(t)\|^2 - \|z(0)\|^2 \leq -2\int_0^t \frac{\langle C^* CBz(s), z(s)\rangle \langle Bz(s), z(s)\rangle}{1 + |\langle C^* CBz(s), z(s)\rangle|} ds.$$

It follows that

$$\|z(t)\| \leq \|z_0\|. \tag{18}$$

From hypothesis 1 of Theorem 1.3, we have

$$\frac{d}{dt}\|Cz(t)\|_Y^2 \leq -2\frac{|\langle C^* CBz(t), z(t)\rangle|^2}{1 + |\langle C^* CBz(t), z(t)\rangle|}.$$

We deduce

$$\|Cz(t)\|_Y^2 - \|Cz(0)\|_Y^2 \leq -2\int_0^t \frac{|\langle C^* CBz(s), z(s)\rangle|^2}{1 + |\langle C^* CBz(s), z(s)\rangle|} ds. \tag{19}$$

Using (17) and Schwartz inequality, we get

$$\|z(t) - S(t)z_0\| \leq \|B\|\|z_0\| \left(T\int_0^t \frac{|\langle C^* CBz(s), z(s)\rangle|^2}{1 + |\langle C^* CBz(s), z(s)\rangle|} ds \right)^{\frac{1}{2}}, \quad \forall t \in [0, T]. \tag{20}$$

Since B is bounded and C continuous, we have

$$|\langle C^* CBS(s)z_0, S(s)z_0\rangle| \leq 2K\|B\|\|z(s) - S(s)z_0\|\|z_0\| + |\langle C^* CBz(s), z(s)\rangle|, \tag{21}$$

where K is a positive constant.
Replacing z_0 by $z(t)$ in (20) and (21), we get

$$|\langle C^* CBS(s)z(t), S(s)z(t)\rangle| \leq 2K\|B\|^2\|z_0\|^2 \left(T\int_t^{t+T} \frac{|\langle C^* CBz(s), z(s)\rangle|^2}{1 + |\langle C^* CBz(s), z(s)\rangle|} ds \right)^{\frac{1}{2}}$$
$$+ |\langle C^* CBz(t+s), z(t+s)\rangle|, \qquad \forall t \geq s \geq 0.$$

Integrating this relation over $[0, T]$ and using Cauchy-Schwartz, we deduce

$$\int_0^T |\langle C^* CBS(s)z(t), S(s)z(t)\rangle| ds \leq \left(2K\|B\|^2 T^{\frac{3}{2}} + T\left(\left(1 + K\|B\|\|z_0\|^2\right)\right)\right.$$

$$\left. \times \left(\int_t^{t+T} \frac{|\langle C^* CBz(s), z(s)\rangle|^2}{1+|\langle C^* CBz(s), z(s)\rangle|} ds\right)^{\frac{1}{2}},\right.$$

which achieves the proof.

The following result gives sufficient conditions for strong stabilization of the output (5).

Theorem 1.4 Let A generate a semigroup $(S(t))_{t\geq 0}$ of contractions on H, B is a bounded linear operator. If the assumptions 1, 2 of Theorem 1.3 and

$$\int_0^T |\langle C^* CBS(t)\psi, S(t)\psi\rangle| dt \geq \gamma \|C\psi\|_Y^2, \quad \forall \psi \in H, \quad (for\ some\ \ T, \gamma > 0), \tag{22}$$

holds, then control (14) strongly stabilizes the output (5) with decay estimate

$$\|Cz(t)\|_Y = O\left(\frac{1}{\sqrt{t}}\right), \quad \text{as} \quad t \to +\infty. \tag{23}$$

Proof: Using (19), we deduce

$$\|Cz(kT)\|_Y^2 - \|Cz((k+1)T)\|_Y^2 \geq 2\int_{kT}^{k(T+1)} \frac{|\langle C^* CBz(t), z(t)\rangle|^2}{1 + |\langle C^* CBz(t), z(t)\rangle|} dt, \quad k \geq 0.$$

From (15) and (22), we have

$$\|Cz(kT)\|_Y^2 - \|Cz((k+1)T)\|_Y^2 \geq \beta \|Cz(kT)\|_Y^4, \tag{24}$$

where $\beta = \frac{\gamma^2}{2\left(2K\|B\|^2 T^{\frac{3}{2}} + T\left(1 + K\|B\|\|z_0\|^2\right)\right)^2}$.

Taking $s_k = \|Cz(kT)\|_Y^2$, the inequality (24) can be written as

$$\beta s_k^2 + s_{k+1} \leq s_k, \quad \forall k \geq 0.$$

Since $s_{k+1} \leq s_k$, we obtain

$$\beta s_{k+1}^2 + s_{k+1} \leq s_k, \quad \forall k \geq 0.$$

Taking $p(s) = \beta s^2$ and $q(s) = s - (I + p)^{-1}(s)$ in Lemma 3.3, page 531 in [11], we deduce

$$s_k \leq x(k), \quad k \geq 0,$$

where $x(t)$ is the solution of equation $x'(t) + q(x(t)) = 0, \quad x(0) = s_0$.

Since $x(k) \geq s_k$ and $x(t)$ decreases give $x(t) \geq 0$, $\forall t \geq 0$. Furthermore, it is easy to see that $q(s)$ is an increasing function such that

$$0 \leq q(s) \leq p(s), \forall s \geq 0.$$

We obtain $-\beta x(t)^2 \leq x'(t) \leq 0$, which implies that

$$x(t) = O\left(t^{-1}\right), \quad as\ \ t \to +\infty.$$

Finally the inequality $s_k \leq x(k)$, together with the fact that $\|Cz(t)\|_Y$ decreases, we deduce the estimate (23).

Example 2 Let us consider a system defined on $\Omega =]0,1[$ by

$$\begin{cases} \dfrac{\partial z(x,t)}{\partial t} = Az(x,t) + v(t)a(x)z(x,t) & \Omega \times]0, +\infty[\\ z(x,0) = z_0(x) & \Omega \\ z(0,t) = z(1,t) = 0 & t > 0, \end{cases} \tag{25}$$

where $H = L^2(\Omega)$, $Az = -z$, and $a \in L^\infty(]0,1[)$ such that $a(x) \geq 0$ a.e on $]0,1[$ and $a(x) \geq c > 0$ on subregion ω of Ω and $v(.) \in L^\infty(0,+\infty)$ the control function. System (25) is augmented with the output

$$w(t) = \chi_\omega z(t). \tag{26}$$

The operator A generates a semigroup of contractions on $L^2(\Omega)$ given by $S(t)z_0 = e^{-t}z_0$. For $z_0 \in L^2(\Omega)$ and $T = 2$, we obtain

$$\int_0^2 \langle \chi_\omega^* \chi_\omega BS(t)z_0, S(t)z_0 \rangle dt = \int_0^2 e^{-2t} dt \int_\omega a(x)|z_0|^2 dx \geq \beta \|\chi_\omega z_0\|^2_{L^2(\omega)},$$

with $\beta = c\int_0^2 e^{-2t}dt > 0$.

Applying Theorem 1.4, we conclude that the control

$$v(t) = -\frac{\int_\omega a(x)|z(x,t)|^2 dx}{1 + \int_\omega a(x)|z(x,t)|^2 dx}$$

strongly stabilizes the output (26) with decay estimate

$$\|\chi_\omega z(t)\|_{L^2(\omega)} = O\left(\frac{1}{\sqrt{t}}\right), \quad \text{as} \quad t \to +\infty.$$

2.3 Weak stabilization

The following result provides sufficient conditions for weak stabilization of the output (5).

Theorem 1.5 Let A generate a semigroup $(S(t))_{t\geq 0}$ of contractions on H and B is a compact operator. If the conditions:

1. $\mathcal{R}e(\langle C^*CA\psi, \psi \rangle) \leq 0, \quad \forall \psi \in D(A)$,

2. $\mathcal{R}e(\langle C^*CB\psi, \psi \rangle \langle B\psi, \psi \rangle) \geq 0, \quad \forall \psi \in H$,

3. $\langle C^*CBS(t)\psi, S(t)\psi \rangle = 0, \quad \forall t \geq 0 \Rightarrow C\psi = 0$ hold, then control (14) weakly stabilizes the output (5).

Proof: Let us consider the nonlinear semigroup $\Gamma(t)z_0 := z(t)$ and let (t_n) be a sequence of real numbers such that $t_n \to +\infty$ as $n \to +\infty$.

From (18), $\Gamma(t_n)z_0$ is bounded in H, then there exists a subsequence $(t_{\phi(n)})$ of (t_n) such that

$$\Gamma(t_{\phi(n)})z_0 \rightharpoonup \psi, \quad as\ n \to \infty.$$

Since B is compact and C continuous, we have

$$\lim_{n\to+\infty} \langle C^* CBS(t)\Gamma(t_{\phi(n)})z_0, S(t)\Gamma(t_{\phi(n)})z_0\rangle = \langle C^* CBS(t)\psi, S(t)\psi\rangle.$$

For all $n \geq$, we set

$$\Lambda_n(t) := \int_{\phi(n)}^{\phi(n)+t} \frac{|\langle C^* CB\Gamma(s)z_0, \Gamma(s)z_0\rangle|^2}{1 + |\langle C^* CB\Gamma(s)z_0, \Gamma(s)z_0\rangle|} ds.$$

It follows that $\forall t \geq 0$, $\Lambda_n(t) \to 0$ as $n \to +\infty$.
Using (15), we get

$$\lim_{n\to+\infty} \int_0^t |\langle C^* CBS(s)\Gamma(t_{\phi(n)})z_0, S(s)\Gamma(t_{\phi(n)})z_0\rangle| ds = 0.$$

Hence, by the dominated convergence Theorem, we have

$$\int_0^t |\langle C^* CBS(s)\psi, S(s)\psi\rangle| ds = 0.$$

We conclude that

$$\langle C^* CBS(s)\psi, S(s)\psi\rangle = 0, \quad \forall s \in [0, t].$$

Using condition 3 of Theorem 1.5, we deduce that

$$C\Gamma(t_{\phi(n)})z_0 \rightharpoonup 0, \quad \text{as} \quad n \to +\infty. \tag{27}$$

On the other hand, it is clear that (27) holds for each subsequence $(t_{\phi(n)})$ of (t_n) such that $C\Gamma(t_{\phi(n)})z_0$ weakly converges in Y. This implies that $\forall \varphi \in Y$, we have $\langle C\Gamma(t_n)z_0, \varphi\rangle \to 0$ as $n \to +\infty$ and hence

$$C\Gamma(t)z_0 \rightharpoonup 0, \quad \text{as} \quad t \to +\infty.$$

Example 3 Consider a system defined in $\Omega =]0, +\infty[$, and described by

$$\begin{cases} \frac{\partial z(x,t)}{\partial t} = -\frac{\partial z(x,t)}{\partial x} + v(t)Bz(x,t) & x \in \Omega,\ t > 0 \\ z(x,0) = z_0(x) & x \in \Omega \\ z(0,t) = z(\infty,t) = 0 & t > 0, \end{cases} \tag{28}$$

where $Az = -\frac{\partial z}{\partial x}$ with domain $D(A) = \{z \in H^1(\Omega) \mid z(0) = 0,\ z(x) \to 0\ as\ x \to +\infty\}$ and $Bz(.) = \int_0^1 z(x)dx(.)$ is the control operator. The operator A generates a semigroup of contractions

$$(S(t)z_0)(x) = \begin{cases} z_0(x-t) & \text{if } x \geq t \\ 0 & \text{if } x < t. \end{cases}$$

Let $\omega =]0, 1[$ be a subregion of Ω and system (28) is augmented with the output

$$w(t) = \chi_\omega z(t). \tag{29}$$

We have

$$\langle \chi_\omega^* \chi_\omega Az, z\rangle = -\int_0^1 z'(x)z(x)dx = -\frac{z^2(1)}{2} \leq 0,$$

so, the assumption 1 of Theorem 1.5 holds. The operator B is compact and verifies

$$\langle \chi_\omega^* \chi_\omega BS(t)z_0, S(t)z_0\rangle = \left(\int_0^{1-t} z_0(x)dx\right)^2, \quad 0 \leq t \leq 1.$$

Thus

$$\langle \chi_\omega^* \chi_\omega BS(t)z_0, S(t)z_0\rangle = 0, \quad \forall t \geq 0 \Rightarrow z_0(x) = 0, \ a.e \text{ on } \omega.$$

Then, the control

$$v(t) = -\frac{\left(\int_0^1 z(x,t)dx\right)^2}{1 + \left(\int_0^1 z(x,t)dx\right)^2}, \tag{30}$$

weakly stabilizes the output (29).

3. Stabilization for semilinear systems

In this section, we give sufficient conditions for exponential, strong and weak stabilization of the output (5). Consider the semilinear system (1) augmented with the output (5).

4. Exponential stabilization

In this section, we develop sufficient conditions for exponential stabilization of the output (5).

The following result concerns the exponential stabilization of (5).

Theorem 1.6 Let A generate a semigroup $(S(t))_{t\geq 0}$ of contractions on H and B be locally Lipschitz. If the conditions:

1. $\mathcal{Re}(\langle C^* CAy, y\rangle) \leq 0, \quad \forall y \in D(A)$,

2. $\mathcal{Re}(\langle C^* CBy, y\rangle \langle By, y\rangle) \geq 0, \quad \forall y \in H$,

3. there exist $T, \gamma > 0$, such that

$$\int_0^T |\langle C^* CBS(t)y, S(t)y\rangle| dt \geq \gamma \|Cy\|_Y^2, \quad \forall y \in H, \tag{31}$$

hold, then the control

$$v(t) = \begin{cases} -\dfrac{\langle C^* CBz(t), z(t)\rangle}{\|z(t)\|^2}, & \text{if} \quad z(t) \neq 0, \\ 0, & \text{if} \quad z(t) = 0, \end{cases} \tag{32}$$

exponentially stabilizes the output (5).

Proof: Since $(S(t))_{t\geq 0}$ is a semigroup of contractions, we have

$$\frac{d}{dt}\|z(t)\|^2 \leq 2\mathcal{R}e(v(t)\langle Bz(t), z(t)\rangle).$$

Integrating this inequality, and using hypothesis 2 of Theorem 1.6, it follows that

$$\|z(t)\| \leq \|z_0\|. \tag{33}$$

For all $z_0 \in H$ and $t \geq 0$, we have

$$\langle C^* CBS(t)z_0, S(t)z_0\rangle = \langle C^* CBz(t), z(t)\rangle - \langle C^* CBz(t), z(t) - S(t)z_0\rangle$$
$$+ \langle C^* CBS(t)z_0 - C^* CBz(t), S(t)z_0\rangle.$$

Since B is locally Lipschitz, there exists a constant positive L that depends on $\|z_0\|$ such that

$$|\langle C^* CBS(t)z_0, S(t)z_0\rangle| \leq |\langle C^* CBz(t), z(t)\rangle| + 2\alpha L\|z(t) - S(t)z_0\|\|z_0\|, \tag{34}$$

where α is a positive constant.

Using (33), we deduce

$$|\langle C^* CBz(t), z(t)\rangle| \leq |v(z(t))|\|z(t)\|\|z_0\|, \quad \forall t \in [0, T]. \tag{35}$$

While from the variation of constants formula and using Schwartz's inequality, we obtain

$$\|z(t) - S(t)z_0\| \leq L\left(T\int_0^T |v(z(t))|^2\|z(t)\|^2 dt\right)^{\frac{1}{2}}. \tag{36}$$

Integrating (34) over the interval $[0, T]$ and taking into account (35) and (36), we get

$$\int_0^T |\langle C^* CBS(t)z_0, S(t)z_0\rangle| dt \leq 2\alpha T^{\frac{3}{2}} L^2\|z_0\|\left(\int_0^T |v(z(t))|^2\|z(t)\|^2 dt\right)^{\frac{1}{2}}$$
$$+ T^{\frac{1}{2}}\|z_0\|\left(\int_0^T |v(z(t))|^2\|z(t)\|^2 dt\right)^{\frac{1}{2}}.$$

Now, let us consider the nonlinear semigroup $U(t)z_0 := z(t)$ [1].

Replacing z_0 by $U(t)z_0$ in (37), and using the superposition properties of the semigroup $(U(t))_{t\geq 0}$, we deduce that

$$\int_0^T |\langle C^* CBS(s)U(t)z_0, S(s)U(t)z_0\rangle| ds \leq 2\alpha T^{\frac{3}{2}} L^2\|U(t)z_0\|$$
$$\times \left(\int_t^{t+T} |v(U(s)z_0)|^2\|U(s)z_0\|^2 ds\right)^{\frac{1}{2}} \tag{37}$$
$$+ T^{\frac{1}{2}}\|U(t)z_0\|\left(\int_t^{t+T} |v(U(s)z_0)|^2\|U(s)z_0\|^2 ds\right)^{\frac{1}{2}}$$

Thus, by using (31) and (37), it follows that

$$\gamma\|CU(t)z_0\|_Y \le M\left(\int_t^{t+T} |v(U(s)z_0)|^2\|U(s)z_0\|^2 ds\right)^{\frac{1}{2}}, \tag{38}$$

where $M = \left(2\alpha TL^2 + 1\right)T^{\frac{1}{2}}$ is a non-negative constant depending on $\|z_0\|$ and T. From hypothesis 1 of Theorem 1.6, we have

$$\frac{d}{dt}\|CU(t)z_0\|_Y^2 \le -2|v(U(t)z_0)|^2\|U(t)z_0\|^2. \tag{39}$$

Integrating (39) from nT and $(n+1)T$, $(n \in \mathbb{N})$, we obtain

$$\|CU(nT)z_0\|_Y^2 - \|CU((n+1)T)z_0\|_Y^2 \ge 2\int_{nT}^{(n+1)T} |v(U(s)z_0)|^2\|U(s)z_0\|^2 ds.$$

Using (38), (39) and the fact that $\|CU(t)z_0\|_Y$ decreases, it follows

$$\left(1 + 2\left(\frac{\gamma}{M}\right)^2\right)\|CU((n+1)T)z_0\|_Y^2 \le \|CU(nT)z_0\|_Y^2.$$

Then

$$\|CU((n+1)T)z_0\|_Y \le \beta\|CU(nT)z_0\|_Y,$$

where $\beta = \frac{1}{\left(1+2\left(\frac{\gamma}{M}\right)^2\right)^{\frac{1}{2}}}$.

By recurrence, we show that $\|CU(nT)z_0\|_Y \le \beta^n\|Cz_0\|_Y$.

Taking $n = E\left(\frac{t}{T}\right)$ the integer part of $\frac{t}{T}$, we deduce that

$$\|CU(t)z_0\|_Y \le \mathrm{Re}^{-\sigma t}\|z_0\|,$$

where $R = \alpha\left(1 + 2\left(\frac{\gamma}{M}\right)^2\right)^{\frac{1}{2}}$, with $\alpha > 0$ and $\sigma = \frac{\ln\left(1+2\left(\frac{\gamma}{M}\right)^2\right)}{2T} > 0$, which achieves the proof.

4.1 Strong stabilization

The following result provides sufficient conditions for strong stabilization of the output (5).

Theorem 1.7 Let A generate a semigroup $(S(t))_{t\ge 0}$ of contractions on H and B be locally Lipschitz. If the conditions:

1. $\mathcal{R}e(\langle C^*CAy, y\rangle) \le 0, \quad \forall y \in D(A)$,

2. $\mathcal{R}e(\langle C^*CBy, y\rangle\langle By, y\rangle) \ge 0, \quad \forall y \in H$,

3. there exist $T, \gamma > 0$, such that

$$\int_0^T |\langle C^*CBS(t)y, S(t)y\rangle| dt \ge \gamma\|Cy\|_Y^2, \quad \forall y \in H, \tag{40}$$

hold, then the control

$$v(t) = -\langle C^* CBz(t), z(t) \rangle, \tag{41}$$

strongly stabilizes the output (5).

Proof: From hypothesis 1 of Theorem 1.7, we obtain

$$\frac{d}{dt}\|Cz(t)\|_Y^2 \leq -2|\langle C^* CBz(t), z(t)\rangle|^2. \tag{42}$$

Integrating this inequality, gives

$$2\int_0^t |\langle C^* CBz(s), z(s)\rangle|^2 ds \leq \|Cz(0)\|_Y^2.$$

Thus

$$\int_0^{+\infty} |\langle C^* CBz(s), z(s)\rangle|^2 ds < +\infty, \tag{43}$$

From the variation of constants formula and using Schwartz's inequality, we deduce

$$\|z(t) - S(t)z_0\| \leq LT^{\frac{1}{2}}\left(\int_0^T |\langle C^* CBz(s), z(s)\rangle|^2 ds\right)^{\frac{1}{2}}. \tag{44}$$

Integrating (34) over the interval $[0, T]$ and taking into account (44), we obtain

$$\int_0^T |\langle C^* CBS(s)z_0, S(s)z_0\rangle| ds \leq 2\alpha L^2 T^{\frac{3}{2}}\|z_0\|^2\left(\int_0^T |\langle C^* CBz(s), z(s)\rangle|^2 ds\right)^{\frac{1}{2}}$$
$$+ T^{\frac{1}{2}}\left(\int_0^T |\langle C^* CBz(s), z(s)\rangle|^2 ds\right)^{\frac{1}{2}}.$$

Replacing z_0 by $z(t)$ and using the superposition property of the solution, we get

$$\begin{aligned}\int_0^T |\langle C^* CBS(s)z(t), S(s)z(t)\rangle| ds \leq 2\alpha L^2 T^{\frac{3}{2}}\|z_0\|^2\left(\int_t^{t+T} |\langle C^* CBz(s), z(s)\rangle|^2 ds\right)^{\frac{1}{2}} \\ + T^{\frac{1}{2}}\left(\int_t^{t+T} |\langle C^* CBz(s), z(s)\rangle|^2 ds\right)^{\frac{1}{2}}.\end{aligned} \tag{45}$$

By (43), we get

$$\int_t^{t+T} |\langle C^* CBS(s)z(t), S(s)z(t)\rangle| ds \to 0, \text{ as } t \to +\infty. \tag{46}$$

From (40) and (46), we deduce that $\|Cz(t)\|_Y \to 0$, as $t \to +\infty$, which completes the proof.

Proposition 1.8 Let A generate a semigroup $(S(t))_{t\geq 0}$ of contractions on H, B be locally Lipschitz and the assumptions 1, 2 and 3 of Theorem 1.7 hold, then the control (41) strongly stabilizes the output (5) with decay estimate

$$\|Cz(t)\|_Y = O\left(t^{-\frac{1}{2}}\right), \quad as\ t \to +\infty. \tag{47}$$

Proof: Using (45), we get

$$\int_0^T |\langle C^* CBS(s)U(t)z_0, S(s)U(t)z_0\rangle| ds \leq \theta\sqrt{\xi(t)}, \tag{48}$$

where $\theta = \left(2\alpha TL^2\|z_0\|^2 + 1\right)T^{\frac{1}{2}}$ and $\xi(t) = \left(\int_t^{t+T} |\langle C^* CBU(s)z_0, U(s)z_0\rangle|^2 ds\right)$.
From (40) and (48), we deduce that

$$\varrho\sqrt{\xi(nT)} \geq \|CU(nT)z_0\|_Y^2, \quad \forall n \geq 0, \tag{49}$$

where $\varrho = \frac{1}{\gamma}\theta$.
Integrating the above inequality gives

$$\frac{d}{dt}\|CU(t)z_0\|_Y^2 \leq -2|\langle C^* CBU(t)z_0, U(t)z_0\rangle|^2,$$

from nT to $(n+1)T$, $(n \in \mathbb{N})$ and using (49), we obtain

$$\|CU(nT)z_0\|_Y^2 - \|CU(nT+T)z_0\|_Y^2 \geq 2\xi(nT), \quad \forall n \geq 0.$$

We obtain

$$\varrho^2\|CU(nT+T)z_0\|_Y^2 - \varrho^2\|CU(nT)z_0\|_Y^2 \leq -2\|CU(nT)z_0\|_Y^4, \quad \forall n \geq 0. \tag{50}$$

Let us introduce the sequence $r_n = \|CU(nT)z_0\|_Y^2, \quad \forall n \geq 0$.
Using (50), we deduce that

$$\frac{r_n - r_{n+1}}{r_n^2} \geq \frac{2}{\varrho^2}, \quad \forall n \geq 0.$$

Since the sequence (r_n) decreases, we get

$$\frac{r_n - r_{n+1}}{r_n . r_{n+1}} \geq \frac{2}{\varrho^2}, \quad \forall n \geq 0,$$

and also

$$\frac{1}{r_{n+1}} - \frac{1}{r_n} \geq \frac{2}{\varrho^2}, \quad \forall n \geq 0.$$

We deduce that

$$r_n \leq \frac{r_0}{\frac{2r_0}{\varrho^2}n + 1}, \quad \forall n \geq 0.$$

Finally, introducing the integer part $n = E\left(\frac{t}{T}\right)$ and from (42), the function $t \to \|CU(t)z_0\|_Y$ decreases. We deduce the estimate

$$\|Cz(t)\|_Y = O\left(t^{-1/2}\right), \quad as\ t \to +\infty.$$

4.2 Weak stabilization

The following result discusses the weak stabilization of the output (5).

Theorem 1.9 Let A generate a semigroup $(S(t))_{t\geq 0}$ of contractions on H, B be locally Lipschitz and weakly sequentially continuous. If assumptions 1, 2 of Theorem 1.7 and

$$\langle C^*CBS(t)y, S(t)y\rangle = 0, \quad \forall t\geq 0 \Rightarrow Cy = 0, \tag{51}$$

hold, then the control

$$v(t) = -\langle C^*CBz(t), z(t)\rangle, \tag{52}$$

weakly stabilizes the output (5).

Proof: Let us consider $\psi \in Y$ and $(t_n)\geq 0$ be a sequence of real numbers such that $t_n \to +\infty$, as $n \to +\infty$.

Using (42), we deduce that the sequence $h_n = \langle Cz(t_n), \psi\rangle_Y$ is bounded.

Let $h_{\gamma(n)}$ be an arbitrary convergent subsequence of h_n.

From (33), the subsequence $z\left(t_{\gamma(n)}\right)$ is bounded in H, so we can extract a subsequence still denoted by $z\left(t_{\gamma(n)}\right)$ such that $z\left(t_{\gamma(n)}\right) \rightharpoonup \varphi \in H$, as $n \to +\infty$.

Since C is continuous, B is weakly sequentially continuous and $S(t)$ is continuous $\forall t\geq 0$, we get

$$\lim_{n\to+\infty} \left\langle C^*CBS(t)z\left(t_{\gamma(n)}\right), S(t)z\left(t_{\gamma(n)}\right)\right\rangle = \langle C^*CBS(t)\varphi, S(t)\varphi\rangle.$$

From (46), we have

$$\int_0^T \left\langle C^*CBS(s)z\left(t_{\gamma(n)}\right), S(s)z\left(t_{\gamma(n)}\right)\right\rangle ds \to 0, \quad \text{as} \quad n \to +\infty.$$

Using the dominated convergence Theorem, we deduce that

$$\langle C^*CBS(t)\varphi, S(t)\varphi\rangle = 0, \quad \text{for all} \quad t\geq 0,$$

which implies, according to (51), that $C\varphi = 0$, and hence $h_n \to 0$, as $t \to +\infty$.

We deduce that $\langle Cz(t), \psi\rangle_Y \to 0$, as $t \to +\infty$. In other words $Cz(t) \rightharpoonup 0$, as $t \to +\infty$, which achieves the proof.

Example 4 Let us consider the system defined in $\Omega =]0, +\infty[$ by

$$\begin{cases} \dfrac{\partial z(x,t)}{\partial t} = -\dfrac{\partial z(x,t)}{\partial x} + v(t)Bz(x,t), & x\in\Omega, \ t>0, \\ z(x,0) = z_0(x), & x\in\Omega, \end{cases} \tag{53}$$

where $H = L^2(\Omega)$, $Az = -\frac{\partial z}{\partial x}$ with domain $D(A) = \left\{z\in H^1(\Omega) \mid z(0) = 0, \ z(x) \to 0, \ \text{as} \ x \to +\infty\right\}$, $Bz = |\int_0^1 z(x)dx|$ the control operator and $v(.)\in L^2(0,+\infty)$. The operator A generates a semigroup of contractions

$$(S(t)z_0)(x) = \begin{cases} z_0(x-t), & \text{if} \ x\geq t, \\ 0, & \text{if} \ x<t. \end{cases}$$

Let $\omega =]0, 1[$ be a subregion of Ω and system (53) is augmented with the output

$$w(t) = \chi_\omega z(t). \tag{54}$$

The operator B is sequentially continuous and verifies

$$\langle \chi_\omega^* \chi_\omega BS(t)z_0, S(t)z_0 \rangle = |\int_0^{1-t} z_0(x)dx| \int_0^{1-t} z_0(x)dx, \quad 0 \le t \le 1.$$

Thus

$$\langle \chi_\omega^* \chi_\omega BS(t)z_0, S(t)z_0 \rangle = 0, \quad \forall t \ge 0 \quad \Rightarrow z_0(x) = 0 \text{ a.e } x \in]0,1[, \quad \text{i.e} \quad \chi]0,1[z_0 = 0.$$

Then, the control

$$v(t) = -|\int_0^1 z(x,t)dx| \int_0^1 z(x,t)dx, \tag{55}$$

weakly stabilizes the output (54).

5. Simulations

Consider system (53) with $z(x,0) = \sin(\pi x)$, and augmented with the output (54).

For $\omega =]0,2[$, we have

Figure 1 shows that the output (54) is weakly stabilized on ω with error equals 6.8×10^{-4} and the evolution of control is given by **Figure 2**.

For $\omega =]0,3[$, we have

Figure 3 shows that the output (54) is weakly stabilized on ω with error equals 9.88×10^{-4} and the evolution of control is given by **Figure 4**.

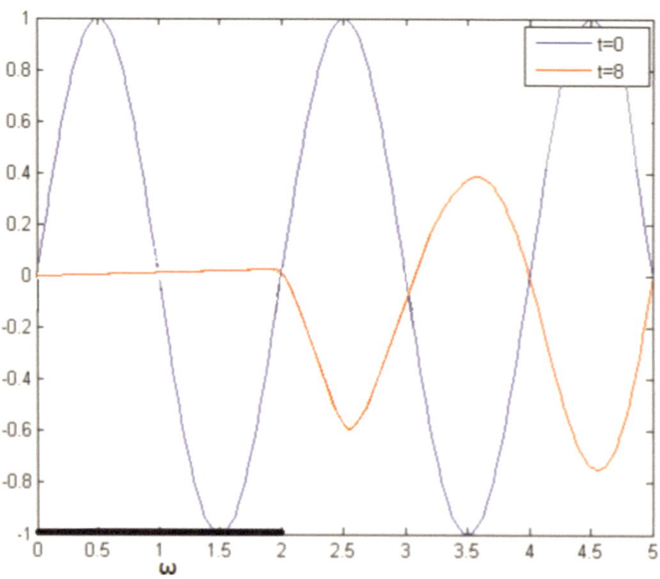

Figure 1.
The stabilization on $\omega =]0,2[$.

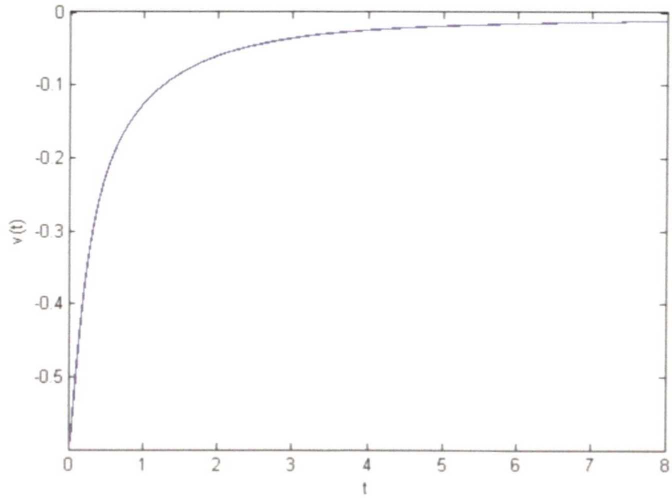

Figure 2.
The evolution control in the interval $]0, 8]$.

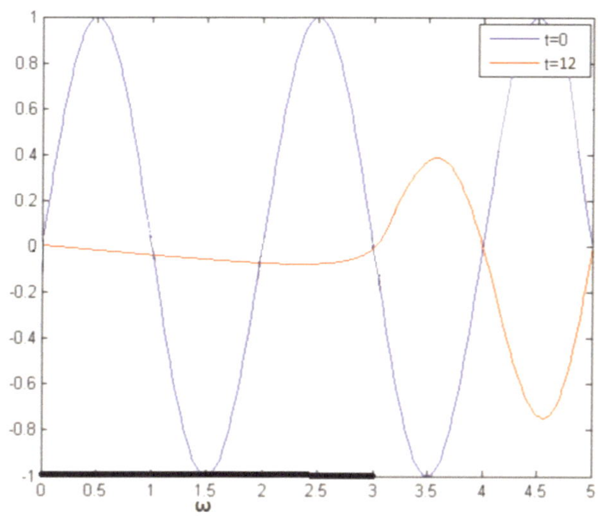

Figure 3.
The stabilization on $\omega =]0, 3[$.

Remark 2. It is clear that the control (55) stabilizes the state on ω, but do not take into account the residual part $\Omega\ \omega$.

6. Conclusions

In this work, we discuss the question of output stabilization for a class of semilinear systems. Under sufficient conditions, we obtain controls depending on the output operator that strongly and weakly stabilizes the output of such systems. This work gives an opening to others questions; this is the case of output stabilization for hyperbolic semilinear systems. This will be the purpose of a future research paper.

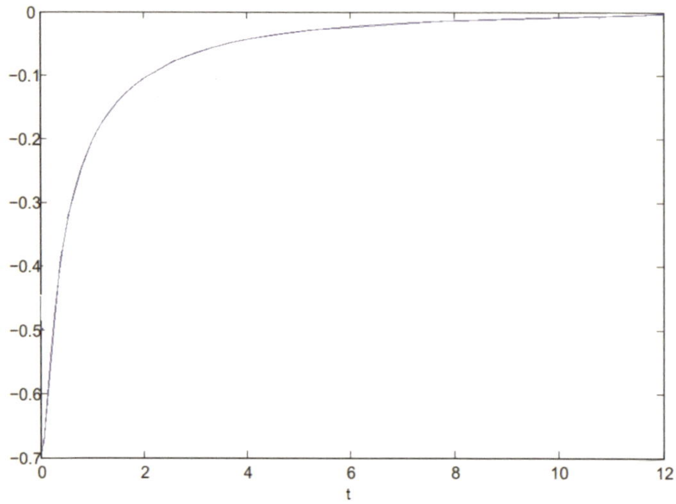

Figure 4.
The evolution control in the interval $]0, 12]$.

Author details

El Hassan Zerrik† and Abderrahman Ait Aadi*†
MACS Team, Department of Mathematics, Moulay Ismail University, Meknes, Morocco

*Address all correspondence to: abderrahman.aitaadi@gmail.com

†These authors contributed equally.

References

[1] Ball JM, Slemrod M. Feedback stabilization of distributed semilinear control systems. Journal of Applied Mathematics and Optimization. 1979;**5**: 169-179

[2] Berrahmoune L. Stabilization and decay estimate for distributed bilinear systems. Journal of Systems Control Letters. 1999;**36**:167-171

[3] Bounit H, Hammouri H. Feedback stabilization for a class of distributed semilinear control systems. Journal of Nonlinear Analysis. 1999;**37**:953-969

[4] Zerrik E, Ait Aadi A, Larhrissi R. Regional stabilization for a class of bilinear systems. IFAC-PapersOnLine. 2017;**50**:4540-4545

[5] Zerrik E, Ait Aadi A, Larhrissi R. On the stabilization of infinite dimensional bilinear systems with unbounded control operator. Journal of Nonlinear Dynamics and Systems Theory. 2018;**18**: 418-425

[6] Zerrik E, Ait Aadi A, Larhrissi R. On the output feedback stabilization for distributed semilinear systems. Asian Journal of Control. 2019. DOI: 10.1002/asjc.2081

[7] Zerrik E, Ouzahra M. Regional stabilization for infinite-dimensional systems. International Journal of Control. 2003;**76**:73-81

[8] Zerrik E, Ouzahra M, Ztot K. Regional stabilization for infinite bilinear systems. IET Proceeding of Control Theory and Applications. 2004; **151**:109-116

[9] Ouzahra M. Partial stabilization of semilinear systems using bounded controls. International Journal of Control. 2013;**86**:2253-2262

[10] Pazy A. Semi-Groups of Linear Operators and Applications to Partial Differential Equations. New York: Springer Verlag; 1983

[11] Lasiecka I, Tataru D. Uniform boundary stabilisation of semilinear wave equation with nonlinear boundary damping. Journal of Differential and Integral Equations. 1993;**6**:507-533

Chapter 7

Existence, Regularity, and Compactness Properties in the α-Norm for Some Partial Functional Integrodifferential Equations with Finite Delay

Boubacar Diao, Khalil Ezzinbi and Mamadou Sy

Abstract

The objective, in this work, is to study the alpha-norm, the existence, the continuity dependence in initial data, the regularity, and the compactness of solutions of mild solution for some semi-linear partial functional integrodifferential equations in abstract Banach space. Our main tools are the fractional power of linear operator theory and the operator resolvent theory. We suppose that the linear part has a resolvent operator in the sense of Grimmer. The nonlinear part is assumed to be continuous with respect to a fractional power of the linear part in the second variable. An application is provided to illustrate our results.

Keywords: integrodifferential, mild solution, resolvent operator, fractional power operator

1. Introduction

We consider, in this manuscript, partial functional equations of retarded type with deviating arguments in terms of involving spatial partial derivatives in the following form [1]:

$$\begin{cases} \dfrac{du(t)}{dt} = -Au(t) + \displaystyle\int_0^t B(t-s)u(s)ds + F(t, u_t) \text{ for } t \geq 0, \\ u_0 = \varphi \in \mathcal{C}_\alpha = C([-r, 0], D(A^\alpha)]), \end{cases} \tag{1}$$

where $-A$ is the infinitesimal generator of an analytic semigroup $(T(t))_{t\geq 0}$ on a Banach space $\mathbb{X}$. $B(t)$ is a closed linear operator with domain $D(B(t)) \supset D(A)$ time-independent. For $0 < \alpha < 1$, A^α is the fractional power of A which will be precise in the sequel. The domain $D(A^\alpha)$ is endowed with the norm $\|x\|_\alpha = \|A^\alpha x\|$ called $\alpha-$ norm. $\mathcal{C}_\alpha$ is the Banach space $C([-r, 0], D(A^\alpha))$ of continuous functions from $[-r, 0]$ to $D(A^\alpha)$ endowed with the following norm:

$$\|\phi\|_\alpha = \sup_{-r\leq\theta\leq 0} \|\phi(\theta)\|_\alpha \text{ for } \phi \in \mathcal{C}_\alpha.$$

$F : \mathbb{R}_+ \times \mathcal{C}_\alpha \to \mathbb{X}$ is a continuous function, and as usual, the history function $u_t \in \mathcal{C}_\alpha$ is defined by

$$u_t(\theta) = u(t+\theta) \text{ for } \theta \in [-r, 0].$$

As a model for this class, one may take the following Lotka-Volterra equation:

$$\begin{cases} \dfrac{\partial u(t,x)}{\partial t} = \dfrac{\partial^2 u(t,x)}{\partial x^2} + \displaystyle\int_0^t h(t-s)\dfrac{\partial^2 u(s,x)}{\partial x^2} ds \\ \qquad + \displaystyle\int_{-r}^0 g\left(t, \dfrac{\partial u(t+\theta,x)}{\partial x}\right) d\theta \text{ for } t \geq 0 \text{ and } x \in [0,\pi], \\ u(t,0) = u(t,\pi) = 0 \text{ for } t \geq 0, \\ u(\theta,x) = u_0(\theta,x) \text{ for } \theta \in [-r,0] \text{ and } x \in [0,\pi]. \end{cases} \tag{2}$$

Here $u_0 : [-r, 0] \times [0, \pi] \to \mathbb{R}, g : \mathbb{R}_+ \times \mathbb{R} \to \mathbb{R}$ and $h : \mathbb{R}_+ \to \mathbb{R}$ are appropriate functions.

In the particular case where $\alpha = 0$, many results are obtained in the literature under various hypotheses concerning A, B, and F (see, for instance, [2–6] and the references therein). For example, in [7], Ezzinbi et al. investigated the existence and regularity of solutions of the following equation:

$$\begin{cases} \dfrac{du(t)}{dt} = -Au(t) + \displaystyle\int_0^t B(t-s)u(s)ds + F(t, u_t) \text{ for } t \geq 0, \\ u_0 = \varphi \in C([-r, 0]; \mathbb{X}), \end{cases} \tag{3}$$

The authors obtained also the uniqueness and the representation of solutions via a variation of constant formula, and other properties of the resolvent operator were studied. In [8], Ezzinbi et al. studied a local existence and regularity of Eq. (3). To achieve their goal, the authors used the variation of constant formula, the theory of resolvent operator, and the principle contraction method. Ezzinbi et al. in [9] studied the local existence and global continuation for Eq. (3). Recall that the resolvent operator plays an important role in solving Eq. (3); in the weak and strict sense, it replaces the role of the c_0 semigroup theory. For more details in this topic, here are the papers of Chen and Grimmer [2], Hannsgen [10], Smart [11], Miller [12, 13], and Miller and Wheeler [14, 15]. In the case where the nonlinear part involves spatial derivative, the above obtained results become invalid. To overcome this difficulty, we shall restrict our problem in a Banach space $\mathbb{Y}_\alpha \subset \mathbb{X}$, to obtain our main results for Eq. (1).

Considering the case where $B = 0$, Travis and Webb in [16] obtained results on the existence, stability, regularity, and compactness of Eq. (1). To achieve their goal, the authors assumed that $-A$ is the infinitesimal generator of a compact analytic semigroup and F is only continuous with respect to a fractional power of A in the second variable. The present paper is motivated by the paper of Travis and Webb in [16].

The paper is organized as follows. In Section 2, we recall some fundamental properties of the resolvent operator and fractional powers of closed operators. The global existence, uniqueness, and continuous dependence with respect to the initial data are studied in Section 3. In Section 4, we study the local existence and bowing up phenomena. In Section 5 we prove, under some conditions, the regularity of the mild solutions. And finally, we illustrate our main results in Section 6 by examining an example.

2. Fractional power of closed operators and resolvent operator for integrodifferential equations

We shall write $\mathbb{Y}$ for $D(A)$ endowed with the graph norm $\|x\|_{\mathbb{Y}} = \|x\| + \|Ax\|$, $\mathbb{Y}_\alpha$ for $D(A^\alpha)$ and $\mathcal{L}(\mathbb{Y}_\alpha, \mathbb{X})$ will denote the space of bounded linear operators from $\mathbb{Y}_\alpha$ to $\mathbb{X}$, and for $\mathbb{Y}_0 = \mathbb{X}$, we write $\mathcal{L}(\mathbb{X})$ with norm $\|.\|_{\mathcal{L}(\mathbb{X})}$. We also frequently use the Laplace transform of f which is denoted by f^*. If we assume that $-A$ generates an analytic semigroup and, without loss of generality, that $0 \in \varrho(A)$, then one can define the fractional power A^α for $0 < \alpha < 1$, as a closed linear operator on its domain $\mathbb{Y}_\alpha$ with its inverse $A^{-\alpha}$ given by

$$A^{-\alpha} = \frac{1}{\Gamma(\alpha)} \int_0^\infty t^{\alpha-1} T(t)\, dt,$$

where Γ is the gamma function

$$\Gamma(\alpha) = \int_0^\infty t^{\alpha-1} e^{-t}\, dt.$$

We have the following known results.

Theorem 2.1. [17] The following properties are true.

i. $\mathbb{Y}_\alpha = D(A^\alpha)$ is a Banach space with the norm $|x|_\alpha = \|A^\alpha x\|$ for $x \in \mathbb{Y}_\alpha$.

ii. A^α is a closed linear operator with domain $\mathbb{Y}_\alpha = \mathrm{Im}(A^{-\alpha})$ and $A^\alpha = (A^{-\alpha})^{-1}$.

iii. $A^{-\alpha}$ is a bounded linear operator in $\mathbb{X}$.

iv. If $0 < \alpha \leq \beta$ then $D(A^\beta) \rightarrowtail D(A^\alpha)$. Moreover the injection is compact if $T(t)$ is compact for $t > 0$.

Definition 2.2. [18] A family of bounded linear operators $(R(t))_{t \geq 0}$ in $\mathbb{X}$ is called resolvent operator for the homogeneous equation of Eq. (3) if:

a. $R(0) = I$ and $\|R(t)\| \leq M_1 \exp(\sigma t)$ for some $M_1 \geq 1$ and $\sigma \in \mathbb{R}$.

b. For all $x \in \mathbb{X}$, $t \to R(t)x$ is continuous for $t \geq 0$.

c. $R(t) \in \mathcal{L}(\mathbb{Y})$ for $t \geq 0$. For $x \in \mathbb{Y}$, $R(.)x \in C^1(\mathbb{R}_+, \mathbb{X}) \cap C(\mathbb{R}_+, \mathbb{Y})$, and for $t \geq 0$ we have

$$\begin{aligned} R'(t)x &= -AR(t)x + \int_0^t B(t-s)R(s)x\, ds \\ &= -R(t)Ax + \int_0^t R(t-s)B(s)x\, ds. \end{aligned} \tag{4}$$

What follows is we assume the hypothesis taken from [1] which implies the existence of an analytic resolvent operator $(R(t))_{t \geq 0}$.

(V1) $-A$ generates an analytic semigroup on $\mathbb{X}$. $(B(t))_{t \geq 0}$ is a closed operator on $\mathbb{X}$ with domain at least $D(A)$ a.e $t \geq 0$ with $B(t)x$ strongly measurable for each $x \in \mathbb{Y}$ and $\|B(t)x\| \leq b(t)\|x\|_{\mathbb{Y}}$, for $b \in L^1_{loc}(0, \infty)$ with $b^*(\lambda)$ absolutely convergent for $Re\lambda > 0$.

(V2) $\rho(\lambda) = (\lambda I + A - B^*(\lambda))^{-1}$ exists as a bounded operator on $\mathbb{X}$ which is analytic for $\lambda \in \Lambda = \{\lambda \in \mathbb{C} : |arg\lambda| < \pi/2 + \delta\}$, where $0 < \delta < \pi/2$. In Λ if $|\lambda| \geq \epsilon > 0$, there exists $M = M(\epsilon) > 0$ so that $\|\rho(\lambda)\| \leq M/|\lambda|$.

(V3) $A\rho(\lambda) \in \mathcal{L}(\mathbb{X})$ for $\lambda \in \Lambda$ and is analytic from Λ to $\mathcal{L}(\mathbb{X})$. $B^*(\lambda) \in \mathcal{L}(\mathbb{Y}, \mathbb{X})$ and $B^*(\lambda)\rho(\lambda) \in \mathcal{L}(\mathbb{Y}, \mathbb{X})$ for $\lambda \in \Lambda$. Given $\epsilon > 0$, there exists a positive constant $M = M(\varepsilon)$ so that $\|A\rho(\lambda)x\| + \|B^*(\lambda)\rho(\lambda)x\| \leq (M/|\lambda|)\|x\|_{\mathbb{Y}}$ for $x \in \mathbb{Y}$ and $\lambda \in \Lambda$ with $\lambda \geq \varepsilon$ and $\|B^*(\lambda)\| \mapsto 0$ as $|\lambda| \mapsto \infty$ in Λ. In addition, $\|A\rho(\lambda)x\| \leq M|\lambda|^n \|x\|$ for some $n > 0$, $\lambda \in \Lambda$ with $\lambda \geq \varepsilon$. Further, there exists $D \subset D(A^2)$ which is dense in $\mathbb{Y}$ such that $A(D)$ and $B^*(\lambda)(D)$ are contained in $\mathbb{Y}$ and $\|B^*(\lambda)x\|_{\mathbb{Y}}$ is bounded for each $x \in D$ and $\lambda \in \Lambda$ with $|\lambda| \geq \epsilon$.

Theorem 2.3. [1] Assume that conditions **(V1)–(V3)** are satisfied. Then there exists an analytic resolvent operator $(R(t))_{t \geq 0}$. Moreover, there exist positive constants N, N_α such that $\|R(t)\| \leq N$and$\|A^\alpha R(t)\| \leq \frac{N_\alpha}{t^\alpha}$ for $t > 0$ and $0 \leq \alpha < 1$.

We take the following hypothesis.

(H0) The semigroup $(T(t))_{t \geq 0}$ is compact for $t > 0$.

Theorem 2.4. [19] Under the conditions **(V1)–(V3)** and **(H0)**, the corresponding resolvent operator $(R(t))_{t \geq 0}$ is compact for $t > 0$.

3. Global existence, uniqueness, and continuous dependence with respect to the initial data

Definition 3.1. A function $u : [0, b] \to \mathbb{Y}_\alpha$ is called a strict solution of Eq. (1), if:

i. $t \to u(t)$ is continuously differentiable on $[0, b]$.

ii. $u(t) \in \mathbb{Y}$ for $t \in [0, b]$.

iii. u satisfies Eq. (1) on $[0, b]$.

Definition 3.2. A continuous function $u : [0, b] \to \mathbb{Y}_\alpha$ is called a mild solution of Eq. (1) if

$$\begin{cases} u(t) = R(t)\varphi(0) + \int_0^t R(t-s)F(s, u_s)ds \text{ for } t \in [0, b], \\ u_0 = \varphi \in \mathcal{C}_\alpha. \end{cases} \tag{5}$$

Now to obtain our first result, we take the following assumption.

(H1) There exists a constant $L_F > 0$ such that

$$\|F(t, \varphi_1) - F(t, \varphi_2)\| \leq L_F \|\varphi_1 - \varphi_2\|_\alpha \text{ for } t \geq 0 \text{ and } \varphi_1, \varphi_2 \in \mathcal{C}_\alpha.$$

Theorem 3.3. Assume that **(V1)–(V3)** and **(H1)** hold. Then for $\varphi \in \mathcal{C}_\alpha$, Eq. (1) has a unique mild solution which is defined for all $t \geq 0$.

Proof. Let $a > 0$. For $\varphi \in \mathcal{C}_\alpha$, we define the set $\wedge$ by

$$\wedge = \{y \in C([0, a]; \mathbb{Y}_\alpha) : y(0) = \varphi(0)\}.$$

The set $\wedge$ is a closed subset of $C([0, a]; \mathbb{Y}_\alpha)$ where $C([0, a]; \mathbb{Y}_\alpha)$ is the space of continuous functions from $[0, a]$ to $\mathbb{Y}_\alpha$ equipped with the uniform norm topology

$$\|y\|_\alpha = \sup_{0 \le t \le a} \|y(t)\|_\alpha \text{ for } y \in C([0,a]; \mathbb{Y}_\alpha).$$

For $y \in \wedge$, we introduce the extension $\overline{y}$ of y on $[-r,a]$ defined by $\overline{y}(t) = y(t)$ for $t \in [0,a]$ and $\overline{y}(t) = \varphi(t)$ for $t \in [-r,0]$. We consider the operator Γ defined on $\wedge$ by

$$\Gamma y(t) = R(t)\varphi(0) + \int_0^t R(t-s)F(s,\overline{y}_s)\, ds \text{ for } t \in [0,a].$$

We claim that $\Gamma(\wedge) \subset \wedge$. In fact for $y \in \wedge$, we have $(\Gamma y)(0) = \varphi(0)$, and by continuity of F and $R(t)x$ for $x \in \mathbb{X}$, we deduce that $\Gamma y \in \wedge$. In order to obtain our result, we apply the strict contraction principle. In fact, let $u, v \in \wedge$ and $t \in [0,a]$. Then

$$(\Gamma(u) - \Gamma(v))(t) = \int_0^t R(t-s)(F(s,\overline{u}_s) - F(s,\overline{v}_s))\, ds.$$

Using the $\alpha-$ norm, we have

$$\begin{aligned}
\|A^\alpha(\Gamma(u) - \Gamma(v))(t)\| &\le L_F N_\alpha \int_0^t \frac{1}{(t-s)^\alpha} \|\overline{u}_s - \overline{v}_s\|_\alpha\, ds \\
&\le L_F N_\alpha \int_0^t \frac{1}{(t-s)^\alpha} \sup_{0 \le \tau \le a} \|u(\tau) - v(\tau)\|_\alpha\, ds \\
&\le \left(L_F N_\alpha \int_0^a \frac{ds}{s^\alpha} \right) \|u - v\|_\alpha.
\end{aligned}$$

Now we choose a such that

$$L_F N_\alpha \int_0^a \frac{ds}{s^\alpha} < 1.$$

Then Γ is a strict contraction on $\wedge$, and it has a unique fixed point y which is the unique mild solution of Eq. (1) on $[0,a]$. To extend the solution of Eq. (1) in $[a,2a]$, we show that the following equation has a unique mild solution:

$$\begin{cases} \dfrac{d}{dt} z(t) = -Az(t) + \displaystyle\int_a^t B(t-s)\, z(s)\, ds + F(t,z_t) \text{ for } t \in [a,2a], \\ z_a = y_a \in C([-r,a], \mathbb{Y}_\alpha). \end{cases} \tag{6}$$

Notice that the solution of Eq. (6) is given by

$$z(t) = R(t-a)z(a) + \int_a^t R(t-s)F(s,z_s)\, ds \text{ for } t \in [a,2a].$$

Let $\overline{z}$ be the function defined by $\overline{z}(t) = z(t)$ for $t \in [a,2a]$ and $\overline{z}(t) = y(t)$ for $t \in [-r,a]$. Consider now again the set $\wedge$ defined by

$$\wedge = \{z \in C([a,2a]; \mathbb{Y}_\alpha) : z(a) = y(a)\},$$

provided with the induced topological norm. We define the operator Γ_a on $\wedge$ by

$$(\Gamma_a z)(t) = R(t-a)z(a) + \int_a^t R(t-s)F(s,\overline{z}_s)\, ds \text{ for } t \in [a,2a].$$

We have $(\Gamma_a z)(a) = y(a)$ and $\Gamma_a z$ is continuous. Then it follows that $\Gamma_a \wedge \subset \wedge$. Moreover, for $u, v \in \wedge$, one has

$$\|A^\alpha(\Gamma_a(u) - \Gamma_a(v))(t)\| \leq L_F N_\alpha \int_a^t \frac{1}{(t-s)^\alpha} \|\overline{u}_s - \overline{v}_s\|_\alpha ds.$$

Since $\overline{u} = \overline{v} = \varphi$ in $[-r, 0]$, we deduce that

$$\|A^\alpha(\Gamma_a(u) - \Gamma_a(v))\| \leq \left(L_F N_\alpha \int_0^a \frac{ds}{s^\alpha}\right) \|u - v\|_\alpha.$$

Then we deduce that Γ_a has a unique fixed point in $\wedge$ which extends the solution y in $[a, 2a]$. Proceeding inductively, y is uniquely and continuously extended to $[na, (n+1)a]$ for all $n \geq 1$, and this ends the proof.

Now we show the continuous dependence of the mild solutions with respect to the initial data.

Theorem 3.4. Assume that **(V1)**–**(V3)** and **(H1)** hold. Then the mild solution $u(.,\varphi)$ of Eq. (1) defines a continuous Lipschitz operator $U(t), t \geq 0$ in $\mathcal{C}_\alpha$ by $U(t)\varphi = u_t(.,\varphi)$. That is, $U(t)\varphi$ is continuous from $[0;\infty)$ to $\mathcal{C}_\alpha$ for each fixed $\varphi \in \mathcal{C}_\alpha$. Moreover there exist a real number δ and a scalar function P such that for $t \geq 0$ and $\varphi_1, \varphi_2 \in \mathcal{C}_\alpha$ we have

$$\|U(t)\varphi_1 - U(t)\varphi_2\| \leq P(\delta) e^{\delta t} \|\varphi_1 - \varphi_2\|_\alpha. \tag{7}$$

Proof. We use the gamma formula

$$\Gamma(1-\alpha) k^{\alpha-1} = \int_0^\infty e^{-ks} s^{-\alpha} ds,$$

where $k > 0$ (see [20], p. 265). The continuity is obvious that the map $t \to u_t(.,\varphi)$ is continuous. Now, let $\varphi_1, \varphi_2 \in \mathcal{C}_\alpha$. If we pose $w(t) = u(t,\varphi_1) - u(t,\varphi_2)$, then we have

$$\|w(t)\|_\alpha \leq M_1 e^{\sigma t} \|\varphi_1 - \varphi_2\|_\alpha + L_F N_\alpha \int_0^t \frac{e^{\sigma(t-s)}}{(t-s)^\alpha} \|w_s\|_\alpha ds. \tag{8}$$

Let δ a real number be such that

$$\sigma - \delta < 0 \text{ and } M_1 \max\{e^{-\delta r}, 1\} L_F \Gamma(1-\alpha)(\delta - \sigma)^{\alpha-1} < 1.$$

We define the function P by

$$P(\delta) = M_1 M_3 M_4 \left(1 - M_1 M_4 L_F \Gamma(1-\alpha)(\delta - \sigma)^{\alpha-1}\right)^{-1}$$

where

$$M_3 = \max\{e^{\delta r}, 1\}, M_4 = \max\{e^{-\delta r}, 1\}.$$

Fix $\bar{t} > 0$ and let $E = \sup_{0 \leq s \leq \bar{t}} e^{-\delta s} \|w_s\|$. If $0 \leq \tau \leq \bar{t}$, then from Eq. (8), we have

$$\begin{aligned} e^{-\delta\tau} \|w(\tau)\|_\alpha &\leq M_1 e^{(\sigma-\delta)\tau} \|\varphi_1 - \varphi_2\|_\alpha + L_F N_\alpha \int_0^\tau \frac{e^{(\sigma-\delta)(\tau-s)}}{(\tau-s)^\alpha} e^{-\delta s} \|w_s\|_\alpha ds \\ &\leq M_1 \|\varphi_1 - \varphi_2\|_\alpha + L_F M_1 E \Gamma(1-\alpha)(\delta - \sigma)^{\alpha-1}. \end{aligned} \tag{9}$$

If $-r \leq \tau \leq 0$, we have

$$e^{-\delta\tau}\|w(\tau)\|_\alpha \leq \|\varphi_1 - \varphi_2\|_\alpha M_3. \tag{10}$$

Therefore, Eqs. (9) and (10) imply that

$$\sup_{-r\leq\tau\leq\bar{t}} e^{-\delta\tau}\|w(\tau)\|_\alpha \leq M_1M_3\|\varphi_1 - \varphi_2\|_{C_\alpha} + L_FM_1E\Gamma(1-\alpha)(\delta-\sigma)^{\alpha-1}. \tag{11}$$

For $0 \leq t \leq \bar{t}$, we have

$$\begin{aligned} e^{-\delta t}\|w_t\|_\alpha &= \sup_{-r\leq\theta\leq 0} e^{\delta\theta}e^{-\delta(t+\theta)}\|w(t+\theta)\|_\alpha \\ &\leq M_4 \sup_{-r\leq\theta\leq 0} e^{-\delta(t+\theta)}\|w(t+\theta)\|_\alpha \\ &\leq M_4 \sup_{-r\leq\tau\leq\bar{t}} e^{-\delta\tau}\|w(\tau)\|_\alpha. \end{aligned} \tag{12}$$

Then from Eqs. (11) and (12), we deduce that for $0 \leq t \leq \bar{t}$

$$e^{-\delta t}\|w_t\|_\alpha \leq M_1M_3M_4\|\varphi_1 - \varphi_2\|_\alpha + L_FM_1M_3E\Gamma(1-\alpha)(\delta-\sigma)^{\alpha-1},$$

which implies that

$$E \leq M_1M_3M_4\|\varphi_1 - \varphi_2\|_\alpha + L_FM_1M_4E\Gamma(1-\alpha)(\delta-\sigma)^{\alpha-1}.$$

Then the result follows.

4. Local existence, blowing up phenomena, and the compactness of the flow

We start by generalizing a result, obtained in [19] in the case of the usual norm on $\mathbb{X}$ $(\alpha = 0)$, in the case where $\alpha \neq 0$. We take the following assumption.

(H2) $B(t) \in \mathcal{L}(\mathbb{X}_\beta, \mathbb{X})$ for some $0 < \beta < 1$, a.e $t \geq 0$ and $\|B(t)x\| \leq b(t)\|x\|_\beta$ for $x \in \mathbb{X}_\beta$, with $b \in \mathbf{L}^q_{loc}(0,\infty)$ where $q > 1/(1-\beta)$.

Theorem 4.1. Assume that **(V1)**–**(V3)** and **(H2)** hold. Then for any $a > 0$, there exists a positive constant $M = M(a)$ such that for $x \in \mathbb{X}$ we have

$$\|A^\alpha(R(t+h)x - R(h)R(t)x)\| \leq M\int_0^h \frac{ds}{s^\alpha}\|x\| \text{ for } 0 \leq h < t \leq a.$$

Proof. Let $a > 0$ and $x \in \mathbb{X}$. Then

$$\begin{aligned} \frac{d}{dt}R(t+h)x &= -AR(t+h)x + \int_0^{t+h} B(t+h-s)R(s)x\,ds \\ &= -AR(t+h)x + \int_0^t B(t-s)R(s+h)x\,ds \\ &\quad + \int_t^{t+h} B(t+h-s)R(s)x\,ds. \end{aligned}$$

We deduce that $R(t+h)x$ satisfies the equation of the form

$$\frac{d}{dt}y(t) = -Ay(t) + \int_0^t B(t-s)y(s)\,ds + f(t).$$

Then by the variation o constante formula, it follows that

$$\begin{aligned} R(t+h)x &= R(t)R(h)x + \int_0^t R(t-s)\int_s^{s+h} B(s+h-u)R(u)x\,du\,ds \\ &= R(h)R(t)x + \int_0^h R(h-s)\int_0^t B(u)R(s+t-u)x\,du\,ds. \end{aligned}$$

Which yields that

$$R(t+h)x - R(h)R(t)x = \int_0^h R(h-s)\int_0^t B(u)R(s+t-u)x\,du\,ds.$$

Taking the $\alpha-$norm, we obtain that

$$\begin{aligned} \|A^\alpha(R(t+h)x - R(h)R(t)x)\| &\le N_\alpha \int_0^h \frac{1}{(h-s)^\alpha}\left\|\int_0^t B(u)R(s+t-u)x\,du\right\| ds \\ &\le N_\alpha \int_0^h \frac{1}{(h-s)^\alpha}\int_0^t b(u)\|A^\beta R(t+s-u)x\|\,du\,ds \\ &\le N_\alpha N_\beta \int_0^h \frac{ds}{(h-s)^\alpha}\int_0^t \frac{b(u)}{(t-u)^\beta}\|x\|\,du. \end{aligned}$$

Let p be such that $1/q + 1/p = 1$, so $p < 1/\beta$. Then it follows that

$$\|A^\alpha(R(t+h)x - R(h)R(t)x)\| \le N_\alpha N_\beta \|b\|_{\mathbf{L}^q(0,a)} \left\|u^{-\beta}\right\|_{\mathbf{L}^p(0,a)} \int_0^h \frac{ds}{s^\alpha}\|x\|.$$

And the proof is complete.

The local existence result is given by the following Theorem.

Theorem 4.2. Suppose that **(V1)**–**(V3)**, **(H0)**, and **(H2)** hold. Moreover, assume that F defined from $J \times \Omega$ into $\mathbb{X}$ is continuous where $J \times \Omega$ is an open set in $\mathbb{R}_+ \times \mathcal{C}_\alpha$. Then for each $\varphi \in \Omega$, Eq. (1) has at least one mild solution which is defined on some interval $[0, b]$.

Proof. Let $\varphi \in \Omega$. For any real $\zeta \in J$ and $p > 0$, we define the following sets:

$$I_\zeta = \{t : 0 \le t \le \zeta\} \quad \text{and} \quad H_p = \{\phi \in \mathcal{C}_\alpha : \|\phi\|_\alpha \le p\}.$$

For $\phi \in H_p$, we choose ζ and p such that $(t, \phi + \varphi) \in I_\zeta \times H_p$ and $H_p \subseteq \Omega$. By continuity of F, there exists $N_1 \ge 0$ such that $\|F(t, \phi + \varphi)\| \le N_1$ for (t, ϕ) in $I_\zeta \times H_p$. We consider $\overline{\varphi} \in C([-r, \zeta]; \mathbb{Y}_\alpha)$ as the function defined by $\overline{\varphi}(t) = R(t)\varphi(0)$ for $t \in I_\zeta$ and $\overline{\varphi}_0 = \varphi$. Suppose that $\overline{p} < p$ and choose $0 < b < \zeta$ such that

$$N_\alpha N_1 \int_0^b \frac{ds}{s^\alpha} < \overline{p} \quad \text{and} \quad \|\overline{\varphi}_t - \varphi\|_\alpha \le p - \overline{p} \text{ for } t \in I_b.$$

Let $K_0 = \{\eta \in C([-r, b]; \mathbb{Y}_\alpha) : \eta_0 = 0 \text{ and } \|\eta_t\|_\alpha \le \overline{p} \text{ for } 0 \le t \le b\}$. Then we have $\|F(t, \overline{\varphi}_t + \eta_t)\| \le N_1$ for $0 \le t \le b$ and $\eta \in K_0$, since $\|\eta_t + \overline{\varphi}_t - \varphi\|_\alpha \le p$. Consider the mapping S from K_0 to $C([-r, b]; \mathbb{Y}_\alpha)$ defined by $(S\eta)(0) = 0$

$$(S\eta)(t) = \int_0^t R(t-s)F(s,\eta_s+\overline{\varphi}_s)\,ds \text{ for } 0 \le t \le b. \tag{13}$$

Notice that finding a fixed point of S in K_0 is equivalent to finding a mild solution of Eq. (1) in K_0. Furthermore, S is a mapping from K_0 to K_0, since if $\eta \in K_0$ we have $(S\eta)_0 = 0$ and

$$\|(S\eta)(t)\|_\alpha \le \int_0^t \|A^\alpha R(t-s)F(s,\eta_s+\overline{\varphi}_s)\|ds.$$

Then

$$\begin{aligned}\|(S\eta)(t)\|_\alpha &\le N_\alpha N_1 \int_0^t \frac{ds}{(t-s)^\alpha}\\ &\le N_\alpha N_1 \int_0^b \frac{ds}{s^\alpha} < \overline{p}\end{aligned}$$

which implies that $S(K_0) \subset K_0$. We claim that $\{(S\eta)(t)) : \eta \in K_0\}$ is compact in $\mathbb{Y}_\alpha$ for fixed $t \in [-r,b]$. In fact, let β be such that $0<\alpha\le\beta<1$. The above estimate show that $\{A^\beta(S\eta)(t) : \eta \in K_0\}$ is bounded in $\mathbb{X}$. Since $A^{\alpha-\beta}$ is compact operator, we infer that $\{A^{\alpha-\beta}A^\beta(S\eta)(t) : \eta \in K_0\}$ is compact in $\mathbb{X}$, hence $\{(S\eta)(t) : \eta \in K_0\}$ is compact in $\mathbb{Y}_\alpha$. Next, we show that $\{(S\eta)(t) : \eta \in K_0\}$ is equicontinuous. The equicontinuity of $\{(S\eta)(t) : \eta \in K_0\}$ at $t=0$ follows from the above estimation of $(S\eta)(t)$. Now let $0<t_0<t\le b$ with t_0 be fixed. Then we have

$$\begin{aligned}\|A^\alpha((S\eta)(t)-(S\eta)(t_0))\| \le& \int_0^{t_0} \|A^\alpha(R(t-s)-R(t-t_0)R(t_0-s))F(s,\eta_s+\overline{\varphi}_s)\|\,ds\\ &+\left\|A^\alpha(R(t-t_0)-I)\int_0^{t_0} R(t_0-s)F(s,\eta_s+\overline{\varphi}_s)ds\right\|\\ &+\int_{t_0}^t \|A^\alpha R(t-s)F(s,\eta_s+\overline{\varphi}_s)\|\,ds.\end{aligned} \tag{14}$$

Using Theorem 4.1, it follows that

$$\begin{aligned}&\|A^\alpha((S\eta)(t)-(S\eta)(t_0))\|\\ &\le t_0N_1M\int_0^{t-t_0}\frac{ds}{s^\alpha}+\left\|(R(t-t_0)-I)A^\alpha\int_0^{t_0}R(t_0-s)F(s,\eta_s+\overline{\varphi}_s)ds\right\|\\ &+N_\alpha N_1\int_0^{t-t_0}\frac{1}{s^\alpha}ds.\end{aligned}$$

As the set $\{(S\eta)(t_0) : \eta \in K_0\}$ is compact in $\mathbb{Y}_\alpha$, we have

$$\lim_{t\to t_0^+}\|(S\eta)(t)-(S\eta)(t_0)\|_\alpha = 0 \ \text{ uniformly in } \eta \in K_0.$$

We obtain the same results by taking t_0 be fixed with $0<t<t_0\le b$. Then we claim that $\lim_{t\to t_0}\|(S\eta)(t)-(S\eta)(t_0)\|_\alpha = 0$ uniformly in $\eta \in K_0$ which means that $\{(S\eta)(t) : \eta \in K_0\}$ is equicontinuous. Then by Ascoli-Arzela theorem, $\{S\eta : \eta \in K_0\}$

is relatively compact in K_0. Finally, we prove that S is continuous. Since F is continuous, given $\varepsilon > 0$, there exists $\delta > 0$, such that

$$\sup_{0\le s\le b} \|\eta(s) - \hat{\eta}(s)\|_\alpha < \delta \text{ implies that } \|F(s,\eta_s + \overline{\varphi}_s) - F(s,\hat{\eta}(s) + \overline{\varphi}_s)\| < \varepsilon.$$

Then for $0 \le t \le b$, we have

$$\begin{aligned}\|(S\eta)(t) - (S\hat{\eta})(t)\|_\alpha &\le N_\alpha \int_0^t \frac{1}{(t-s)^\alpha} \|F(s,\eta_s + \overline{\varphi}_s) - F(s,\hat{\eta}(s) + \overline{\varphi}_s)\| ds \\ &\le N_\alpha \varepsilon \int_0^t \frac{ds}{s^\alpha}.\end{aligned}$$

This yields the continuity of S, and using Schauder's fixed point theorem, we deduce that S has a fixed point. Then the proof of the theorem is complete.

The following result gives the blowing up phenomena of the mild solution in finite times.

Theorem 4.3. Assume that **(V1)**–**(V3)**, **(H0)**, and **(H2)** hold and F is a continuous and bounded mapping. Then for each $\varphi \in \mathcal{C}_\alpha$, Eq. (1) has a mild solution $u(.,\varphi)$ on a maximal interval of existence $[-r, b_\varphi)$. Moreover if $b_\varphi < \infty$, then $\overline{\lim}_{t\to b_\varphi^-} \|u(t,\varphi)\|_\alpha = +\infty$.

Proof. Let $u(.,\varphi)$ be the mild solution of Eq. (1) defined on $[0,b]$. Similar arguments used in the local existence result can be used for the existence of $b_1 > b$ and a function $u(.,u_b)$ defined from $[b,b_1]$ to $\mathbb{Y}_\alpha$ satisfying

$$u(t,u_b(.,\varphi)) = R(t)u(b,\varphi) + \int_b^t R(t-s)F(s,u_s)ds \text{ for } t \in [b,b_1].$$

By a similar proceeding, we show that the mild solution $u(.,\varphi)$ can be extended to a maximal interval of existence $[-r, b_\varphi)$. Assume that $b_\varphi < +\infty$ and $\overline{\lim}_{t\to b_\varphi^-} \|u(t,\varphi)\|_\alpha < +\infty$. There exists $N_2 > 0$ such that $\|F(s,u_s)\| \le N_2$, for $s \in [0, b_\varphi)$. We claim that $u(.,\varphi)$ is uniformly continuous. In fact, let $0 < h \le t \le t + h < b_\varphi$. Then

$$\begin{aligned}u(t+h) - u(t) &= (R(t+h) - R(t))\varphi(0) + \int_0^t (R(t+h-s) - R(t-s))F(s,u_s)ds \\ &+ \int_t^{t+h} R(t+h-s)F(s,u_s)ds.\end{aligned}$$

By continuity of $A^\alpha R(t)$, we claim that $A^\alpha(R(t+h) - R(t))\varphi(0)$ is uniformly continuous on each compact set. Moreover, Theorem 4.1 implies that $A^\alpha(R(t+h-s) - R(t-s))F(s,u_s) \to 0$ uniformly in t when $h \to 0$. In fact we have

$$\begin{aligned}&\int_0^t \|(R(t+h-s) - R(t-s))F(s,u_s)\|_\alpha ds \\ &\le \int_0^t \|(R(t+h-s) - R(h)R(t-s))F(s,u_s)\|_\alpha ds \\ &+ \left\|(R(h) - I)A^\alpha \int_0^t R(t-s)F(s,u_s)ds\right\|\end{aligned}$$

Then using Theorem 4.1, we obtain that

$$\int_0^t \|(R(t+h-s)-R(t-s))F(s,u_s)\|_\alpha ds$$
$$\leq b_\varphi N_2 M \int_0^h \frac{ds}{s^\alpha} + \left\| (R(h)-I)A^\alpha \int_0^t R(t-s)F(s,u_s)ds \right\|.$$

We claim that the set $\left\{A^\alpha \int_0^t R(t-s)F(s,u_s)ds : t \in [0,b_\varphi)\right\}$ is relatively compact. In fact, let $(t_n)_{n\geq 0}$ be a sequence of $[0,b_\varphi)$. Then there exist a subsequence $(t_{n_k})_k$ and a real number t_0 such that $t_{n_k} \to t_0$. Using the dominated convergence theorem, we deduce that

$$\int_0^{t_{nk}} A^\alpha R(t_{n_k}-s)F(s,u_s)ds \to \int_0^{t_0} A^\alpha R(t_0-s)F(s,u_s)ds.$$

This implies that $\left\{A^\alpha \int_0^t R(t-s)F(s,u_s)ds : t \in [0,b_\varphi)\right\}$ is relatively compact. Now using Banach-Steinhaus' theorem, we deduce that

$$(R(h)-I)A^\alpha \int_0^t R(t-s)F(s,u_s)ds \to 0$$

uniformly when $h \to 0$ with respect to $t \in [0,b_\varphi)$. Moreover we have

$$\| \int_t^{t+h} R(t+h-s)F(s,u_s)ds\|_\alpha \leq N_2 N_\alpha \int_0^h \frac{ds}{s^\alpha}.$$

Consequently $\|u(t+h)-u(t)\|_\alpha \to 0$ *as* $h \to 0$ uniformly in $t \in [0,b_\varphi)$. If $h \leq 0$, that is, for $t \leq t_0$, we have

$$u(t)-u(t_0) = (R(t)-R(t_0))\varphi(0) - \int_0^t (R(t_0-s)-R(t_0-t)R(t-s))F(s,u_s)ds$$
$$-(R(t_0-t)-I)\int_0^t R(t-s)F(s,u_s)ds - \int_t^{t_0} R(t_0-s)F(s,u_s)ds,$$

one can show similar results by using the same reasoning. This implies that $u(.,\varphi)$ is uniformly continuous. Therefore $\lim_{t\to b_\varphi^-} u(t,\varphi)$ *exists in* $\mathbb{Y}_\alpha$. And consequently, $u(.,\varphi)$ can be extended to b_φ which contradicts the maximality of $[0,b_\varphi)$.

The next result gives the global existence of the mild solutions under weak conditions of F. To achieve our goal, we introduce a following necessary result which is a consequence of Lemma 7.1.1 given in ([21], p. 197, Exo 4).

Lemma 4.4. [21] Let $\alpha, a, b \geq 0, \beta < 1$ and $0 < d < \infty$. Also assume that v is nonnegative and locally integrable on $[0,d)$ with

$$v(t) \leq \frac{a}{t^\alpha} + b\int_0^t \frac{v(s)}{(t-s)^\beta} ds \text{ for } t \in (0,d).$$

Then there exists a constant $M_2 = M_2(a,b,\alpha,\beta,d) < \infty$ such that $v(t) \leq M_2/t^\alpha$ on $(0,d)$.

Theorem 4.5. Assume that **(V1)–(V3)**, **(H0)**, and **(H2)** hold and F is a completely continuous function on $\mathbb{R}_+ \times \mathcal{C}_\alpha$. Moreover suppose that there exist continuous nonnegative functions f_1 and f_2 such that $\|F(t,\varphi)\| \leq f_1(t)\|\varphi\|_\alpha + f_2(t)$ for $\varphi \in \mathcal{C}_\alpha$ and $t \geq 0$. Then Eq. (1) has a mild solution which is defined for $t \geq 0$.

Proof. Let $[0, b_\varphi)$ be the maximal interval of existence of a mild solution $u(., \varphi)$. Assume that $b_\varphi < +\infty$. By Theorem 4.3 we have $\overline{\lim}_{t \to t_\varphi^-} \|u(t, \varphi)\|_\alpha = +\infty$. Recall that the solution of Eq. (1) is given by $u_0 = \varphi$ and

$$u(t, \varphi) = R(t)\varphi(0) + \int_0^t R(t-s)F(s, u_s(., \varphi))\, ds \text{ for } t \in [0, b_\varphi).$$

Then taking the α−norm, we obtain

$$\|u(t, \varphi)\|_\alpha \leq \|R(t)\|\|\varphi(0)\|_\alpha + k_2 N_\alpha \int_0^{b_\varphi} \frac{ds}{s^\alpha}\, ds + k_1 N_\alpha \int_0^t \frac{1}{(t-s)^\alpha} \|u_s(., \varphi)\|_\alpha ds,$$

where $k_1 = \max_{0 \leq t \leq b_\varphi} |f_1(t)|$ and $k_2 = \max_{0 \leq t \leq b_\varphi} |f_2(t)|$. Then we deduce that

$$\|u(t, \varphi)\|_\alpha \leq N\|\varphi(0)\|_\alpha + k_1 N_\alpha \int_0^t \frac{1}{(t-s)^\alpha} \sup_{-r \leq \tau \leq s} \|u(\tau, \varphi)\|_\alpha ds + k_2 N_\alpha \int_0^{b_\varphi} \frac{ds}{s^\alpha}. \quad (15)$$

Now we claim that the function

$$t \to \int_0^t \frac{1}{(t-s)^\alpha} \sup_{-r \leq \tau \leq s} \|u(\tau, \varphi)\|_\alpha ds,$$

is nondecreasing. In fact, let $0 \leq t_1 \leq t_2$. Then

$$\begin{aligned}
\int_0^{t_1} \frac{1}{(t_1-s)^\alpha} \sup_{-r \leq \tau \leq s} \|u(\tau, \varphi)\|_\alpha ds &= \int_0^{t_1} \frac{1}{s^\alpha} \sup_{-r \leq \tau \leq t_1 - s} \|u(\tau, \varphi)\|_\alpha ds \\
&\leq \int_0^{t_2} \frac{1}{s^\alpha} \sup_{-r \leq \tau \leq t_2 - s} \|u(\tau, \varphi)\|_\alpha ds \\
&= \int_0^{t_2} \frac{1}{(t_2-s)^\alpha} \sup_{-r \leq \tau \leq s} \|u(\tau, \varphi)\|_\alpha ds
\end{aligned}$$

which yields the result. Then it follows from Eq. (15) that

$$\sup_{-r \leq s \leq t} \|u(s, \varphi)\|_\alpha \leq N\|\varphi(0)\|_\alpha + k_2 N_\alpha \int_0^{b_\varphi} \frac{ds}{s^\alpha}\, ds + k_1 N_\alpha \int_0^t \frac{1}{(t-s)^\alpha} \sup_{-r \leq \tau \leq s} \|u(\tau, \varphi)\|_\alpha ds.$$

Then using Lemma 4.4, we deduce that $u(., \varphi)$ is bounded in $[0, b_\varphi)$. Then we obtain that $\overline{\lim}_{t \to b_\varphi^-} \|u(t, \varphi)\|_\alpha < \infty$, which contradicts our hypothesis. Then the mild solution is global.

We focus now to the compactness of the flow defined by the mild solutions.

Theorem 4.6. Assume that **(V1)**–**(V3)** and **(H0)**–**(H2)** hold. Then the flow $U(t)$ defined from $\mathcal{C}_\alpha$ to $\mathcal{C}_\alpha$ by $U(t)\varphi = u_t(., \varphi)$ is compact for $t > r$, where $u_t(., \varphi)$ denotes the mild solution starting from φ.

Proof. We use Ascoli-Arzela's theorem. Let $E = \{\varphi_\gamma : \gamma \in \Gamma\}$ be a bounded subset of $\mathcal{C}_\alpha$ and let $t > r$ be fixed, but arbitrary. We will prove that $\overline{U(t)E}$ is compact. It follows from **(H1)** and inequality Eq. (7) that there exists N_5 such that

$$\|F(t, u_t(\varphi_\gamma))\| \leq N_2 \|u_t(\varphi_\gamma))\| + \|F(t, 0)\| = N_5 \text{ for } \gamma \in \Gamma.$$

For each $\gamma \in \Gamma$, we define $f_\gamma \in \mathcal{C}_\alpha$ by $f_\gamma = u_t(., \varphi_\gamma)$. We show now that for fixed $\theta \in [-r, 0]$, the set $\{f_\gamma(\theta) : \gamma \in \Gamma\}$ is precompact in $\mathbb{Y}_\alpha$. For any $\gamma \in \Gamma$, we have

$$f_\gamma(\theta) = R(t+\theta)\varphi_\gamma(0) + \int_0^{t+\theta} R(t+\theta-s)F(s, u_s(., \varphi))\, ds.$$

As $R(t)$ is compact for $t > 0$, we need only to prove that the set

$$\left\{\int_0^{t+\theta} R(t+\theta-s)F\left(s, u_s\left(., \varphi_\gamma\right)\right) ds : \gamma \in \Gamma\right\}$$

is compact. Also we have

$$\mu\left(\left\{R(\varepsilon)\int_0^{t+\theta-\varepsilon} R(t+\theta-\varepsilon-s)F\left(s, u_s\left(., \varphi_\gamma\right)\right) ds : \gamma \in \Gamma\right\}\right) = 0,$$

where μ is the measure of non-compactness. Moreover, using Theorem 4.1, we have

$$\begin{aligned}
&\left\|A^\alpha\left(\int_0^{t+\theta-\varepsilon} R(t+\theta-\varepsilon-s) - R(\varepsilon)R(t+\theta-\varepsilon-s)F(s, u_s(., \varphi_\gamma))\, ds\right)\right\| \\
&\qquad \leq \int_0^{t+\theta-\varepsilon} \|(R(t+\theta-s) - R(\varepsilon)R(t+\theta-\varepsilon-s))F(s, u_s(., \varphi))\|_\alpha\, ds \\
&\qquad \leq N_5 M \int_0^{\varepsilon} \frac{ds}{s^\alpha} \to 0 \quad \text{as} \quad \varepsilon \to 0.
\end{aligned}$$

We deduce that

$$\mu\left(\left\{\int_0^{t+\theta-\varepsilon} R(t+\theta-s)F\left(s, u_s\left(., \varphi_\gamma\right)\right) ds : \gamma \in \Gamma\right\}\right) = 0.$$

On the other hand, for $0 < \alpha \leq \beta < 1$, we have

$$\begin{aligned}
\left\|A^\beta \int_{t+\theta-\varepsilon}^{t+\theta} R(t+\theta-s)F(s, u_s(., \varphi_\gamma))ds\right\| &\leq \int_{t+\theta-\varepsilon}^{t+\theta} \|R(t+\theta-s)F(s, u_s(., \varphi_\gamma))\|_\beta ds \\
&\leq N_\beta N_5 \int_{t+\theta-\varepsilon}^{t+\theta} \frac{ds}{(t+\theta-s)^\beta} \\
&= N_\beta N_5 \int_0^{\varepsilon} \frac{ds}{s^\beta} \to 0 \quad \text{as} \quad \varepsilon \to 0.
\end{aligned}$$

Thus $\left\{A^\beta \int_{t+\theta-\varepsilon}^{t+\theta} R(t+\theta-s)F\left(s, u_s\left(., \varphi_\gamma\right)\right)ds : \gamma \in \Gamma\right\}$ is a bounded subset of $\mathbb{X}$. The precompactness in $\mathbb{Y}_\alpha$ now follows from the compactness of $A^{-\beta} : \mathbb{X} \to \mathbb{Y}_\alpha$. Then the set $\{(U(t)E)(\theta) : -r \leq \theta \leq 0\}$ is precompacted in $\mathbb{Y}_\alpha$. We prove that the family $\{f_\gamma : \gamma \in \Gamma\}$ is equicontinuous. Let γ in Γ, $0 < \varepsilon < t - r$, and $-r \leq \hat{\theta} \leq \theta \leq 0$ with $\hat{\theta}$ be fixed and $h = \theta - \hat{\theta}$. Then

$$\begin{aligned}\left\|A^\alpha\left(f_\gamma(h+\hat{\theta})-f_\gamma(\hat{\theta})\right)\right\| &\le \left\|R(t+\hat{\theta}+h)-R(t+\hat{\theta})\varphi_\gamma(0)\right\|_\alpha \\ &+\int_0^{t+\hat{\theta}}\left\|A^\alpha(R(t+\hat{\theta}+h-s)-R(h)R(t+\hat{\theta}-s))F(s,u_s(.,\varphi_\gamma))\right\|ds \\ &+\left\|(R(h)-I)A^\alpha\int_0^{t+\hat{\theta}}R(t+\hat{\theta}-s)F(s,u_s(.,\varphi_\gamma))\,ds\right\| \\ &+\int_{t+\hat{\theta}}^{t+\hat{\theta}+h}\left\|A^\alpha R(t+\hat{\theta}+h-s)F(s,u_s(.,\varphi_\gamma))\right\|ds.\end{aligned}$$

Then it follows that

$$\begin{aligned}\left\|A^\alpha\left(f_\gamma(h+\hat{\theta})-f_\gamma(\hat{\theta})\right)\right\| &\le \left\|(R(t+\hat{\theta}+h)-R(t+\hat{\theta}))A^\alpha\varphi_\gamma(0)\right\|+MN_5(t+\hat{\theta})\int_0^h\frac{ds}{s^\alpha} \\ &+\left\|(R(h)-I)A^\alpha\int_0^{t+\hat{\theta}}R(t+\hat{\theta}-s)F(s,u_s(.,\varphi_\gamma))\,ds\right\| \\ &+N_5N_\alpha\int_0^h\frac{ds}{s^\alpha}.\end{aligned}$$

Using the compactness of the set $\left\{A^\alpha\int_0^{t+\theta}R(t+\theta-s)F\left(s,u_s\left(.,\varphi_\gamma\right)\right)ds:\gamma\in\Gamma\right\}$ and the continuity of $t\to R(t)x$ for $x\in\mathbb{X}$, the right side of the above inequality can be made sufficiently small for $h>0$ small enough. Then we conclude that $\left\{f_\gamma:\gamma\in\Gamma\right\}$ is equicontinuous. Consequently, by Ascoli-Arzela's theorem, we conclude that the set $\{U(t)\varphi:\varphi\in E\}$ is compact, which means that the operator $U(t)$ is compact for $t>r$.

5. Regularity of the mild solutions

We define the set C_α^1 by $C_\alpha^1=C^1([-r,0];\mathbb{Y}_\alpha)$ as the set of continuously differentiable functions from $[-r,0]$ to $\mathbb{Y}_\alpha$. We assume the following hypothesis.

(H3) F is continuously differentiable, and the partial derivatives D_tF and $D_\varphi F$ are locally Lipschitz in the classical sense with respect to the second argument.

Theorem 5.1. Assume that **(V1)–(V3)**, **(H1)**, and **(H3)** hold. Let φ in C_α^1 be such that $\varphi(0)\in\mathbb{Y}$ and $\dot{\varphi}(0)=-A\varphi(0)+F(0,\varphi)$. Then the corresponding mild solution u becomes a strict solution of Eq. (1).

Proof. Let $a>0$. Take $\varphi\in C_\alpha^1$ such that $\varphi(0)\in\mathbb{Y}$ and $\dot{\varphi}(0)=-A\varphi(0)+F(0,\varphi)$, and let u be the mild solution of Eq. (1) which is defined on $[0,+\infty[$. Consider the following equation:

$$\begin{cases}v(t)=R(t)\dot{\varphi}(0)+\displaystyle\int_0^t R(t-s)\left[D_tF(s,u_s)+D_\varphi F(s,u_s)v_s\right]ds \\ \qquad+\displaystyle\int_0^t R(t-s)B(s)\varphi(0)\,ds \text{ for } t\ge 0, \\ v_0=\dot{\varphi}.\end{cases}\tag{16}$$

Using the strict contraction principle, we can show that there exists a unique continuous function v solution in $[0,a]$ of Eq. (16). We introduce the function w defined by

$$w(t)=\begin{cases}\varphi(0)+\displaystyle\int_0^t v(s)\,ds\,\text{if}\;\;t\geq 0,\\ \varphi(t)\qquad\qquad\quad\text{if}\;\;-r\leq t\leq 0.\end{cases}$$

Then it follows

$$w_t=\varphi+\int_0^t v_s\,ds\text{ for }t\in[0,a].$$

Consequently, the maps $t\to w_t$ and $t\to\int_0^t R(t-s)F(s,w_s)ds$ are continuously differentiable, and the following formula holds

$$\begin{aligned}\frac{d}{dt}\int_0^t R(t-s)F(s,w_s)\,ds&=R(t)F(0,w_0)+\int_0^t R(t-s)\left[D_tF(s,w_s)+D_\varphi F(s,w_s)v_s\right]ds\\&=R(t)F(0,\varphi)+\int_0^t R(t-s)\left[D_tF(s,w_s)+D_\varphi F(s,w_s)v_s\right]ds.\end{aligned}$$

This implies that

$$\int_0^t R(s)F(0,\varphi)\,ds=\int_0^t R(t-s)F(s,w_s)\,ds-\int_0^t\int_0^s R(s-\tau)\left[D_tF(\tau,w_\tau)+D_\varphi F(\tau,w_\tau)v_\tau\right]d\tau ds.$$

On the other hand, from equality Eq. (4), we have

$$-\int_0^t R(s)A\varphi(0)\,ds=R(t)\varphi(0)-\varphi(0)-\int_0^t\int_0^s R(s-\tau)B(\tau)\varphi(0)\,d\tau ds.$$

We rewrite w as follows:

$$\begin{aligned}w(t)=\varphi(0)&-\int_0^t R(s)A\varphi(0)\,ds+\int_0^t R(s)F(0,\varphi)\,ds\\&+\int_0^t\int_0^s R(s-\tau)\left[D_tF(\tau,u_\tau)+D_\varphi F(\tau,u_\tau)v_\tau\right]d\tau ds\\&+\int_0^t\int_0^s R(s-\tau)B(\tau)\varphi(0)\,d\tau ds.\end{aligned}$$

Then it follows that

$$\begin{aligned}w(t)=R(t)\varphi(0)&+\int_0^t R(t-s)F(s,w_s)\,ds\\&+\int_0^t\int_0^s R(s-\tau)[(D_\tau F(\tau,u_\tau)-D_\tau F(\tau,w_\tau))]\,d\tau ds\\&+\int_0^t\int_0^s \left(D_\varphi F(\tau,u_\tau)v_\tau-D_\varphi F(\tau,w_\tau)v_\tau\right)d\tau ds.\end{aligned}$$

We deduce, for $t\in[0,a]$, that

$$\begin{aligned}\|u(t)-w(t)\|_\alpha\leq&\int_0^t\|A^\alpha R(t-s)(F(s,u_s)-F(s,w_s))\|\,ds\\&+\int_0^t\int_0^s\|A^\alpha R(s-\tau)(D_\tau F(\tau,u_\tau)-D_\tau F(\tau,w_\tau))\|d\tau ds\\&+\int_0^t\int_0^s\left\|A^\alpha R(s-\tau)(D_\varphi F(\tau,u_\tau)-D_\varphi F(\tau,w_\tau))v_\tau\right\|d\tau ds.\end{aligned}\tag{17}$$

The set $H = \{u_s, w_s : s \in [0,a]\}$ is compact in $\mathcal{C}_\alpha$. Since the partial derivatives of F are locally Lipschitz with respect to the second argument, it is well-known that they are globally Lipschitz on H. Then we deduce that

$$\begin{aligned}\|u(t) - w(t)\|_\alpha &\leq N_\alpha h(a) \int_0^t \frac{1}{(t-s)^\alpha} \|u_s - w_s\|_\alpha \, ds \\ &\leq N_\alpha h(a) \int_0^t \frac{1}{(t-s)^\alpha} \sup_{0 \leq \tau \leq a} \|u(\tau) - w(\tau)\|_\alpha ds,\end{aligned}$$

where $h(a) = L_F N_\alpha + a N_\alpha \mathrm{Lip}(D_t F) + a N_\alpha \mathrm{Lip}(D_\varphi F)$, with $\mathrm{Lip}(D_\varphi F)$ and $\mathrm{Lip}(D_t F)$ the Lipschitz constant of $D_\varphi F$ and $D_t F$, respectively, which implies that

$$\|u - w\|_\alpha \leq \left(N_\alpha h(a) \int_0^a \frac{ds}{s^\alpha} \right) \|u - w\|_\alpha.$$

If we choose a such that

$$N_\alpha h(a) \int_0^a \frac{ds}{s^\alpha} < 1,$$

then $u = w$ in $[0,a]$. Now we will prove that $u = w$ in $[0,+\infty)$. Assume that there exists $t_0 > 0$ such that $u(t_0) \neq w(t_0)$. Let $t_1 = \inf\{t > 0 : \|u(t) - w(t)\| > 0\}$. By continuity, one has $u(t) = w(t)$ for $t \leq t_1$, and there exists $\varepsilon > 0$ such that $\|u(t) - w(t)\| > 0$ for $t \in (t_1, t_1 + \varepsilon)$. Then it follows that for $t \in (t_1, t_1 + \varepsilon)$,

$$\|u(t) - w(t)\|_\alpha \leq N_\alpha h(\varepsilon) \int_0^\varepsilon \frac{ds}{s^\alpha} \sup_{\varepsilon \leq \tau \leq t_1 + \varepsilon} \|u(\tau) - w(\tau)\|_\alpha.$$

Now choosing ε such that

$$N_\alpha h(\varepsilon) \int_0^\varepsilon \frac{ds}{s^\alpha} < 1,$$

then $u = w$ in $[t_1, t_1 + \varepsilon]$ which gives a contradiction. Consequently, $u(t) = w(t)$ for $t \geq 0$. We conclude that $t \to u_t$ from $[0,+\infty)$ to $\mathbb{Y}_\alpha$ and $t \to F(t, u_t)$ from $[0,+\infty) \times \mathcal{C}_\alpha$ to $\mathbb{X}$ are continuously differentiable. Thus, we claim that u is a strict solution of Eq. (1) on $[0,+\infty)$ [22–31].

6. Application

For illustration, we propose to study the model Eq. (2) given in the Introduction. We recall that this is defined by

$$\begin{cases} \dfrac{\partial}{\partial t} w(t,x) = \dfrac{\partial^2}{\partial x^2} w(t,x) + \displaystyle\int_0^t h(t-s) \frac{\partial^2}{\partial x^2} w(s,x) ds \\ \qquad + \displaystyle\int_{-r}^0 g\left(t, \frac{\partial}{\partial x} w(t+\theta, x)\right) d\theta \text{ for } t \geq 0 \text{ and } x \in [0,\pi], \\ w(t,0) = w(t,\pi) = 0 \text{ for } t \geq 0, \\ w(\theta, x) = w_0(\theta, x) \text{ for } \theta \in [-r, 0] \text{ and } x \in [0,\pi], \end{cases} \tag{18}$$

where $w_0 : [-r, 0] \times [0, \pi] \to \mathbb{R}$, $g : \mathbb{R}_+ \times \mathbb{R} \to \mathbb{R}$ and $h : \mathbb{R}_+ \to \mathbb{R}+$ are appropriate functions. To study this equation, we choose $\mathbb{X} = L^2([0, \pi])$, with its usual norm $\|.\|$. We define the operator $A : \mathbb{Y} = D(A) \subset \mathbb{X} \to \mathbb{X}$ by

$$Aw = -w'' \text{ with domain } D(A) = H^2(0, \pi) \cap H_0^1(0, \pi),$$

and $B(t)x = h(t)Ax \in \mathbb{X}$, *for* $\geq 0, x \in \mathbb{Y}$. For $\alpha = 1/2$, we define $\mathbb{Y}_{1/2} = \left(D\left(A^{1/2}\right), |\cdot|_{1/2}\right)$ where $|x|_{1/2} = \|A^{1/2}x\|$ for each $x \in \mathbb{Y}_{1/2}$. We define $\mathcal{C}_{1/2} = C([-r, 0], \mathbb{Y}_{1/2})$ equipped with norm $|\cdot|_\infty$ and the functions u and φ and F by $u(t) = w(t, x)$, $\varphi(\theta)(x) = w_0(\theta, x)$ for a.e $x \in [0, \pi]$ and $\theta \in [-r, 0]$, $t \geq 0$, and finally

$$F(t, \varphi)(x) = \int_{-r}^{0} g\left(t, \frac{\partial}{\partial x}\varphi(\theta)(x)\right) d\theta \text{ for a.e } x \in [0, \pi] \text{ and } \varphi \in \mathcal{C}_{1/2}.$$

Then Eq. (18) takes the abstract form

$$\begin{cases} \dfrac{du(t)}{dt} = -Au(t) + \displaystyle\int_0^t B(t-s)u(s)\,ds + F(t, u_t) \text{ for } t \geq 0, \\ u_0 = \varphi \in \mathcal{C}_{1/2} = C\left([-r, 0], D\left(A^{1/2}\right)\right), \end{cases} \tag{19}$$

The $-A$ is a closed operator and generates an analytic compact semigroup $(T(t))_{t\geq 0}$ on $\mathbb{X}$. Thus, there exists δ in $(0, \pi/2)$ and $M \geq 0$ such that $\Lambda = \left\{\lambda \in \mathbb{C} : |arg\lambda| < \frac{\pi}{2} + \delta\right\} \cup \{0\}$ is contained in $\rho(-A)$, the resolvent set of $-A$, and $\|R(\lambda, -A)\| < M/|\lambda|$ for $\lambda \in \Lambda$. The operator $B(t)$ is closed and for $x \in \mathbb{Y}$, $\|B(t)x\| \leq h(t)\|x\|_{\mathbb{Y}}$. The operator A has a discrete spectrum, the eigenvalues are n^2, and the corresponding normalized eigenvectors are $e_n(x) = \sqrt{\frac{2}{\pi}} \sin(nx)$, $n = 1, 2, \cdots$. Moreover the following formula holds:

i. $Au = \sum_{n=1}^{\infty} n^2 \langle u, e_n\rangle e_n \; u \in D(A)$.

ii. $A^{-1/2}u = \sum_{n=1}^{\infty} \frac{1}{n} \langle u, e_n\rangle e_n$ for $u \in \mathbb{X}$.

iii. $A^{1/2}u = \sum_{n=1}^{\infty} n\langle u, e_n\rangle e_n$ for $u \in D\left(A^{1/2}\right) = \left\{u \in \mathbb{X} : \sum_{n=1}^{\infty} \frac{1}{n} \langle u, e_n\rangle e_n \in \mathbb{X}\right\}$.

One also has the following result.

Lemma 6.1 [16] *Let* $\varphi \in \mathbb{Y}_{1/2}$. *Then* φ *is absolutely continuous,* $\varphi' \in \mathbb{X}$ *and*

$$\|\varphi'\| = \|A^{\frac{1}{2}}\varphi\|.$$

We assume the following assumptions.

(H4) The scalar function $h(.) \in L^1(0, \infty)$ and satisfies $g_1(\lambda) = 1 + h^*(\lambda) \neq 0$ (h^* the Laplace transform of h) and $\lambda g_1^{-1}(\lambda) \in \Lambda$ for $\lambda \in \Lambda$. Further, $h^*(\lambda) \to 0$ as $|\lambda| \to \infty$, for $\lambda \in \Lambda$ and $(h^*(\lambda))^{-1} = \circ(|\lambda|^n)$.

(H5) The function $g : \mathbb{R}_+ \times \mathbb{R} \to \mathbb{R}$ is continuous and Lipschitz with respect to the second variable.

By assumption **(H4)**, the operator $\rho(\lambda) = (\lambda I + g_1(\lambda)A)^{-1} = g_1^{-1}(\lambda)\left(\lambda g_1^{-1}(\lambda)I + A\right)^{-1}$ exists as a bounded operator on $\mathbb{X}$, which is analytic in Λ and satisfies $\|\rho(\lambda)\| < M/|\lambda|$. On the other hand, for $x \in \mathbb{X}$, we have

$$\begin{aligned} A\rho(\lambda)x &= A\left(\lambda I + g_1(\lambda)A\right)^{-1}x \\ &= \left(A + \lambda g_1^{-1}(\lambda)I - \lambda g_1^{-1}(\lambda)I\right)\left(\lambda I + g_1(\lambda)A\right)^{-1}x \\ &= g_1^{-1}(\lambda)\left[I - \lambda g_1^{-1}(\lambda)\left(\lambda g_1^{-1}(\lambda)I + A\right)^{-1}\right]x. \end{aligned}$$

Since $\lambda g_1^{-1}(\lambda)\left(\lambda g_1^{-1}(\lambda)I + A\right)^{-1}$ is bounded because $g_1^{-1}(\lambda) \in \Lambda$, then $\|A\rho(\lambda)x\|$ has the growth properties of $g_1^{-1}(\lambda)$ which tends to 1 if $|\lambda|$ goes to infinity. Then we deduce that $A\rho(\lambda) \in \mathcal{L}(\mathbb{X})$. Moreover, it is analytic from Λ to $\mathcal{L}(\mathbb{X})$. Now, for $x \in \mathbb{Y}$, one has

$$A\rho(\lambda)x = g_1^{-1}(\lambda)\left(\lambda g_1^{-1}(\lambda)I + A\right)^{-1}Ax \text{ and } B^*(\lambda)\rho(\lambda)x = h^*(\lambda)\rho(\lambda)Ax.$$

Then it follows that

$$\|A\rho(\lambda)x\| \le M/|\lambda|\|x\|_{\mathbb{Y}} \text{ and } \|B^*(\lambda)\rho(\lambda)\| \le M/|\lambda|\|x\|_{\mathbb{Y}}.$$

We deduce that $A\rho(\lambda) \in \mathcal{L}(\mathbb{Y},\mathbb{X})$, $B^*(\lambda) = h^*(\lambda)A \in \mathcal{L}(\mathbb{Y},\mathbb{X})$, and $B^*(\lambda)\rho(\lambda) \in \mathcal{L}(\mathbb{Y},\mathbb{X})$. Considering $D = C_0^\infty([0,\pi])$, we see that the conditions **(V1)**–**(V3)** and **(H0)** are verified. Hence the homogeneous linear equation of Eq. (18) has an analytic compact resolvent operator $(R(t))_{t\ge 0}$. The function F is continuous in the first variable from the fact that g is continuous in the first variable. Moreover from Lemma 6.1 and the continuity of g, we deduce that F is continuous with respect to the second argument. This yields the continuity of F in $\mathbb{R}_+ \times C_{1/2}$. In addition, by assumption **(H5)** we deduce that

$$\|F(t,\varphi_1) - F(t,\varphi_2)\| \le rL_f\|\varphi_1 - \varphi_2\|_{C_{1/2}}.$$

Then F is a continuous globally Lipschitz function with respect to the second argument. We obtain the following important result.

Proposition 6.2. Suppose that the assumptions **(H4)**–**(H5)** hold. Then Eq. (19) has a mild solution which is defined for $t \ge 0$.

Author details

Boubacar Diao[1*], Khalil Ezzinbi[2] and Mamadou Sy[1]

1 Laboratoire L.A.N.I, Université Gaston Berger de Saint-Louis, Saint-Louis, Senegal

2 Département de Mathématiques, Faculté des Sciences Semlalia, Université Cadi Ayyad, Marrakesh, Moroco

*Address all correspondence to: diaobacar@yahoo.fr

References

[1] Grimmer R, Pritchard AJ. Analytic resolvent operators for integral equations in a Banach space. Journal of Differential Equations. 1983;**50**(2):234-259

[2] Chen G, Grimmer R. Semigroup and integral equations. Journal of Integral Equations. 1980;**2**(2):133-154

[3] Hale JK, Lunel V. Introduction to functional differential equations. In: Applied Mathematical Sciences. Vol. 99. New York: Springer-Verlag; 1993

[4] Pruss J. Evolutionary Integral Equations and Applications. In: Lecture Notes in Mathematics. New York: Springer, Birkhauser; 1993

[5] Travis CC, Webb GF. Existence and stability for partial functional differential equations. Transactions of the American Mathematical Society. 1974;**200**:395-418

[6] Wu J. Theory and applications of partial functional differential equations. In: Applied Mathematical Sciences. Vol. 119. New York: Springer-Verlag; 1996

[7] Ezzinbi K, Touré H, Zabsonre I. Existence and regularity of solutions for some partial functional integrodifferential equations in Banach spaces. Nonlinear Analysis: Theory Methods & Applications. 2009;**70**(7):2761-2771

[8] Ezzinbi K, Touré H, Zabsonre I. Local existence and regularity of solutions for some partial functional integrodifferential equations with infinite delay in Banach spaces. Nonlinear Analysis. 2009;**70**(9):3378-3389

[9] Ezzinbi K, Ghnimi S. Local existence and global continuation for some partial functional integrodifferential equations. Afr. Diaspora J. Math. 2011;**12**(1):34-45

[10] Hannsgen KB. The resolvent kernel of an integrodifferential equation in Hilbert space. SIAM Journal on Mathematical Analysis. 1976;7(4):481-490

[11] Smart DR. Uniform L^1 behavior for an integrodifferential equation with parameter. SIAM Journal on Mathematical Analysis. 1977;**8**:626-639

[12] Miller RK. Volterra integral equations in a Banach space. Fako de l'Funkcialaj Ekvacioj Japana Matematika Societo. 1975;**18**:2163-2193

[13] Miller RK. An integrodifferential equation for rigid heat conductors with memory. Journal of Mathematical Analysis and Applications. 1978;**66**:313-332

[14] Miller RK, Wheeler RL. Asymptotic behavior for a linear Volterra integral equation in Hilbert space. Journal of Differential Equations. 1977;**23**:270-284

[15] Miller RK. Well-posedness and stability of linear Volterra integrodifferential equations in abstract spaces. Fako de l'Funkcialaj Ekvacioj Japana Matematika Societo. 1978;**21**:279-305

[16] Travis CC, Webb GF. Existence, stability and compactness in the α-norm for partial functional differential equations. Transactions of the American Mathematical Society. 1978;**240**:129-143

[17] A. Pazy, Semigroups of linear operators and application to partial differential equations, Applied Mathematical Sciences Vol. 44, Springer-Verlag, New York, (2001).

[18] Grimmer R. Resolvent operators for integral equations in a Banach space. Transactions of the American Mathematical Society. 1982;**273**:333-349

[19] Desch W, Grimmer R, Schappacher W. Some considerations for linear integrodifferential equations. Journal of Mathematical Analysis and Applications. 1984;**104**:219-234

[20] Hale JK. Functional Differential Equation. New York: Springer-Verlag; 1971

[21] Henry D. Geometric Theory of Semilinear Parabolic Equations. Berlin/Heidelberg/New York: Springer-Verlag; 1981

[22] Adimy M, Ezzinbi K. Existence and linearized stability for partial neutral functional differential equations. Differential Equations and Dynamical Systems. 1999;**7**(4):371-417

[23] Engel KJ, Nagel R. One parameter semigroups of linear evolution equations. In: Brendle S, Campiti M, Hahn T, Metafune G, Nickel G, Pallara D, Perazzoli C, Rhandi A, Romanelli S, Schnaubelt R, editors. Graduate Texts in Mathematics. Vol. 194. New York: Springer-Verlag, Elsevier; 2001

[24] Ezzinbi K, Benkhalti R. Existence and stability in the $\alpha-$ norm for some partial functional differential equations with infinite delay. Differential and Integral Equations. 2006;**19**(5):545-572

[25] Ezzinbi K, N'Guérékata GM. Almost automorphic solutions for some partial functional differential equations. Journal of Mathematical Analysis and Applications. 2007;**328**:344-358

[26] Goldstein JA. Some remarks on infinitesimal generators of analytic semigroups. American Mathematical Society. 1969;**22**(1):91-93

[27] Grimmer R, Pruss R. On linear Voltera equations in Banach spaces, hyperbolic partial differential equations, II. Computers and Mathematics with Applications; **11**(1–3):189-205

[28] Haraux A, Cazenave T. An Introduction to Semilinear Evolution Equations, Oxford Lecture Series in Mathematics and its Applications. Vol. 131998

[29] Lunardi A. Analytic Semigroup and Optimal Regularity in Parabolic Problems, Progress in Nonlinear Differential Equations and their Applications. Vol. 16. Basel: Birkhäuser Verlag; 1995

[30] Smart DR. Fixed Point Theorems, Cambridge Tracts in Mathematics. Vol. 66. London/New York: Cambridge University Press; 1974

[31] Travis CC, Webb GF. Partial differential equations with deviating arguments in the time variable. Journal of Mathematical Analysis and Applications. 1976;**56**:397-409

Section 2

Recent Applications in Nonlinear Systems

Chapter 8

Nonlinear Resonances in 3D Printed Structures

Astitva Tripathi and Anil K. Bajaj

Abstract

Nonlinear resonators can have advantages over linear designs including increased sensitivity towards changes in their physical properties and environment, and high quality factors which make them attractive in applications such as mass/chemical sensors or signal filters. Designing nonlinear structures, however, requires much understanding of nonlinear behavior characteristics of structures. Similarly, the proliferation of 3D or additive manufacturing/printing capabilities has opened the doors to deploying nonlinear resonators on scales not possible earlier. However, to obtain consistent nonlinear dynamic performance the designer must perform a careful analysis to explore the existence and repeatability of desired nonlinear behavior. Also, the use of 3D printing with the associated substrate material properties poses its own challenges in regards to device simulation in view of the fact that most of the traditional literature on nonlinear resonators assumes linear material stiffness. In this chapter, the authors discuss computational design methods for structural design, and specifically study the case of 1:2 internal resonances in resonators made of nonlinear (hyperelastic) materials. The design methods allow for development of large number of candidate resonator designs without a required significant nonlinear structural design experience, and the study of the dynamic response of the resonators provides a glimpse in to the 1:2 nonlinear internal resonance exhibited by the candidate resonators.

Keywords: nonlinear dynamics, internal resonances, hyperelastic materials, 3D printing, topology optimization

1. Introduction

The development of the micro- and nano-electronics industry coupled with the advances in semiconductor manufacturing techniques led to an interest in developing and applying micro- and nano-electromechanical systems (MEMS and NEMS) [1, 2]. Due to the small length scales of such devices and the popular modes of actuation employed by designers for MEMS or NEMS, such as electrostatic actuation, it led to the observations that nonlinear effects in many of these devices were the norm rather than the exception. This realization led to the study of effects of nonlinearities on the dynamic response of MEMS and NEMS in their various modes of operation, the effects in some cases being detrimental to linear design performance, and in some cases being beneficial to performance [3]. The study of nonlinear dynamics is an area of research with long history in structural and mechanical systems [4]. Several attempts have been made to incorporate the nonlinear dynamic effects and use the plethora of associated phenomena into operating mechanisms of MEMS devices particularly as mass and/or chemical sensors and filters [3]. While such nonlinear devices have several advantages over linear ones with the same functionality in terms of measurement resolution, there have been challenges in making sure that the designed topologies are "in-tune" with the semiconductor manufacturing processes. At a conceptual level, researchers in nonlinear dynamics often work with lumped-parameter models [3] and many interesting applications have been considered [5, 6] for systems characterized by a single degree of freedom. For systems with more than one degree of freedom, one of the most compact representations is a system exhibiting a nonlinear 1:2 internal resonance, the spring-pendulum system [4, 7]. While this system lends itself very conveniently to a systematic analytical study [8], it is a relatively hard task to replicate it physically at micro- or nano-scales. This is even more so when the structural components fabricated involve two- or three-dimensional elastic structures.

3D printing or additive manufacturing offers several appealing advantages in terms of building devices based on nonlinear dynamic principles. The fabrication processes and dimensions are such that there is better repeatability with complex mechanical designs, as well as initial prototyping and low volume production costs come without the need for significant capital expenditure [9, 10]. Recently studies have been reported for prototyping nonlinear vibratory components made with 3D printers having a feature size of less than 1 mm [11]. While the dimensions of polymeric resonators created by 3D printers may limit their measurement resolution and frequency range of operations, the manufacturing process itself is more repeatable for certain kinds of resonators. However, using 3D printing for producing nonlinear resonators also comes with its own challenges. While micro- and nano-resonators operate in an environment with many kinds of nonlinearities, 3D printed structures have to largely rely on only two sources of nonlinearities, namely, geometric and material nonlinearities unless controlled material variations or composite structures are explicitly introduced in fabrication. Fortunately, as was demonstrated by Tripathi and Bajaj [12, 13], both geometric nonlinearities due to finite deformations and material nonlinearities due to nonlinear hyperelastic properties of the 3D printed material can produce nonlinear dynamic effects such as 1:2 internal resonance.

A 1:2 internal resonance is a popular mechanism exhibited and employed by many nonlinear dynamics based resonators [14]. Internal resonance in a structure refers to the energy transfer that occurs between two modes of the structure when their natural frequencies are almost commensurable and the structure has some appropriate nonlinearity. For example, for 1:2 internal resonance, if the two modes of a structure have their natural frequencies close to the ratio 1:2 and harmonic excitation of the higher mode is above a certain threshold, energy can be transferred from the resonant response of the higher mode to the lower frequency mode in the presence of quadratic nonlinearities. The mathematical description of a 1:2 internal resonance and the dynamics is well established [4, 7, 8]. For the purposes of evaluating the suitability of using 3D printing to produce resonators exhibiting 1:2 internal resonances, it is important to demonstrate that the dynamic response equations for the resonators exhibit the same mathematical characteristics as the cardinal examples.

In the context of elastic structures exhibiting various internal resonances, the present work focuses on elastic plate-type structures [4, 15]. A few representative works on different aspects of nonlinear vibrations of rectangular plates with internal resonances are [16–18]. In general, isotropic plates with different simple boundary conditions do not exhibit any commensurate frequencies unless there exists some type of symmetry of the structure. Some works have considered optimization of geometry and material distribution to affect frequency distributions as well as internal resonances [19, 20]. A systematic approach is based on the concepts in "topology optimization" [21]. Applications of topology or shape optimization have now appeared in the literation on nonlinear dynamics as well, with the works in [22–24] focusing on general one-dimensional elastic systems whereas the works in [12, 13] focusing on plate structures with internal resonances. The overall goal is to tailor the system's dynamic response to some desired form for appropriate external excitations.

In the present study, particular classes of resonator designs consisting of rectangular plates with cutouts which can be easily fabricated using 3D printing are analyzed for their nonlinear dynamic response. To obtain a suitable resonator design with commensurable (1:2) natural frequencies, a parametric optimization process which varies the sizes of the cutouts is employed. The natural frequencies themselves are computed using linear finite element analysis (FEA). The resonators are assumed to be made of a Mooney-Rivlin hyperelastic material [25] which is anticipated to provide the material nonlinearity necessary to produce 1:2 internal resonances. Once the optimization process is able to provide a candidate structure, the mode shapes obtained by the finite element analysis are used to build a reduced order model of the resonator displacements. This displacement field can then be used to derive the kinetic and strain energies of the structure which can provide the system Lagrangian. This Lagrangian is then averaged and subjected to the Euler-Lagrange conditions to derive the slow-amplitude equations of motion of the structure that provide the dynamic steady state response.

This work has two following sections: Section 2 describes the design and optimization process which leads to a desired candidate structure. It discusses the aspects of the hyperelastic material model as well as the use of mode shapes to construct the reduced order model. Section 3 describes the development of the structure's Lagrangian, the extraction of the nonlinear equations of motion for the modal amplitudes, and the steady state dynamic response of the system under harmonic excitation of the higher frequency mode. Section 4 contains some concluding remarks for this work.

2. Candidate structure synthesis

The principal objective of the structural synthesis proves is to obtain a resonator design with commensurable natural frequencies. As it is difficult to come up with such a structure by just relying on the researcher's experience, a computational optimization method is proposed to design the candidate resonators. For 1:2 internal resonances, the desired frequency relation between the two modes taking part in the energy transfer can be expressed as:

$$\frac{\omega_1}{\omega_2} = \frac{1}{2} \tag{1}$$

where ω_1 and ω_2 are the natural frequencies of the lower and the higher mode, respectively. To obtain such a candidate resonator, the natural frequency requirement represented by Eq. (1) can be formulated as an optimization problem. An example optimization problem for such a task can be represented by

$$minimize, c(\omega) = \left(\frac{1}{2} - \frac{\omega_1}{\omega_2}\right) \tag{2}$$

Thus, the optimization process attempts to minimize the deviation of the two natural frequencies from the perfect 1:2 natural frequency ratio. Solving the optimization problem posed by Eq. (2) would lead to a structure with two of its natural frequencies close to the ratio of 1:2 which is a major requirement for resonators exhibiting 1:2 internal resonance. In this study, two methods for solving the optimization problem are discussed. The first method is a topology optimization method based on simple isotropic material with penalization (SIMP) model and the method of moving asymptotes (MMA) [21]. The second method is a parametric optimization method in which a starting parameterized base structure is chosen whose topology is similar to the final desired candidate structure. Then this base structure is optimized by a nonlinear quadratic programming process to produce a viable candidate structure.

2.1 Topology optimization with SIMP

Topology optimization techniques have been widely used to solve a broad range of structural optimization problems. While quite versatile, an occasional drawback against topology optimization generated optimal topologies has been the difficulty of their reproduction using conventional manufacturing processes. In this regard 3D printing is eminently suited to produce topologically optimized design as both techniques are adept at producing extruded structures with complex planar patterns. Topology optimization methods are based on finite element discretization of the design spaces. In the context of designing candidate hyperelastic resonators for 1:2 internal resonance, the design space can be discretized with finite elements and the density and material stiffness of the ith element in the design space can be written using the SIMP formulation as

$$\rho_i = \rho_{min} + x_i^{n1} \rho_0 \tag{3}$$

$$E_i = E_{min} + x_i^{n1} E_0 \tag{4}$$

where, ρ_0 is the material density, E_0 is the material stiffness and x_i is the design variable which varies between 0 and 1. ρ_{min} and E_{min} are infinitesimal constants to prevent numerical singularities in case x_i becomes equal to zero. The exponents, *n1* and *n2* are usually chosen larger than three so as to penalize any intermediate values of the design variable. As can be observed from Eqs. (3) and (4), a value of $x_i = 1$, implies that material is present and $x_i = 0$ implies presence of a void. Any intermediate value of x_i would produce non-physical results which is why a high value of exponents *n1* and *n2* are chosen to initially penalize intermediate values followed by filtering process at the end of optimization in which intermediate properties are taken to one extreme or the other depending on the filter design.

As an example of the topology optimization based resonator generation process, consider the structure shown in **Figure 1**. This structure is a rectangular plate which is constrained at its bottom edge. This plate is assigned Mooney-Rivlin material properties and is meshed with four node planar elements as it is assumed that the plate is undergoing vibrations in its plane.

The ratio of the first two planar natural frequencies of the base structure shown in **Figure 1** was 3.3. The aim of the topology optimization process is to fill the central cavity of the starting structure so that its first two planar natural frequencies are in the ratio close to 1:2. Thus, the design space is discretized with finite elements and the optimization problem posed by Eq. (2) is solved using the method moving asymptotes (MMA) which yields the optimized structure shown in **Figure 2**.

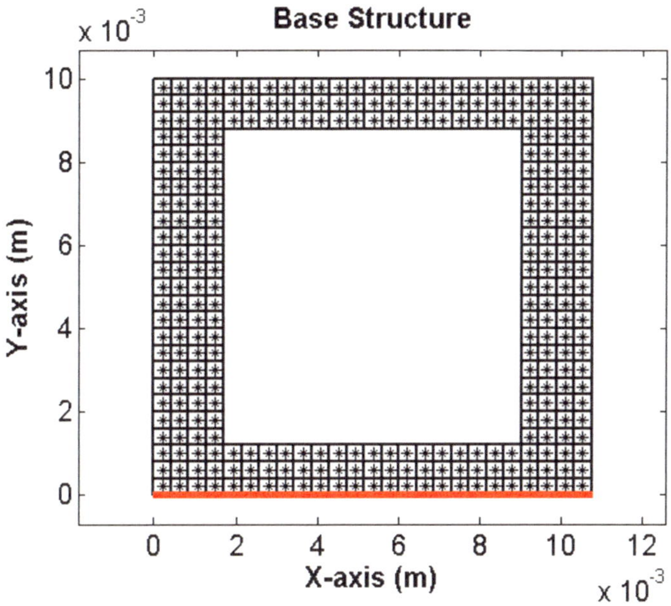

Figure 1.
The base structure used as a starting point for the topology optimization process. The red line indicates the edge that is fixed.

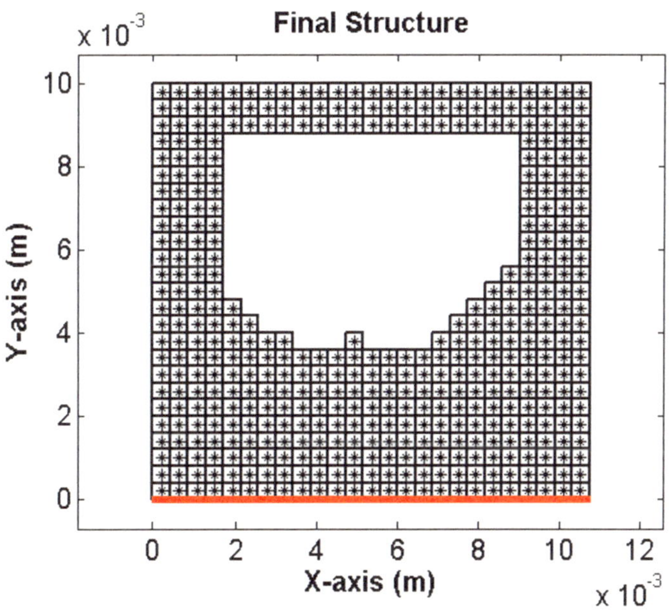

Figure 2.
The optimized structure obtained after applying the topology optimization process to the base structure shown in ***Figure 1****.*

The ratio of the first two planar natural frequencies of the optimized structure shown in **Figure 2** was 1.99. Thus, the topology optimization process was successful in bringing the natural frequencies of interest close to the ratio of 1:2. As the optimization process uses finite elements to compute the natural frequencies of the structure, the mode shapes of the optimized structure also become available and are shown in **Figure 3**.

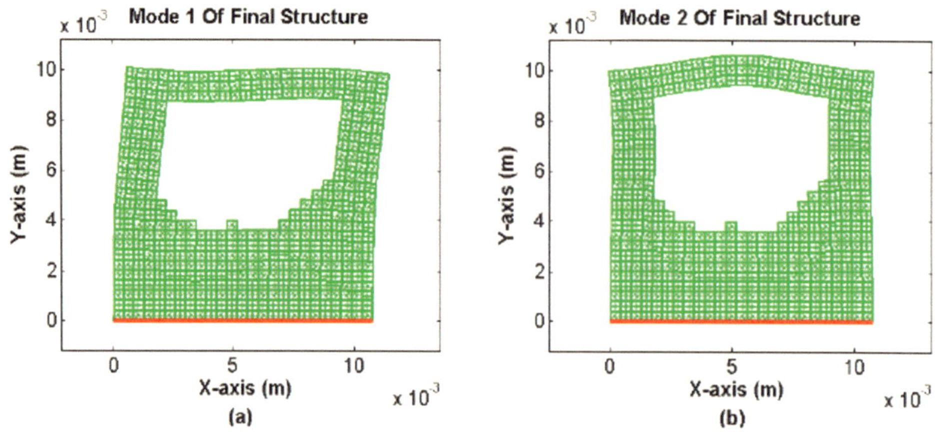

Figure 3.
Mode shapes of the optimized structure shown in ***Figure*** *2. (a) Lower Mode (Mode 1 of the structure) (b) Upper Mode (Mode 2 of the structure).*

2.2 Parametric optimization

Parametric optimization process is a simple but powerful tool which can also be used to generate various candidate structures for 1:2 internal resonance. As an example to illustrate the essential aspects of this procedure, consider the base structure shown in **Figure 4**. This base structure consists of a rectangular cantilever plate with two cutouts.

This plate can be assigned Mooney-Rivlin material properties and meshed with four node shell elements. In this study, Abaqus software is used to compute the natural frequencies of the base structure with the frequencies of interest being the second and third natural frequencies respectively. For the base structure shown in **Figure 4**, the ratio between the natural frequencies of the higher and lower mode of interest (third and second natural frequencies, respectively) was computed as 2.4. This base structure was then subjected to an optimization process with the objective

Figure 4.
The base structure used as a starting point for the optimization process in parametric optimization. The red line indicates the edge that is fixed.

function being described by Eq. (2). The design parameters for this optimization were the cut-out size and positions on the cantilever plate and the optimization was performed by a sequential quadratic programming method. The optimized structure obtained by applying the optimization process on the base structure is shown in **Figure 5**.

The ratio between the third and second natural frequencies of the optimized structure shown in **Figure 5** was 2.0. Thus, the optimization method was able to successfully bring the natural frequencies of the structure close to the desired ratio of 1:2. The mode shapes of the optimized structure also become available from the finite element model and are shown in **Figure 6**. These mode shapes can then be used to construct a reduced order model for the system which will be used to develop the nonlinear dynamic response for the candidate structure.

The method of parametric optimization allows for development a wide range of topologies which can each potentially exhibit 1:2 internal response. The optimal topology obtained depends on the starting structure, reflecting the local optimal nature of the solution. For example, consider the starting structure shown in **Figure 7**, the ratio between the natural frequencies of the higher and lower mode of interest (third and second natural frequencies, respectively) was computed as 1.6. Also note that the boundary conditions in this case involve fixing the resonator

Figure 5.
*The optimized structure obtained after applying the parametric optimization process to the base structure shown in **Figure 4**.*

Figure 6.
*Mode shapes of the optimized structure shown in **Figure 5**. (a) Lower Mode (Mode 2 of the structure) (b) Upper Mode (Mode 3 of the structure).*

Figure 7.
The base structure used as a starting point for the parameter optimization process. The red lines indicate the edges that are fixed.

Figure 8.
*The optimal structure obtained after implementing the parametric optimization process on the base structure shown in **Figure 7**.*

along both of its vertical sides. In this particular case, the optimization parameters were the size and location of the three circular cutouts. After performing the shape optimization process again using the sequential quadratic programming method, the optimized structure obtained is shown in **Figure 8**. The ratio between the third and second natural frequencies of the optimized structure shown in **Figure 8** was 2.0. Thus, the examples of **Figures 5** and **8** illustrate the possibilities of generating a large number of examples with different topologies as candidate resonators for 1:2 internal resonance. The mode shapes of the optimized structure shown in **Figure 8** are shown in **Figure 9**.

Figure 9.
*Mode shapes of the optimized structure shown in **Figure 8**. (a) Lower Mode (Mode 2 of the structure) (b) Upper Mode (Mode 3 of the structure).*

3. Nonlinear dynamic response

For the development of the nonlinear dynamic response of a 3D printed structure, consider the structure shown in **Figure 10**. This structure was designed using the simple iterative optimization procedure detailed in Section 2, and the resulting structure's modes 2 and 3 are in near internal resonance of 1:2. Thus, the frequency ratio achieved was 2:0005. The resonator was then fabricated using 3D printing machine Stratsys Dimension 1200es. This structure has the ratio of its second and third natural frequencies as ~2.0. The two transverse modes of interest for this candidate structure are shown in **Figure 11**. Using the mode shapes shown in **Figure 11**, assuming that the structure is subjected to a base excitation, and using the Kirchhoff plate theory [15], the displacement at any point on this rectangular plate can we written as:

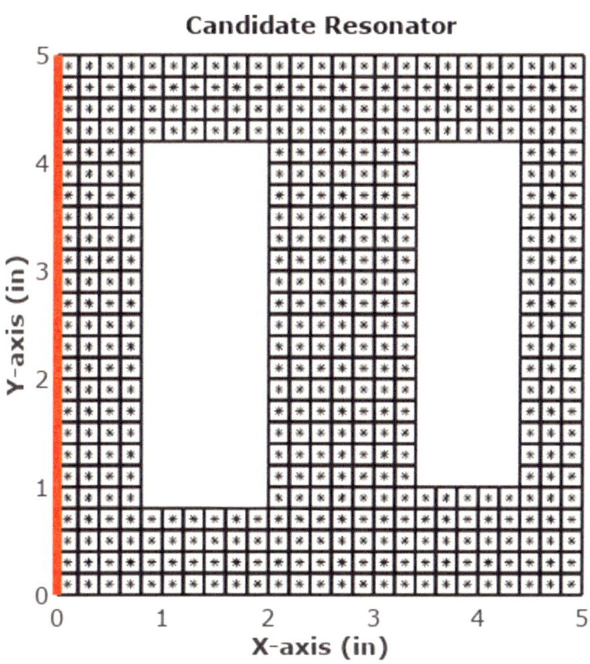

Figure 10.
The candidate structure for which the nonlinear dynamic transverse response is to be developed subject to a base excitation.

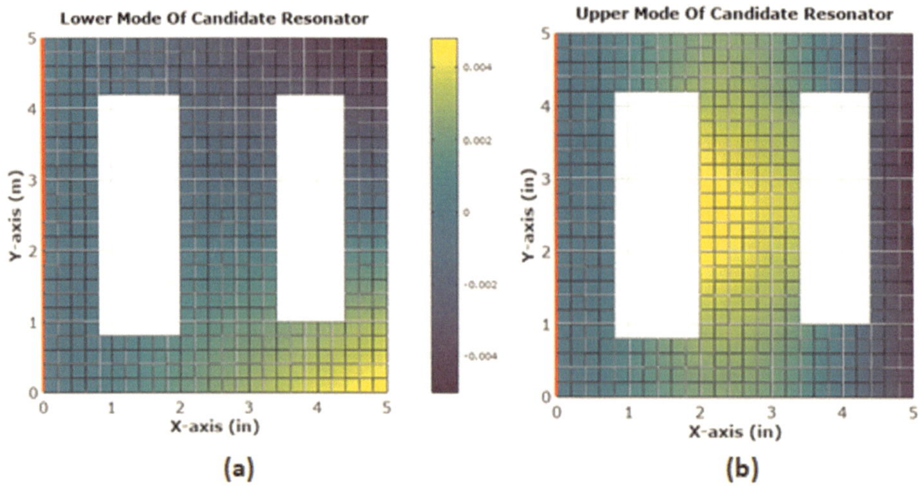

Figure 11.
The mode shapes of the candidate structure shown in ***Figure 10****. The modes are (a) mode 2 and (b) mode 3 respectively of the candidate resonator.*

$$u(X,Y,Z,t) = \varepsilon\left(A_1(t)\left(u_{01}(X,Y) - Z\frac{\partial w_1(X,Y)}{\partial X}\right) + A_2(t)\left(u_{02}(X,Y) - Z\frac{\partial w_2(X,Y)}{\partial X}\right)\right) \tag{5}$$

$$v(X,Y,Z,t) = \varepsilon\left(A_1(t)\left(v_{01}(X,Y) - Z\frac{\partial w_1(X,Y)}{\partial Y}\right) + A_2(t)\left(v_{02}(X,Y) - Z\frac{\partial w_2(X,Y)}{\partial Y}\right)\right) \tag{6}$$

$$w(X,Y,Z,t) = \varepsilon(A_1(t)w_1(X,Y) + A_2(t)w_2(X,Y)) + w_f(t) \tag{7}$$

where, u, v and w are the displacements in the X-, Y- and Z-directions, respectively, A_1 and A_2 are the modal amplitudes, u_{01}; u_{02} and v_{01}; v_{02} are the independent in-plane modal displacements in the X and Y directions, and w_1 and w_2 are the corresponding modal displacements (the mode shapes) in the Z- or the transverse direction. The base excitation is applied in the transverse direction (Z-direction) and is denoted by w_f which only depends on time, and ε is a small dimensionless parameter to keep track of the significant terms in the system response. It is assumed that the displacement field can be written with linear superposition of the two modes because in the canonical 1:2 internal resonance form, the higher mode is directly excited by external means and the system nonlinearities can cause an energy transfer between the higher mode and the lower mode. All other modes, if present, will see their contribution to the displacement field decay over time in the presence of damping as they are neither directly excited, nor excited by internal energy transfer.

The displacement field given in Eqs. (5)–(7) can now be used to write the kinetic and strain energy of the candidate structure. The kinetic energy, T is given by

$$T = \int_0^V \frac{1}{2}\rho\left(\dot{u}^2 + \dot{v}^2 + \dot{w}^2\right) dXdYdZ \tag{8}$$

where the dot ($\dot{}$) represents the derivative of the displacements with respect to time and ρ is the material density. For the time derivatives of the displacement, it must be noted that the modal amplitudes will be differentiated and the mode shapes can be treated as constants as they do not depend on time. In a similar manner, for a Mooney-Rivlin material, the strain energy, U can be written as,

$$U = \int_0^V \left\{ C_{10}\left(\bar{I}_1 - 3\right) + C_{01}\left(\bar{I}_2 - 3\right) + \frac{1}{d}(J-1)^2 \right\} dXdYdZ \tag{9}$$

where, $\bar{I}_1$ and $\bar{I}_2$ are the first and second deviatoric invariants, respectively, of the Left Cauchy Green deformation tensor B, J is the determinant of the deformation gradient, F, and C_{10}, C_{01} and d are the material constitutive parameters. The deformation gradient F is derived from the displacement field of the structure. The relationship between the original coordinates of a point on the plate and the deformed coordinates can be written as,

$$x = X + u \tag{10}$$

$$y = Y + v \tag{11}$$

$$z = Z + w \tag{12}$$

Then the deformation gradient F, can we written as,

$$F = \begin{bmatrix} 1 + \frac{\partial u}{\partial X} & \frac{\partial u}{\partial Y} & \frac{\partial u}{\partial Z} \\ \frac{\partial v}{\partial X} & 1 + \frac{\partial v}{\partial Y} & \frac{\partial v}{\partial Z} \\ \frac{\partial w}{\partial X} & \frac{\partial w}{\partial Y} & 1 \end{bmatrix} \tag{13}$$

The left Cauchy Green deformation tensor is given by

$$B = FF^T \tag{14}$$

The deviatoric strain invariants of the left Cauchy Green deformation tensor B are written as

$$\bar{I}_1 = J^{-\frac{2}{3}} I_1 \tag{15}$$

$$\bar{I}_2 = J^{-\frac{4}{3}} I_2 \tag{16}$$

where J is the determinant of the deformation tensor given by Eq. (13), and the strain invariants I_1 and I_2 are given by

$$I_1 = tr(B) \tag{17}$$

$$I_2 = \frac{1}{2}\left(tr(B)^2 - tr\left(B^2\right)\right) \tag{18}$$

where *tr(B)* refers to the trace of the matrix B. Note that using Eqs. (5)–(7) and (9)–(14), the strain energy of the structure can also be computed. Once the expressions of both the kinetic energy and the strain energy are available, the Lagrangian of the resonator can be written as

$$L = T - U \tag{19}$$

This Lagrangian from Eq. (19) will be a nonlinear function of the modal amplitudes owing to the nonlinear nature of the strain energy potential given in Eq. (9). The base excitation of the structure is now assumed to be of the form

$$\dot{w}_f = \varepsilon^2 V_B \sin(\Omega t) \tag{20}$$

where V_B is the amplitude of the base excitation velocity and Ω is the excitation frequency. For 1:2 internal resonance, the excitation frequency can be near the lower or the upper natural frequency. In case of subharmonic external resonance, the external frequency is assumed to be close to the upper natural frequency so that the higher mode is resonantly excited [4]. The difference between the excitation frequency and the second natural frequency is known as the external mistuning σ_2 defined as

$$\Omega = \omega_2 + \varepsilon\sigma_2 \tag{21}$$

Similarly, another mistuning parameter, the internal mistuning σ_1, is introduced to take into account the deviation of the two interacting natural frequencies from the perfect 1:2 ratio, that is,

$$\omega_2 = 2\omega_1 + \varepsilon\sigma_1 \tag{22}$$

To further study the nonlinear dynamic response of the structure for small nonlinear motions, and to formulate the application of the method of averaging [4], the modal amplitudes (for Eqs. (5)–(7)) can be written as

$$\boldsymbol{A_1(t) = p_1(\varepsilon t) \cos\left(\frac{\Omega}{2} t\right) + q_1(\varepsilon t) \sin\left(\frac{\Omega}{2} t\right)} \tag{23}$$

$$\boldsymbol{A_2(t) = p_2(\varepsilon t) \cos(\Omega t) + q_2(\varepsilon t) \sin(\Omega t)} \tag{24}$$

where p_i and q_i are amplitude components which vary on a slow time scale $\tau = \varepsilon t$, as defined in [4, 8]. Using the expressions for amplitudes in Eqs. (23) and (24), the time derivatives of the amplitudes can be written as

$$\begin{aligned}\dot{A}_1 = {} & \varepsilon\left(p_1'(\varepsilon t) \cos\left(\frac{\Omega}{2} t\right) + q_1'(\varepsilon t) \sin\left(\frac{\Omega}{2} t\right)\right) \\ & + \frac{\Omega}{2}\left(-p_1(\varepsilon t) \sin\left(\frac{\Omega}{2} t\right) + q_1(\varepsilon t) \cos\left(\frac{\Omega}{2} t\right)\right)\end{aligned} \tag{25}$$

$$\dot{A}_2 = \varepsilon\left(p_2'(\varepsilon t) \cos(\Omega t) + q_2'(\varepsilon t) \sin(\Omega t)\right) + \frac{\Omega}{2}\left(-p_2(\varepsilon t) \sin(\Omega t) + q_2(\varepsilon t) \cos(\Omega t)\right) \tag{26}$$

where a prime ($'$) denotes a derivative with respect to the slow time τ. Now the Lagrangian given in Eq. (19) is averaged over the period of oscillation$= \frac{4\pi}{\Omega}$. The slow time amplitudes are treated as constants for this averaging operation and also only terms till $O(\varepsilon^3)$ are retained in the Lagrangian as the cubic nonlinear terms are sufficient to capture the 1:2 internal resonance of the structure. The effects of internal resonance are essentially captured by quadratic nonlinear terms in the equations of motion. The averaged Lagrangian is defined by

$$\langle L \rangle = \int_0^{\frac{4\pi}{\Omega}} (T - U) dt \tag{27}$$

Subjecting the averaged Lagrangian shown in Eq. (27) to the Euler-Lagrange conditions ($\frac{d}{d\tau}\frac{\partial\langle L\rangle}{\partial p_i'} - \frac{\partial\langle L\rangle}{\partial p_i} = 0, \frac{d}{d\tau}\frac{\partial\langle L\rangle}{\partial q_i'} - \frac{\partial\langle L\rangle}{\partial q_i} = 0, i = 1, 2$) provides the following equations of motion for the slow time amplitudes

$$p_1' + \zeta_1 p_1 + \left(\frac{\sigma_1 + \sigma_2}{2}\right) q_1 + \Lambda_1 \omega_1 (p_2 q_1 - p_1 q_2) = 0 \tag{28}$$

$$q_1' + \zeta_1 q_1 - \left(\frac{\sigma_1 + \sigma_2}{2}\right) p_1 + \Lambda_1 \omega_1 (p_1 p_2 + q_1 q_2) = 0 \tag{29}$$

$$p_2' + \zeta_1 p_2 + (\sigma_2) q_2 - 2\Lambda_2 \omega_2 p_1 q_1 = 0 \tag{30}$$

$$q_2' + \zeta_1 q_2 - (\sigma_2) p_2 + \Lambda_2 \omega_2 (p_1^2 - q_1^2) = \Lambda_3 V_B \tag{31}$$

where a prime (′) denotes a derivative with respect to the slow time τ, and. Λ_i, i = 1, 2, 3, are constants which come from the averaged Lagrangian of the structure and depend on the material constitutive parameters (Eq. (9)) and mode shapes. The modal damping terms ζ_1 and ζ_2 were introduced in Eqs. (28)–(31) to represent damping in the system to prevent the solutions from becoming unbounded. The expressions in Eqs. (28)–(31) are the same as the expressions in a standard 1:2 internal resonance system [4, 8], thus demonstrating that 3D printed rectangular plates with cutouts are able to exhibit nonlinear dynamic phenomena such as 1:2 internal resonance provided the constants Λ_i, i = 1, 2, 3, are non-zero. To make the analysis of Eqs. (28)–(31) a little more tractable, the following variable transformations are defined

$$p_1 = a_1 \cos(\beta_1), q_1 = a_1 \sin(\beta_1) \tag{32}$$

$$p_2 = a_2 \cos(\beta_2), q_2 = a_2 \sin(\beta_2) \tag{33}$$

where a_i's are the amplitudes and β_i are the phase angles. With the transformations introduced in Eqs. (32) and (33), the equations of motion for modal amplitudes become

$$a_1' = -\zeta_1 a_1 - \Lambda_1 \omega_1 a_1 a_2 \, sin \, (2\beta_1 - \beta_2) \tag{34}$$

$$a_1 \beta_1' = \left(\frac{\sigma_1 + \sigma_2}{2}\right) a_1 - \Lambda_1 \omega_1 a_1 a_2 \, cos \, (2\beta_1 - \beta_2) \tag{35}$$

$$a_2' = -\zeta_2 a_2 - \Lambda_2 \omega_2 a_1^2 \, sin \, (\beta_2 - 2\beta_1) + \Lambda_3 V_B \sin(\beta_2) \tag{36}$$

$$a_2 \beta_2' = \sigma_2 a_2 - \Lambda_2 \omega_2 a_1^2 \, cos \, (2\beta_1 - \beta_2) + \Lambda_3 V_B \cos(\beta_2) \tag{37}$$

The steady-state solutions for the system of equations described by Eqs. (34)–(37) can be obtained by setting $a_i' = 0$ and $\beta_i' = 0$. These equations can be solved for steady-state solutions to give single-mode (only second modal amplitude a_2 is non-zero) and coupled-mode solutions (both first and second mode amplitudes are non-zero, that is, $a_1 \neq 0$ and $a_2 \neq 0$). The coupled-mode solution is the main nonlinear response as it implies energy transfer from the higher to lower mode when the higher mode is directly excited in the presence of quadratic nonlinearities. Plots in **Figure 12** give a representative set of steady-state solutions for the single and coupled-mode response for the structure shown in **Figure 10** with zero damping. Note that the coupled-mode response is slightly asymmetric about $\sigma_2 = 0$ axis for the structure as some minimal internal mistuning exists ($\sigma_1 \neq 0$) due to the fact that $\frac{\omega_2}{\omega_1} = 2{:}0005$. For perfect internal resonance, the coupled-mode solutions will be completely symmetric around $\sigma_2 = 0$. The non-zero first mode arises as a result of the subcritical pitchfork bifurcations (at P_1 and P_2) from the single mode solution consisting of only the second mode (The lower mode amplitude is zero). A more detailed study of the stability of the solutions for the single and coupled mode results is provided in [13].

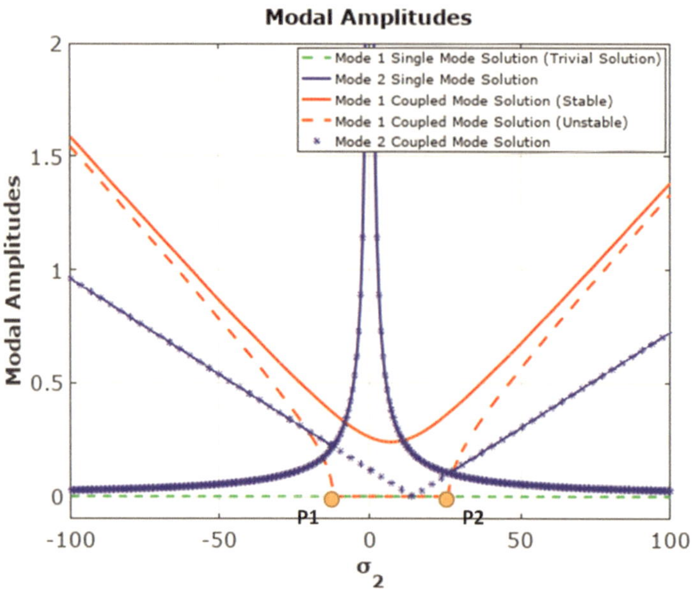

Figure 12.
*Non-linear response of the 3D printed structure shown in **Figure 10** to a transverse harmonic base excitation. The plots are for the amplitudes of the two interacting modes for both the single-mode and coupled-mode response. Note that $\sigma_1 \neq 0$ though very small and the modal damping is low (ζ_1 and ζ_2 equal to 0.05).*

As is clear from Eqs. (34)–(37), the modal amplitudes depend upon many parameters. Some of the more interesting of these are the internal mistuning σ_1 and the modal damping terms ζ_1 and ζ_2. The effect of change in internal mistuning is shown in **Figure 13**. As can be observed from **Figure 13**, changes in internal mistuning can result in the coupled-mode motion to lose existence and disappear, that is, the modal interaction is lost for large internal mistuning. Thus, in actual physical systems deviation of natural frequencies of participating modes from the perfect 1:2 ratio can cause the non-trivial coupled mode response to not manifest itself.

Figure 14 shows the effect of damping coefficients on the nonlinear response curves obtained using Eqs. (34)–(37). Increasing damping coefficients ζ_1 and ζ_2 has interesting effects on the overall nonlinear response. As can be observed from

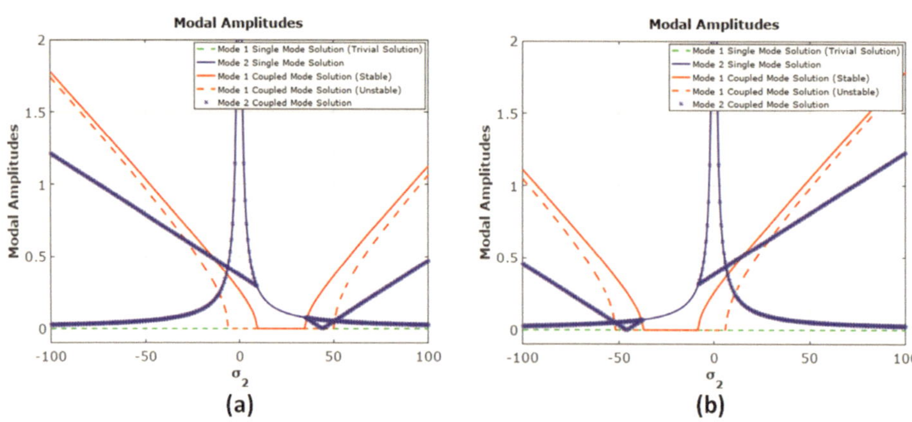

Figure 13.
*Non-linear response curves of the 3D printed (hyperelastic) structure shown in **Figure 10** for (a) large negative internal mistuning, and (b) large positive internal mistuning.*

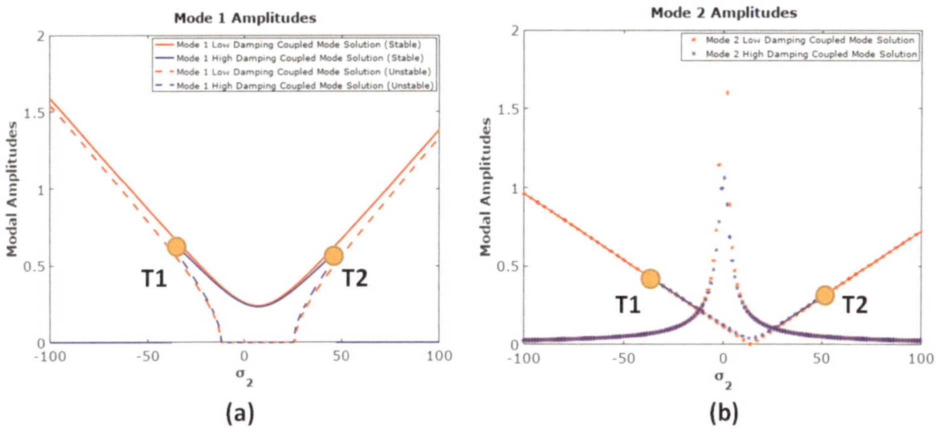

Figure 14.
Non-linear response curves of the hyperelastic structure in ***Figure 10*** *for representative low (red) and higher (blue) damping coefficients. The two figures are for (a) mode 1 amplitudes (b) mode 2 amplitudes. The points T1 and T2 are turning point bifurcations.*

Figure 14a, increasing damping leads to a reduction in the frequency range in which the nonlinear coupled-mode response is observed.

4. Summary and conclusions

This work explored the possibility of synthesizing 3D printed hyperelastic plate structure exhibiting 1:2 internal resonances. 3D printing occupies a potential sweet spot in terms of dimensional capabilities and repeatability to produce nonlinear resonators which can be used as vibration absorbers, sensors, or for signal processing applications. The synthesis methodology allows for designing a large set of designs meeting the desired internal resonance conditions resulting in complex modal coupling and energy transfer between modes of the structure.

While the nonlinear dynamical response studied here was focused on 3D printed cantilever plates that exhibited 1:2 internal resonances on account of material non-linearities, the methodology can be easily applied to other boundary conditions and internal resonances, as well as for structures with geometric nonlinearities caused by finite deformations of plates.

Conflict of interest

The authors declare no conflict of interest.

Author details

Astitva Tripathi[1] and Anil K. Bajaj[2*]

1 Caelynx LLC, Ann Arbor, MI, USA

2 School of Mechanical Engineering, West Lafayette, IN, USA

*Address all correspondence to: bajaj@ecn.purdue.edu

References

[1] Santuria SD. Microsystem Design. New York: Springer; 2001

[2] Lobontiu N. Dynamics of Microelectromechanical Systems. New York: Springer; 2007

[3] Younis MI. MEMS Linear and Nonlinear Statics and Dynamics. New York: Springer; 2011

[4] Nayfeh AH, Mook DT. Nonlinear Oscillations. New York: John Wiley & Sons; 2008

[5] Rhoads JF, Shaw SW, Turner KL. Nonlinear dynamics and its applications in micro- and nanoresonators. Journal of Dynamic Systems, Measurement, and Control. 2010;**132**:034001. DOI: 10.1115/1.4001333

[6] Daqaq MF, Masana R, Erturk A, Quinn DD. On the role of nonlinearities in vibratory energy harvesting: A critical review and discussion. Applied Mechanics Reviews. 2014;**66**(4): 040801. DOI: 10.1115/1.4026278

[7] Bajaj AK, Chang SI, Johnson JM. Amplitude modulated dynamics of a resonantly excited autoparametric two degree-of-freedom system. Nonlinear Dynamics. 1994;**5**:433-457

[8] Nayfeh AH. Nonlinear Interactions: Analytical, Computational, and Experimental Methods. 1st ed. New York: Wiley; 2000

[9] Galeta T, Raos P, Stojšić J, Pakši I. Influence of structure on mechanical properties of 3D printed objects. Procedia Engineering. 2016;**149**:100-104

[10] Kotlinski J. Mechanical properties of commercial rapid prototyping materials. Rapid Prototyping Journal. 2014;**20**: 499-510. DOI: 10.1108/RPJ-06-2012-0052

[11] Grappasonni C, Habib G, Detroux T, Kerschen G. Experimental Demonstration of a 3D-Printed Nonlinear Tuned Vibration Absorber [Internet]. Available from: http://citeseerx.ist.psu.edu/viewdoc/download?doi=10.1.1.1024.23&rep=rep1&type=pdf [Accessed: May 25, 2019]

[12] Tripathi A, Bajaj AK. Design for 1:2 internal resonances in in-plane vibrations of plates with hyperelastic materials. Journal of Vibration and Acoustics. 2014;**136**:061005. DOI: 10.1115/1.4028268

[13] Tripathi A, Bajaj AK. Topology optimization and internal resonances in transverse vibrations of hyperelastic plates. International Journal of Solids and Structures. 2016;**81**:311-328

[14] Vyas A, Peroulis D, Bajaj AK. Dynamics of a nonlinear microresonator based on resonantly interacting flexural-torsional modes. Nonlinear Dynamics. 2008;**54**:31-52

[15] Amabili M. Nonlinear Vibrations and Stability of Shells and Plates. New York: Cambridge University Press; 2008

[16] Ribeiro P, Petyt M. Non-linear free vibration of isotropic plates with internal resonance. International Journal of Nonlinear Mechanics. 2000;**35**(2): 263-278

[17] Amabili M. Nonlinear vibrations of rectangular plates with different boundary conditions: Theory and experiments. Computers & Structures. 2004;**82**(31–32):2587-2605

[18] Chang SI, Bajaj AK, Krousgrill CM. Non-linear vibrations and chaos in harmonically excited rectangular plates with one-to-one internal resonance. Nonlinear Dynamics. 1993;**4**:433-460

[19] Pedersen NL. Optimization of holes in plates for control of eigenfrequencies. Structural and Multidisciplinary Optimization. 2004;**28**:1-10. DOI: 10.1007/s00158-004-0426-8

[20] Pedersen NL. Designing plates for minimum internal resonances. Structural and Multidisciplinary Optimization. 2005;**30**:297-307. DOI: 10.1007/s00158-005-0529-x

[21] Bendsoe M, Sigmund O. Topology Optimization: Theory, Methods and Applications. New York: Springer; 2003

[22] Dai X, Miao X, Sui L, Zhou H, Zhao X, Ding G. Tuning of nonlinear vibration via topology variation and its application in energy harvesting. Applied Physics Letters. 2012;**100**: 031902. DOI: 10.1063/1.3676661

[23] Dou S, Strachan BS, Shaw SW, Jensen JS. Structural optimization for nonlinear dynamic response. Philosophical Transactions of the Royal Society, A. 2015;**373**:20140408

[24] Lily LL, Polunin PM, Dou S, Shoshani O, Strachan BS, Jensen JS, et al. Tailoring the nonlinear response of MEMS resonators using shape optimization. Applied Physics Letters. 2017;**110**:081902

[25] Breslavsky IV, Amabili M, Legrand M. Nonlinear vibrations of thin hyperelastic plates. Journal of Sound and Vibration. 2015;**333**:4668-4681

Chapter 9

Nonlinear Systems in Healthcare towards Intelligent Disease Prediction

Parag Chatterjee, Leandro J. Cymberknop and Ricardo L. Armentano

Abstract

Healthcare is one of the key fields that works quite strongly with advanced analytical techniques for prediction of diseases and risks. Data being the most important asset in recent times, a huge amount of health data is being collected, thanks to the recent advancements of IoT, smart healthcare, etc. But the focal objective lies in making sense of that data and to obtain knowledge, using intelligent analytics. Nonlinear systems find use specifically in this field, working closely with health data. Using advanced methods of machine learning and computational intelligence, nonlinear analysis performs a key role in analyzing the enormous amount of data, aimed at finding important patterns and predicting diseases. Especially in the field of smart healthcare, this chapter explores some aspects of nonlinear systems in predictive analytics, providing a holistic view of the field as well as some examples to illustrate such intelligent systems toward disease prediction.

Keywords: nonlinear systems, healthcare, artificial intelligence, computational intelligence, machine learning, predictive analytics, chronic disease, cancer, cardiometabolic disease, Parkinson's disease

1. Introduction

> *"Prevention is better than cure"*
>
> —*Desiderius Erasmus*

A nonlinear system is a system in which the change of the output is not proportional to the change of the input [1–3]. Especially in the field of healthcare, most of the health systems being inherently nonlinear in nature, nonlinear systems are of special interest to researchers hailing from multidisciplinary areas. Nonlinear dynamical systems, describing changes in variables over time, may appear chaotic, unpredictable, or counterintuitive, contrasting with much simpler linear systems. Nonlinear modeling still has not been able to explain all of the complexity present in human systems, and further models still need to be refined and developed. However, nonlinear modeling is helping to explain some system behaviors that linear systems cannot and thus will augment our understanding of the nature of complex dynamic systems within the human body in health and in disease states [4].

The delivery of healthcare is a complex endeavor at both individual and population levels. At the clinical level, the tailored provision of care to individuals is guided, in part, by medical history, examination, vital signs and evidence. In the twenty-first century these traditional tenets have been supplemented by a focus on learning, metrics and quality improvement. The collection and analysis of data of good quality are critical to improvements in the effectiveness and efficiency of health care delivery [5]. This is also catalyzed by the boost in the field of eHealth across the world. eHealth is emerging as a promising vehicle to address the limited capacity of the health care system to provide health behavior change and chronic disease management interventions. The field of eHealth holds promise for supporting and enabling health behavior change and the prevention and management of chronic disease [6].

With a global increase in the adoption of Electronic Health Records (EHRs) [7–12], the volume and complexity of the data generated increases in all dimensions. In addition to the EHR-sourced patient data, the additional data available from other sources like the data about medical conditions, underlying genetics, medications, and treatment approaches is humongous. But human cognition to learn, understand, and process the data being finite [13], the traditional medical methods of analysis does not stand always to be the most efficient. Thus, computer-assisted methods to organize, interpret, and recognize patterns from these data are needed [14].

In the recent years, the underlying value of data is unfolding like never before and newer systems are being developed concentrating on the data analysis to make sense of the data. Especially in the field of healthcare, the aspect of intelligent data analytics is one of the most trending topics worldwide. One of the focal areas where such analyses have been applied is in the field of chronic diseases. By 2020, chronic diseases are expected to contribute to 73% of all deaths worldwide and 60% of the global burden of disease. Moreover, 79% of the deaths attributed to these diseases occur in the developing countries. Four of the most prominent chronic diseases—cardiovascular diseases (CVD), cancer, chronic obstructive pulmonary disease and type 2 diabetes are linked by common and preventable biological risk factors, notably high blood pressure, high blood cholesterol and overweight, and by related major behavioral risk factors. Action to prevent these major chronic diseases should focus on controlling these and other key risk factors in a well-integrated manner [15]. Apart from the chronic diseases, a key area where nonlinear models are applied from the perspective of intelligent prediction is human movement and locomotion. This leads to topics like fall detection, abnormal gait detection and diseases like Parkinson's. The outreach of intelligent prediction is spread to wider domains like transplantations [16], for example, to predict the success of a liver transplant by analyzing all the relevant health parameters.

Moreover, with the recent trends of smart sensors and eHealth devices powered by the Internet of Things (IoT), the data acquired is more comprehensive and detailed. Both prediction and prevention systems in this case usually use some fundamental steps in common, like collection of data from sensors and its analysis, followed by computing the risk and other possibilities [17]. The entire pool of data originating from this field is mostly nonlinear, invoking the need for the development of nonlinear analysis and predictive models.

Exploring the possible actions toward prevention of the chronic diseases, the key challenge lies in early detection of the diseases. Most of these diseases do not exhibit clearly identifiable signs at the early stage. This leads to harvesting the possibility of early detection of these diseases using artificial intelligence (AI). From the perspective of data science, the fundamental and most valuable resource in this aspect is the health data.

The health data holds immense potential for detailed analyses towards the early detection and prediction of diseases. The prediction of diseases using computational intelligence is multilevel. Most of the health systems being inherently nonlinear in nature, it provides an enormous opportunity to analyze those intricate details of the health systems while searching for the traits or early signals of diseases. On the other hand, given that the importance of health data (mostly the superficially and non-invasively obtained behavioral, physiological and metabolic health data) is quite crucial, a huge opportunity lies on the aspect of analysis of this health data toward the prediction of diseases. The computational aspect of disease prediction is also multifold, including aspects of data analysis, signal and image processing and other fields. However, this chapter is focused to the aspect of data analytics and computational intelligence, highlighting the key aspects of the health data of people pertaining to nonlinear systems, discussing the field of machine learning and intelligence toward the prediction of diseases.

2. Data modeling in healthcare toward predictive analysis

> *"Data is the new oil. It's valuable, but if unrefined it cannot really be used."*
> *—Clive Humby*

The backbone of the intelligent prediction systems in healthcare is the data. Thanks to the skyrocketing advancements in data collection strategies and tools, it estimates a yearly growth of 48% [18] with projected growth rate to more than 2000 exabytes by 2020 [19]. On one hand, this poses an enormous challenge to handle this data. But on the other hand, this also keeps the potential in performing detailed analysis toward interesting insights. To understand the health data from a holistic view, the attributes like volume, variety, velocity, and veracity need to be considered, to decipher its value. This implies that on a larger scale, handling the health data as big data is inevitable. The first part is making this data suitable for analysis. About 80% of the world's healthcare data being unstructured [20], it poses a huge challenge for the data preprocessing. Health data obtained from multiple sources often lack seriously in the aspect of interoperability or uniformity to model and process together. Even with the rise of EHRs, a major challenge lies in normalizing the data and making it suitable for modeling. With a lot of heterogeneous smart eHealth devices and components of IoT, tying the data together stands extremely difficult. For this reason, several EHR management system are being designed recently considering the data models and the aspect of preprocessing; thus, the EHRs are collected in such a way that fits the data models or facilitates the same. An inclusive data preprocessing system holds immense potential to support the aspect of health data modeling from a comprehensive perspective. Especially in the case of nonlinear systems, the aspect of data modeling is critical for processing enormous health data. Badly constructed data models not only skew the results but may also produce erroneous insights appearing to be correct.

The traditional way modeling the health data includes the field of data mining, with the aspects of knowledge discovery in databases (KDD) and exploratory data analysis (EDA). Most of these old KDD techniques visualize data from a perspective of database, looking for interesting knowledge in the data from a high-level point of view. These techniques provide better view to the data and especially in the field of healthcare, is often quite useful in analyzing patterns in large health datasets. However, specifically in the area of prediction of diseases and risks, recent systems are focused toward predictive models to perform a more formal, scalable and more efficient analysis and prediction. Comparing with the traditional data mining models, these predictive models are made in a tailored-approach, mostly dedicated to a specific goal (for example, prediction of cardiovascular diseases). Often, such models also involve machine learning and computational intelligence for the aspect of analysis and prediction.

One of the most crucial tasks in this respect is to identify the mathematical model of a system from measurements of the system inputs and outputs. Especially in the field of disease and risk prediction, the data handled is mostly multidimensional. Keeping the focus toward nonlinear modeling, the first check is of course the identification of the data model, if it is a linear model or a nonlinear one. This is usually done using the superposition principle (properties of additivity and homogeneity). However, in some cases for finding the pattern in the health data, linear models do count useful. A more crude but common approach is to start with a linear model. After the initial tests to check the suitability of the linear model, which if turns out not good enough, leads to the replacement by a nonlinear approach to model the system.

Among several nonlinear models used in analyzing health data especially aimed at intelligent disease prediction, methods like nonlinear regression, clustering, decision trees, nonlinear support vector machines (SVMs), and artificial neural networks (ANNs) are quite profuse in recent times within the field of predictive medical analytics. For example, in nonlinear regression, the observational health data is modeled by a function which is a nonlinear combination of the model parameters and depends on one or more independent variables. The data is fitted by a method of successive approximations. If the health data available is labeled, supervised learning models like SVMs are used often, to analyze the data using classification and regression analysis. Given a set of training examples, each marked as belonging to one or the other of two categories, a SVM training algorithm builds a model that assigns new examples to one category or the other. Training data is usually divided into a training data (70%) and test data (30%). To map nonlinear functions, kernels can be used in SVMs. A kernel is a function that maps the data into a higher dimensional space where the linear mapping is possible. One of the main advantages of SVM with respect to modeling nonlinear systems is the possibility to use kernels, making it possible to represent very complex functions. However, when compared to linear regression, the main drawback is the need of more training and prediction times [21].

Neural networks on the other hand encompass a large class of models and learning methods and are nonlinear statistical models [22]. A recent survey of AI applications in healthcare reported uses in major disease areas such as cancer or cardiology and artificial neural networks as a common machine learning technique [23]. Such networks are organized in layers made of a number of interconnected nodes which contain an activation function. Data is provided to the network via the input layer, following which, the processing is performed in one or more hidden layers using a system of weighted connections. The last hidden layer is linked to the output layer where the result is given [21]. The healthcare domain of intelligent risk prediction is largely governed by the aspect of pattern recognition or finding relationships among several health and behavioral parameters and to study their impacts. One of the principal advantages of ANN (**Figure 1**) [24] is that it can model different types of relationships; systems which otherwise may have been very difficult to represent correctly could be modeled quickly and relatively easily using ANNs. However, compared to other types of networks, ANNs tend to be slower in training. Despite being a system of parallel computation, the slowness of the training step is due to the fact that individual artificial neurons are usually processed sequentially [21].

Ensemble classifiers are constructed from a given training data set and predict the class of a previously unseen object by combining the predictions obtained from these basic classifiers. The importance of different ensemble classifiers has also been at rise attributing to the possibility of determining the risk groups among patient population. For example, the family of simple probabilistic classifiers like naive Bayes classifiers discover application in programmed medicinal analysis [25]. Performing regular analysis of healthcare for a large population makes it possible to act early in the case of health hazards and risks [26]. Clinical decision support systems often count useful in this aspect to assist the medical personnel in designing treatment strategies [27]. Such systems are mostly constructed using decision trees. Decision trees are flowchart structures in which each internal node denotes a test on a characteristic, each branch signifies the result of the test, and each leaf node denotes a class label. The paths from the root to leaf denotes classification rules [25]. Random forest algorithms (**Figure 2**) [21] are specifically suited for decision tree classifiers. In this technique, the basic classifiers are decision trees obtained by manipulating the input features. Basically, random forest builds multiple decision

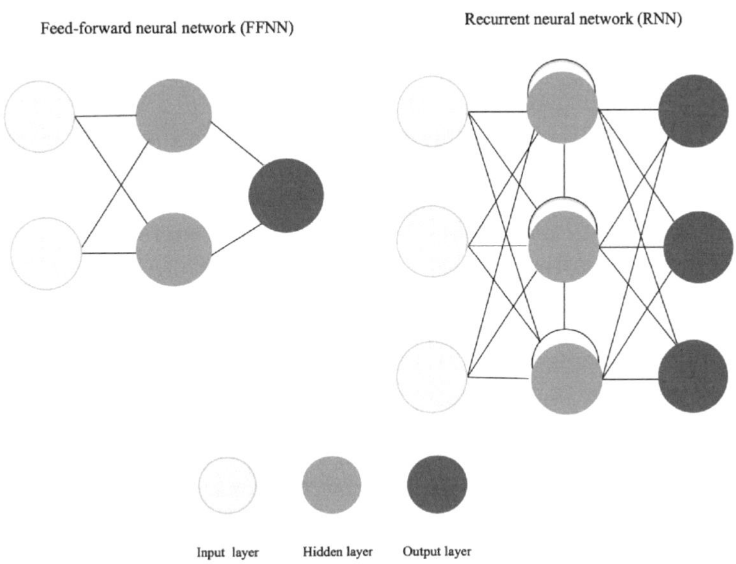

Figure 1.
Artificial neural networks.

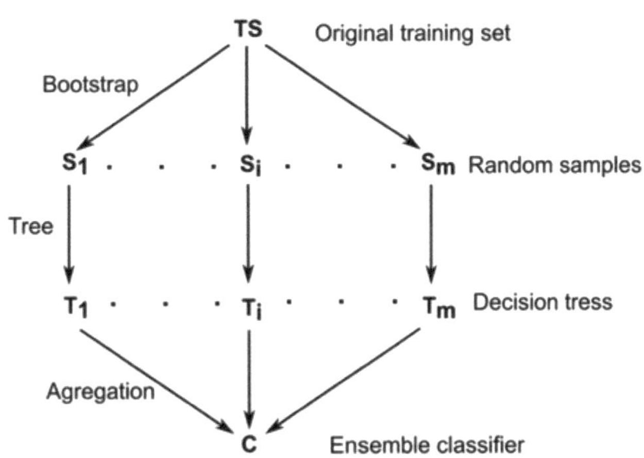

Figure 2.
Working principle of random forest.

trees and merges them together to get a more accurate and stable prediction. A key advantage of random forest is, that it can be used for both classification and regression problems; a huge part of current machine learning systems in healthcare is related to such problems. However, the major limitation of random forest algorithms is that a large number of trees can make the algorithm slow down and ineffective for real-time predictions. In general, these algorithms are fast to train, but quite slow to create predictions once they are trained. A more accurate prediction requires more trees, which results in a slower model [28].

Unsupervised learning techniques are also profuse in health analytics especially in the field of analyzing health risks. Since a considerable part of health-data often arrive unlabeled, unsupervised learning methods help in finding patterns in the data or to analyze the health scenario over a big population. In this aspect, clustering techniques like k-means, gaussian distribution models, and mean-shift clustering often stand very useful in separating a group of patients into different clusters and then to analyze in detail the salient features and distinct characteristics. Among all the unsupervised learning algorithms, clustering via k-means might be one of the simplest and most widely used algorithms. Briefly, k-means clustering aims to find the set of k clusters such that every data point is assigned to the closest center, and the sum of the distances of all such assignments is minimized [29]. Especially when the relationships among different health parameters and their respective impacts are not known well, clustering techniques are used to separate a patient population in order to study the distinguishing features and influencing factors. In this aspect, nonlinear techniques of clustering also count useful.

Another crucial aspect with respect to machine learning algorithms is the aspect of bias and variance. The bias is an error from erroneous assumptions in the learning algorithm. High bias can cause an algorithm to miss the relevant relations between features and target outputs (underfitting). The variance is an error from sensitivity to small fluctuations in the training set. To visualize the degree to which a machine learning algorithm is suffering from bias or variance with respect to a data problem, learning curves are important tools. Learning curves are displays in which the performance of the machine learning algorithms are plotted with respect to the quantity of data used for training where the plotted values are the prediction error measurements [30].

Nevertheless, the choice of algorithm is dependent on multiple factors, the most important being the type of the dataset. Apart from that the aspect of prior knowledge about the data, computational complexity and expected results are also deciding factors, and the correct use of the model is extremely crucial in this regard [31]. Recent research has delved into uniting different techniques to provide hybrid machine learning algorithms [32]. Nevertheless, it is clear that the use of machine learning and computational intelligence takes an active role in predicting the health risks and the probability of diseases using the intelligence hidden in the health data.

3. Applications of computational intelligence toward prediction of diseases

The use of computational intelligence with an objective to predict the health risks and diseases is an extensive and multi-step process. In this section, different scenarios are explained with respect to predictive analytics, aimed at disease and risk prediction. The first case is based on cardiometabolic diseases, and therefore, the most important component of the predictive system stands to be the detailed physiological and health data.

For example, in the field of cardiometabolic diseases, several parameters (age, gender, systolic and diastolic blood pressure, cholesterol, diabetes and smoking habits) are considered for patients registered in the database and their risk scores was calculated. The risk scores give a general idea to classify the entire population into high and low risk groups [33]. Nevertheless, alternative analyses are performed to identify the underlying risk groups for each health parameter in the entire population. But this analysis is scalable for a further detailed technique for prediction of risks and influence of different health parameters on the cardiometabolic disease of a population. For example, Framingham Risk Score is used to predict the hard-coronary heart diseases (myocardial infarction or coronary death) and is calculated based on predictors like age, total cholesterol, high-density lipoproteins (HDL), systolic blood pressure, treatment for hypertension and smoking status [34]. In this case, the Framingham Risk can be expressed as a linear equation considering all the parameters. However, considering sample sets of people which are smaller and more specific, there exists the possibility of nonlinear relationships of several other parameters pertaining to cardiovascular risks, not usually considered in classical risk prediction models.

The traditional approach based on the identification and treatment of risk factors has proven to be insufficient and ignores that the detection of its subclinical stage is valid to define cardiovascular risk strategies. Taking into account that the artery is the main protagonist in this disease, it is necessary to evaluate it directly through a morpho-structural and functional analysis with non-invasive, reliable, reproducible procedures that are applicable in the youngest population. The detection of subclinical disease and the precocity with which it is done defines a safe framework to derive the real individual cardiovascular risk. Because coronary calcifications are an early marker of atherosclerosis detectable non-invasively; a model of cardiovascular risk that incorporates them along with the classic risk factors could have a remarkable interest in clinical practice, having the potential to change the field of preventive cardiology. The traditional approach guides the prevention and treatment of arterial disease, atherosclerotic in particular, based on the detection of cardiovascular risk factors (e.g., hypertension, smoking, dyslipidemia etc.). This approach quantifies the probability (risk) that the subject has a cardiovascular event (accident) in the next 10 years of life. Thus, based on information from large populations and global cardiovascular risk tables, and information obtained regarding the risk factors of each subject (for example, blood pressure, blood lipids (LDL, HDL), etc.), this can be classified in one of three possible categories: low risk, intermediate risk or high risk. However, this method has limitations since it does not take into account the individual cardiovascular risk and to detect early atherosclerosis and other alterations of the arterial wall. It has been demonstrated that a significant number of people considered to be at intermediate risk with the traditional approach, in fact, have a high risk of presenting a cardiovascular event (for example, they have significant atheroma plaques in the coronary arteries). Moreover, quantitatively most deaths due to cardiovascular causes occur in subjects who present low or intermediate risk, evaluated by the traditional approach. This underestimation of individual risk, which shows the traditional risk quantification approach, determines that millions of people do not receive adequate medical treatment every day to reduce their cardiovascular risk. In other words, asymptomatic subjects, but vulnerable to having a cardiovascular or cerebrovascular accident in the short term, are not offered the benefits of available prophylactic therapies, because they have underestimated their real cardiovascular risk. For example, hypertension is considered an asymptotic disease and is easy to detect; however, it has serious and lethal complications if it is not treated in time.

To demonstrate how coronary artery calcium (CAC) can be incorporated into the risk of traditional was calculated in 618 male patients, the Framingham model and the probability that the CAC of each patient falls in every four categories of CAC (0, 1–100, 101–400 and >400) using linear and nonlinear regression models. Then they were adjusted based on a relative risk (RR) that weighted the risk of coronary heart disease in individuals and that are RR = 1.7 (for a CAC of 1–100), RR = 3.0 (for CAC 101–400), and RR = 4.3 (for CAC > 400) obtained from a meta-analysis published by Fletcher. The predictive power was evaluated using ROC curves (receiver operating characteristic). The model included in the CAC has a remarkable predictive value of atherosclerosis of 0.74, which is the area of the ROC curve as a function of the number of sites with extracoronal plates including carotid, femoral and abdominal aorta (coded as 0–1 sites = 0; 2–3 sites = 1). The predictive scale indicated 0.90–1 = excellent, 0.80–0.90 = good, 0.70–0.80 = median, 0.60–0.70 = weak, 0.50–0.60 = zero. The calcium score is a numerical information that allows quantifying the magnitude of coronary atherosclerotic lesions and provides independent predictive information of risk factors in general mortality. The combination of modeling of the CAC with the modeling of conventional risk factors leads to a remarkable improvement in the predictive value of the overall risk assessment of Framingham through the reclassification of the risk of atherosclerosis to a degree that may be clinically important. Adding to this approach, the other indices of subclinical atherosclerosis such as arterial rigidity, intima media thickness, endothelial function, and the presence of plaques will generate an integrative risk that will determine and classify the subjects in relation to short-term risk of suffering a cardiovascular or cerebrovascular accident. It will allow to know in a more precise way the cardiovascular risk of a particular individual. It allows early detection (subclinical stage) of vascular alterations and offer the best current prophylactic therapies available [35–37].

All standard risk assessment models to predict cardiovascular diseases make an implicit assumption that each risk factor is related in a linear fashion to CVD outcomes [38]. Such models may thus oversimplify complex relationships which include large numbers of risk factors with nonlinear interactions [39]. The aspect of computational intelligence comes into play specifically in this situation to decipher the inherent patterns and relationships among the parameters apart from the known and formally specified set, thereby determining more nuanced relationships between risk factors and outcomes. Current approaches to predict cardiovascular risk fail to identify many people who would benefit from preventive treatment, while others receive unnecessary intervention. Machine learning offers the opportunity to improve accuracy by exploiting complex interactions between the risk factors [39]. The established ACC/AHA 10-year risk prediction model used eight core baseline variables (gender, age, smoking status, systolic blood pressure, blood pressure treatment, total cholesterol, HDL cholesterol, and diabetes). However, in [39], additional 22 variables (like Body Mass Index, Triglycerides, C-reactive protein, Serum fibrinogen, etc.) were included in the machine learning algorithms, aimed at finding the influence on cardiovascular diseases, thereby designing the predictive algorithm for the same. Among other machine learning algorithms, neural networks had a 3.6% improved prediction than the existing algorithm. The system of cardiovascular risk prediction varies widely based on geographical factors. Therefore, several risk scores have been developed in different parts of the world like the SCORE by the European Society of Cardiology or the HellenicSCORE in Greece to address more accurately a specific group of population for calculating the cardiovascular risks. The majority of these scores use a common set of the 'classical' CVD risk factors, e.g., age, sex, smoking, blood pressure and lipids levels, whereas others have also incorporated more advanced markers of CVDs.

Most of these risk-prediction tools are based on stochastic models, incorporating variables, based on cohort studies [40]. However, the alternative approaches of machine learning like k-nearest neighbors, random forests and decision tress also generate results quite comparable to the classical risk prediction scores [41], thus demonstrating its possibility as alternative methods of CVD risk prediction along with its added advantages.

With the rise in the prevalence of hypertension globally and its associativity with other parameters of cardiovascular risks, computational techniques like feed-forward ANNs are used to model systolic blood pressure, diastolic blood pressure and pulse pressure variations with biological parameters like age, pulse rate, alcohol addiction, and physical activity level. In this aspect, ANN approaches provided more flexible and nonlinear models for prognosis and prediction of the blood pressure parameters than classical statistical algorithms [42]. Even with the increase of complex cardiovascular diseases, using machine learning models like random forest, ANNs, SVMs and Bayesian Networks to predict the in-hospital length of stay provides a positive impact on healthcare metrics [43]. Nonlinear models of unsupervised learning like clustering are commonly used in stratification of patient population and knowledge extraction from different groups. This is highly relevant in the prediction of risks because individuals with similar characteristics often present a similar risk profile [44].

The aspect of computational intelligence through nonlinear machine learning model even applies to other fields like survival prediction in transplantations and early detection of chronic diseases like cancer. Another interesting example of using computational intelligence and predictive analysis is the prediction of neurodegenerative diseases like Parkinson's disease. Disorders like Parkinson's disease and essential tremor which affect the normal movements of a person share some symptoms or manifestations that make the process of discrimination between them a difficult task. Clinical experience of the medical doctor is crucial at the moment of giving an accurate diagnosis. And still in such a case, that diagnosis is subjective and could be contaminated by several factors beyond the usual capacity of a medical personnel to analyze efficiently [45]. Especially with the use of wearable IoT-based sensors, data obtained about a patient's movement is extensive and complex. Nevertheless, it provides huge scope for using computational intelligence toward the prediction or early detection of such diseases. A major challenge in this aspect is the early detection of such disorders based on the patient's data obtained over a period of time, tracking its changes or finding patterns exhibiting similar trends of having the disease. In this case also, linear models do not count useful always since the parameters are quite dynamic and it needs the provision to continuously analyze other non-formalized parameters to find interesting traits leading to prediction. Thus, the aspect of computational intelligence is not only helpful in designing a better model for analysis but is also useful in prediction of diseases with higher accuracy.

4. Conclusion

Nonlinear systems constitute an important part of the area of predictive analytics aimed at diseases and risks for people. In the new age of data and eHealth, the inherent knowledge of data has turned out to be of immense importance, which needs specific methods with computational intelligence. Especially for chronic diseases, long-term behavioral data stands quite crucial. Data modeling and predictive analytics open a huge avenue toward clinical decision support systems, which is a fundamental tool now-a-days for preventive and personalized healthcare and supports healthcare providers to have deeper insights into patients' data [46] and take clinical decisions [33]. Therefore, the use of intelligent prediction is primarily based in two parts—modeling of the health data and analysis of the knowledge obtained. Computational intelligence and predictive analytics not only help in predicting the risks of diseases, but also supports largely in visualizing the holistic picture of health in a large population, aimed at designing more efficient and robust healthcare strategies across the world. Of course, it deals with growing challenges like the complexity of health data obtained, lack of interoperable systems for extended and unified analysis, intrinsic bias of some machine learning algorithms, and the implementational difficulties. Nonetheless, the application of machine learning and nonlinear methods using computation intelligence have already demonstrated its potential in predicting health risks and diseases, and is expected to reshape the field of health analytics, early detection and prediction of diseases in a global perspective.

Acknowledgements

The authors acknowledge the research facilities and technical assistance of the National Technological University (*Universidad Tecnológica Nacional*), Buenos Aires, Argentina.

The work is supported partly by the National Agency for Research and Innovation (*Agencia Nacional de Investigación e Innovación*) of Uruguay through its grant FSDA_1_2017_1_143653 (*Inteligencia Computacional en Salud. Creando Herramientas para Predecir la Sobrevida en Trasplante Hepático*).

Also it is supported partly by the National Interuniversity Council (CIN) of Argentina through the program '*Programa Estratégico de Formación de Recursos Humanos para la Investigación y Desarrollo (PERHID)*'.

The authors sincerely acknowledge the scientific view and insights of Dr. Josemaría Menéndez Dutra, Dr. Ofelia María Noceti Penza, and Dr. Solange Gerona Sangiovanni, from the Central Hospital of the Armed Forces (*Dirección Nacional de Sanidad de las Fuerzas Armadas*) in Montevideo, Uruguay.

Author details

Parag Chatterjee[1,2]*, Leandro J. Cymberknop[1] and Ricardo L. Armentano[1,2]

1 Universidad Tecnológica Nacional, Buenos Aires, Argentina

2 Universidad de la República, Uruguay

*Address all correspondence to: paragc@ieee.org

References

[1] Boeing G. Visual analysis of nonlinear dynamical systems: Chaos, fractals, self-similarity and the limits of prediction. Systems. 2016;**4**(4):37. DOI: 10.3390/systems4040037

[2] Explained: Linear and Nonlinear Systems. MIT News [Retrieved: 2018-06-30]

[3] Nonlinear Systems, Applied Mathematics. University of Birmingham. Available from: www.birmingham.ac.uk [Retrieved: 2018-06-30]

[4] Higgins JP. Nonlinear systems in medicine. The Yale Journal of Biology and Medicine. 2002;**75**(5-6):247-260

[5] Wyber R, Vaillancourt S, Perry W, Mannava P, Folaranmi T, Celi LA. Big data in global health: Improving health in low- and middle-income countries. Bulletin of the World Health Organization. 2015;**93**:203-208. DOI: 10.2471/BLT.14.139022

[6] Ahern DK, Kreslake JM, Phalen JM. What is ehealth (6): Perspectives on the evolution of ehealth research. Journal of Medical Internet Research. 2006;**8**(1):e4. DOI: 10.2196/jmir.8.1.e4

[7] Charles D, King J Patel V, Furukawa M. Adoption of Electronic Health Record Systems among U.S. Non-Federal Acute Care Hospitals: 2008-2012; 2013. Available from: http://www.healthit.gov/sites/default/files/oncdatabrief9final.pdf [Accessed: 20 April 2014]

[8] Shah NH. Translational bioinformatics embraces big data. Yearbook of Medical Informatics. 2012;**7**(1):130-134

[9] Heinze O, Birkle M, Koster L, Bergh B. Architecture of a consent management suite and integration into IHE-based regional health information networks. BMC Medical Informatics and Decision Making. 2011;**11**:58

[10] Tejero A, de la Torre I. Advances and current state of the security and privacy in electronic health records: Survey from a social perspective. Journal of Medical Systems. 2012;**36**(5):3019-3027

[11] Mense A, Hoheiser-Pfortner F, Schmid M, Wahl H. Concepts for a standard based cross-organisational information security management system in the context of a nationwide EHR. Studies in Health Technology and Informatics. 2013;**192**:548-552

[12] Faxvaag A, Johansen TS, Heimly V, Melby L, Grimsmo A. Healthcare professionals' experiences with EHR-system access control mechanisms. Studies in Health Technology and Informatics. 2011;**169**:601-605

[13] Ross MK, Wei W, Ohno-Machado L. "Big data" and the electronic health record. Yearbook of Medical Informatics. 2014;**9**(1):97-104. DOI: 10.15265/IY-2014-0003

[14] Wagholikar KB, Sundararajan V, Deshpande AW. Modeling paradigms for medical diagnostic decision support: A survey and future directions. Journal of Medical Systems. 2012;**36**(5):3029-3049

[15] WHO. Integrated Chronic Disease Prevention and Control [Internet]. 2019. Available from: https://www.who.int/chp/about/integrated_cd/en/

[16] VanWagner LB, Ning H, Whitsett M, Levitsky J, Uttal S, Wilkins JT, et al. A point-based prediction model for cardiovascular risk in orthotopic liver transplantation: The CAR-OLT score. Hepatology. 2017;**66**:1968-1979. DOI: 10.1002/hep.29329

[17] Hemmatpour M, Ferrero R, Gandino F, Montrucchio B, Rebaudengo M. Nonlinear predictive threshold model for real-time abnormal gait detection. Journal of Healthcare Engineering. 2018;**2018**. Article ID 4750104. 9 p. DOI: 10.1155/2018/4750104

[18] Stanford Medicine. Stanford Medicine 2017 Health Trends Report Harnessing the Power of Data in Health [Internet]. 2017. Available from: med.stanford.edu/content/dam/sm/sm-news/documents/StanfordMedicineHealthTrendsWhitePaper2017.pdf

[19] Corbin, Kenneth. How CIOs Can Prepare for Healthcare 'Data Tsunami'. CIO [Internet]. 2014. Available from: www.cio.com/article/2860072/how-cios-can-prepare-for-healthcare-data-tsunami.html

[20] Health Data Archiver. Health Data Volumes Skyrocket, Legacy Data Archives On The Rise [Internet]. 2017. Available from: https://www.healthdataarchiver.com/health-data-volumes-skyrocket-legacy-data-archives-rise-hie/

[21] Hippolyte T, Adamou M, Blaise N, Pierre C, Olivier M. Linear vs non-linear learning methods—A comparative study for forest above ground biomass, estimation from texture analysis of satellite images. ARIMA Journal. 2014;**18**:114-131

[22] Hastie T et al. The Elements of Statistical Learning. Vol. 2. Heidelberg: Springer; 2009

[23] Jiang F, Jiang Y, Zhi H, Dong Y, Li H, Ma S, et al. Artificial intelligence in healthcare: Past, present and future. Stroke and Vascular Neurology. 2017;**2**(4):230-243

[24] Shahid N, Rappon T, Berta W. Applications of artificial neural networks in health care organizational decision-making: A scoping review. PLoS One. 2019;**14**(2):e0212356. DOI: 10.1371/journal.pone.0212356

[25] Deepa et al. Health care analysis using random Forest algorithm. Journal of Chemical and Pharmaceutical Sciences. 2017;**10**(3):1359-1361. Available from: www.jchps.com/issues/Volume%2010_Issue%203/20171025_075052_0180417.pdf

[26] Chatterjee P, Cymberknop L, Armentano R. IoT-based ehealth toward decision support system for CBRNE events. In: Malizia A, D'Arienzo M, editors. Enhancing CBRNE Safety & Security: Proceedings of the SICC 2017 Conference. Cham: Springer; 2018. DOI: 10.1007/978-3-319-91791-7_21

[27] Chatterjee P, Cymberknop L, Armentano R. IoT-Based Decision Support System Towards Cardiovascular Diseases. Córdoba, Argentina: SABI; 2017

[28] Donges N. The Random Forest Algorithm. Towards Data Science. 2018. Available from: towardsdatascience.com/the-random-forest-algorithm-d457d499ffcd

[29] Healthcare.ai. Step by Step to K-Means Clustering. Data Science Blog. 2017. Available from: healthcare.ai/step-step-k-means-clustering/

[30] Mueller J, Massaron L. Machine Learning for Dummies. Hoboken: John Wiley & Sons, Inc.; 2016

[31] Raschka S. Model evaluation, model selection, and algorithm selection in machine learning. ArXiv. 2018. Available from: arxiv.org/abs/1811.12808v2

[32] Dinesh KG, Arumugaraj K, Santhosh KD, Mareeswari V. Prediction of cardiovascular disease using machine learning algorithms. In: International Conference on Current Trends towards

Converging Technologies (ICCTCT). 2018. DOI: 10.1109/icctct.2018.8550857

[33] Chatterjee P, Armentano RL, Cymberknop LJ. Internet of things and decision support system for eHealth-applied to cardiometabolic diseases. In: International Conference on Machine Learning and Data Science (MLDS); Noida. Piscataway: IEEE; 2017. pp. 75-79. DOI: 10.1109/MLDS.2017.22

[34] Framingham Heart Study. The Adult Treatment Panel III, JAMA. 2001 [Internet]. 2019. Available from: https://www.framinghamheartstudy.org/fhs-risk-functions/hard-coronary-heart-disease-10-year-risk/

[35] Chironi G, Simon A, Megnien JL, Sirieix ME, Mousseaux E, Pessana F, et al. Impact of coronary artery calcium on cardiovascular risk categorization and lipid-lowering drug eligibility in asymptomatic hypercholesterolemic men. International Journal of Cardiology. 2011;**151**(2):200-204. DOI: 10.1016/j.ijcard.2010.05.024

[36] Pessana F, Armentano R, Chironi G, Megnien JL, Mousseaux E, Simon A. Subclinical atherosclerosis modeling: Integration of coronary artery calcium score to Framingham equation. In: Annual International Conference of the IEEE Engineering in Medicine and Biology Society. 2009. DOI: 10.1109/iembs.2009.5334049

[37] Bucci CM, Legnani WE, Armentano RL. Clustering of cardiovascular risk factors highlighted the coronary artery calcium as a strong clinical discriminator. Health and Technology. 2016;**6**(3):159-165. DOI: 10.1007/s12553-016-0139-1

[38] Obermeyer Z, Emanuel EJ. Predicting the future—Big data, machine learning, and clinical medicine. The New England Journal of Medicine. 2016;**375**(13):1216-1219. DOI: 10.1056/NEJMp1606181

[39] Weng SF, Reps J, Kai J, Garibaldi JM, Qureshi N. Can machine-learning improve cardiovascular risk prediction using routine clinical data? PLoS One. 2017;**12**(4):e0174944. DOI: 10.1371/journal.pone.0174944

[40] Panagiotakos D. Health measurement scales: Methodological issues. Open Cardiovascular Medicine Journal. 2009;**3**:160

[41] Dimopoulos AC et al. Machine learning methodologies versus cardiovascular risk scores, in predicting disease risk. BMC Medical Research Methodology. 2018;**18**(1):1-11. DOI: 10.1186/s12874-018-0644-1

[42] Bhaduri A, Bhaduri A, Bhaduri A, Mohapatra PK. Blood pressure modeling using statistical and computational intelligence approaches. In: IEEE International Advance Computing Conference. 2009. DOI: 10.1109/iadcc.2009.4809156

[43] Daghistani TA, Elshawi R, Sakr S, Ahmed AM, Al-Thwayee A, Al-Mallah MH. Predictors of in-hospital length of stay among cardiac patients: A machine learning approach. International Journal of Cardiology. 2019;**288**:140-147. DOI: 10.1016/j.ijcard.2019.01.046

[44] Paredes S, Henriques J, Rochar T, Mendes D, Carvalho P, Moraisl J, et al. A clinical interpretable approach applied to cardiovascular risk assessment. In: 40th Annual International Conference of the IEEE Engineering in Medicine and Biology Society (EMBC). 2018. DOI: 10.1109/embc.2018.8512956

[45] Romero L, Chatterjee P, Armentano R. An IoT approach for integration of computational intelligence and wearable sensors for Parkinson's disease diagnosis and monitoring. Health and Technology. 2016;**6**(3):167-172. DOI: 10.1007/s12553-016-0148-0

[46] Chatterjee P, Cymberknop LJ, Armentano RL. IoT-based decision support system for intelligent healthcare—Applied to cardiovascular diseases. In: 7th International Conference on Communication Systems and Network Technologies (CSNT); Nagpur. 2017. pp. 362-366. DOI: 10.1109/CSNT.2017.8418567

Chapter 10

Mathematical Modeling and Well-Posedness of Three-Dimensional Shell in Disorders of Human Vascular System

Vishakha Jadaun and Nitin Raja Singh

Abstract

Aortic dissection is the most common aortic emergency requiring surgical intervention. Whether the elective endovascular repair of abdominal aortic aneurysm reduces long-term morbidity and mortality, as compared with traditional open repair, remains uncertain. The foundation of shell element based on the Reissner-Mindlin kinematics assumption is widely applicable, but this cannot model applications of shell surface stresses as needed in analysis of shell in human vascular system. The analysis is designed to assess progression of initial lesion in aortic dissection. Using general shell element analysis and tensor calculus, a higher order differential geometry-based model is proposed. Since the shell is thin, a variational formulation for initial lesion is proposed. The variational formulation for initial lesion is well posed. The weak convergence of the solution to initial lesion model is mathematically substantiated. Asymptotic analysis shows that initial lesion is membrane-dominated and bending-dominated when pure bending is inhibited and noninhibited, respectively. At least two observations are to be noted. First, the mathematical analysis of the initial lesion model is distinct from classical shell models. Second, the asymptotic analysis of the initial lesion model is based on degenerating three-dimensional continuum to bending strains in order to assess initial lesion behavior.

Keywords: aortic dissection, higher order kinematical assumptions, initial lesion model, variational formulation, asymptotic analysis

1. Introduction

The shell structure is generally a three-dimensional structure that is elongated in two directions and thinned out in other direction. The shell structures in nature are profusely impressive such as seashells and eggshells. In various industries including aeronautics, naval architecture, and automotive engineering milieu, many engineering designs are analyzed to design shells as thin as possible and optimize the amount of material [1]. Human anatomy develops cyst-related diseases with progressive severity. These disease states involve single to multiple cyst formations in distinct organ systems including the lung, liver, kidney, brain, bone etc. Pathophysiologically, these cysts emanate from either the underlying genetic anomaly or infections such as helminths and mycobacterial, among others. Interestingly, these cysts can be modeled as shells, albeit in higher dimensions.

Different approaches have been formulated for shell elements discretization. One of the approaches [2] evaluated the shell behavior as the superimposition of membrane bending action as well as plate bending action. The discrete construction of shell elements requires a combination of plane stress matrices as well as plate bending stiffness matrices. However, the resultant shell elements are less accurate since curvature effects are not duly incorporated and the membrane behavior and plate bending behavior are coupled at nodal points only. Another approach [3] is based on variational formulation and perusing relevant shell theory wherein a specific shell theory is constituted of higher-order derivatives and required concomitant nodal point variables beyond the conventional nodal point rotations and displacements. Such an approach is applicable and relevant to certain shell geometries and associated pertinent analysis conditions. Thus, it is difficult to model more complex shell structures. Yet another approach [4] is aimed for very general formulation related to three-dimensional continuum degeneration. In this approach, the mid-surface of the shell element that belongs to the three-dimensional continuum is clearly defined and identified. The first assumption is that the fibers are straight and normal to mid-surface prior to the deformation which continues to remain straight during the course of deformation. The second assumption is that the stress normal to the shell mid-surface is zero throughout the shell motion [5, 6]. The shell models based on the aforementioned kinematical assumptions can be interpreted as a truncation of the expansion of displacements in different directions across the thickness of the shell structure. It is to be noted that such truncated expansion contains terms up to degree one and degree zero for the tangential displacements and transverse displacements, respectively. The physiological and pathological states in the human body undergo dynamic transformations. In cardiovascular dynamics, the interaction of blood to the internal vessel lining is associated with large through-the-thickness displacement of local vessel wall surface owing to distension by propulsion of blood and elastic recoil thereafter. Thus, the aforementioned assumptions might not be applicable to shells in human anatomy. In order to make better estimate, higher-order kinematical assumptions are effective. Yet the detailed analysis of biological shell structures frequently presents challenging problems. One of the difficulties is that the shell structure resists applied loads largely along its curvature such that, in case, curvature is changed and the load bearing capacity of shell is transformed. Therefore, analysis of boundary conditions of a shell structure plays a vital role in shell behavior and its response to stress.

The aorta is the largest diameter blood vessel, emerging from the left ventricle to supply oxygenated blood to the human body. Whenever nonlinear degeneration of the tunica media (middle layer of the vessel wall) occurs, the aorta undergoes dynamic dilatation and marginal elongation. Generally, this degeneration is caused by genetic anomaly and prolonged untreated hypertension in young and senile patients, respectively. It is termed as *aortic aneurysm* [7]. Whenever there is structural discontinuity in nonconformal internal vessel wall, the blood surges through the tear causing the inner and middle layers of the aorta to separate. It is termed as *aortic dissection* (AD) [8]. AD is a life-threatening condition [9]. If the blood-filled channel ruptures through the outside aortic wall, AD is often fatal [10]. It is the most common aortic emergency requiring surgical intervention. AD is classified according to the regional involvement of the segment of the aorta with the Stanford type A dissection and the Stanford type B dissection involving the ascending aorta and occurring distal to the left subclavian artery, respectively. According to the international guidelines on clinical therapeutics, uncomplicated type B dissection should receive optimal medical treatment (OMT). However, in spite of adequate hypertension-related treatment, patients may develop a significant aortic enlargement that necessitates operative intervention. These chronic patients will benefit in the long-term from prophylactic intervention.

Currently, there is no consensus on the management of uncomplicated type B dissection that may be liable for rapid progression. Thus, seeking multiple high-risk attributes/features responsible for rapid progression might help to decide when to treat and how to treat. There is a subgroup of patients who progress very rapidly to terminal dilatation liable for rupture and torrential bleed leading to death. Offering early transthoracic endovascular repair to this subgroup seems to be a life-saving proposition. Finding these patients is a challenge. It is not known that a patient at risk for catastrophic events is following a personal trajectory of disease progression. It is also not known that a threshold for disease progression that can predict a high risk of mortality for a specific patient. By modeling the initial lesion of AD, we can potentially avoid rupture by crossing over to transthoracic endovascular repair at a time that minimizes procedural risks. On asymptotic analysis, we evaluate the point of follow-up; we lose the ability to achieve the same desirable aortic remodeling observed with transthoracic endovascular repair in the more acute setting. Therefore, reliable predictors are needed in the early stage of disease. It aids identification of patients at risk of aortic enlargement.

In early stages of AD, subintimal intramural hemorrhage occurs due to tunica media degeneration. In certain situations, when strains are known on a plane, the low degree of expansion of the transverse displacement is to be recovered. It is to be noted that by dispensing away the assumption of plane stress, an arbitrary three-dimensional material law is applicable in three-dimensional formulation of continuum mechanics. The objective of this chapter is to identify higher-order shell model for initial-stage primary tunica intimal lesion of AD by the general shell element approach and to perform mathematical analysis.

This chapter is organized in the following manner. In Section 2, we give certain definitions, conventions, and notations relevant to the shell geometry and its corresponding deformation. Next in Section 3, we derive initial lesions of AD as the higher-order shell model perusing general shell analysis approach. Then in Section 4, we do mathematical analyses of the initial lesions described in the previous section. In Section 5, we assess asymptotic behavior of the model. Finally, in Section 6, we present our conclusions regarding mathematical modeling of shells in human vascular tissues and future scope.

2. Conventions and notations in higher-order shell geometry

We are interested in modeling early stages of AD; the initial subintimal intramural hemorrhage caused by tunica media degeneration undergoes solidification due to clot formation. Thus, this initial lesion closely follows the principles of continuum mechanics. We consider the initial lesion as a solid medium. It is geometrically defined by a mid-surface immersed in the human vascular compartment ϵ (dimensionless thickness parameter) and a parameter representing the thickness of the medium around this surface.

In order to understand the initial lesion of AD, we model the initial lesion using general shell element theory. A shell is defined as a collection of charts. Let us consider the mid-surface of a shell as a collection of two-dimensional charts. These charts are smooth ono-one maps from domains of $\mathbb{R}^2$ into Euclidean (physical) space $\mathscr{E}$. We consider an initial lesion with a mid-surface $\mathcal{S}$ defined by a two-dimensional chart $\vec{\varphi}$, which is a one-one map from the closure of a bounded open subset of $\mathbb{R}^2$, denoted by ω, into $\mathscr{E}$, hence $\mathcal{S} = \vec{\varphi}\left(\vec{\omega}\right)$. At each point of the mid-surface, the vector $\vec{\mathbf{z}}_\alpha$ is assumed as partial derivative of $\vec{\varphi}$ with respect to ξ^α such that

$$\vec{\mathbf{z}}_\alpha = \frac{\partial\varphi(\xi_1, \xi_2)}{\partial\xi^\alpha}. \tag{1}$$

These vectors are linearly independent from each other, so that they form a basis of the plane tangent to the mid-surface at this point. The unit normal vector is given by

$$\vec{\mathbf{z}_3} = \frac{\vec{\mathbf{z}_1} \times \vec{\mathbf{z}_2}}{\|\vec{\mathbf{z}_1} \times \vec{\mathbf{z}_2}\|}.$$

Definition 1. (Geometric definition of initial lesion). *An initial lesion is a solid medium whose domain* Ω *can be defined by a mid-surface whose map is given by*

$$\varphi : \omega \subseteq \mathbb{R}^2 \to \mathbb{R}^3, s.t.\ \varphi\left(\xi^1, \xi^2\right) = \left(\xi^1, \xi^2, \xi^3\right) \in \mathbb{R}^3 \tag{2}$$

The three-dimensional medium corresponding to the initial lesion is then defined by three-dimensional chart given by

$$\varphi\left(\xi^1, \xi^2, \xi^3\right) = \varphi\left(\xi^1, \xi^2\right) + \xi^3\vec{z}_3\left(\xi^1, \xi^2\right), \tag{3}$$

where $\left(\xi^1, \xi^2, \xi^3\right) \in \Omega = \left\{\left(\xi^1, \xi^2, \xi^3\right) \in \mathbb{R}^3 | \left(\xi^1, \xi^2\right) \in \omega, \xi^3 \in \left(-\frac{t\left(\xi^1, \xi^2\right)}{2}, \frac{t\left(\xi^1, \xi^2\right)}{2}\right)\right\}$ *and* $t\left(\xi^1, \xi^2\right)$ *is the thickness of the initial lesion element at* $\left(\xi^1, \xi^2\right)$.

In Eq. (1), we have defined tangent vector to a point on the mid-surface of the initial lesion (2) which lies in the region of the Euclidean space. Since we are interested in higher-order parameterization of the initial lesion of AD, the three-dimensional chart (3) of this lesion can be very helpful. Thus, transition from the Euclidean space to curvilinear coordinate system will aid to model higher-order initial lesion. It is relevant to grasp few basic notions of surface differential geometry.

2.1 Definitions related to surface differential geometry

Definition 2. (Covariant vector). *Let* $r(z)$ *be a position vector; the differentiation of* $r(z)$ *with respect to each of the coordinate is called covariant basis:*

$$\mathbf{z}_i = \frac{\partial r(z)}{\partial z^i} \tag{4}$$

If Eq. (4) defines three vectors $\mathbf{z}_1$, $\mathbf{z}_2$, and $\mathbf{z}_3$

$$\mathbf{z}_1 = \frac{\partial\, r(z^1, z^2, z^3)}{\partial z^1}, \quad \mathbf{z}_2 = \frac{\partial\, r(z^1, z^2, z^3)}{\partial z^2}, \quad \mathbf{z}_3 = \frac{\partial\, r(z^1, z^2, z^3)}{\partial z^3}. \tag{5}$$

Let $\mathbf{V}$ be a vector in $\mathbb{R}^3$, and then its expansion n terms of basis is

$$\mathbf{V} = V^i\mathbf{z}_i = V^1\mathbf{z}_1 + V^2\mathbf{z}_2 + V^3\mathbf{z}_3 \tag{6}$$

The values V^i are called *contravariant components* of vector $\mathbf{V}$.

Interestingly, covariant basis is useful in the modeling of higher-order initial lesions in human vascular system given by

$$\begin{cases} \vec{g}_i = \dfrac{\partial \varphi}{\partial \xi^i} = z_i + \xi^3\, \mathbf{z}_{\overrightarrow{3,i}} = \mathbf{z}_{\vec{i}} - \xi^3\, b_i^k \cdot \vec{\mathbf{z}_k}, \quad \text{where} \quad \mathbf{z}_{\overrightarrow{3,i}} = \dfrac{\partial \mathbf{z}_{\vec{3}}}{\partial \xi^i}, \\ \vec{g}_i = \left(\delta_i^k - \xi^3\, b_i^k\right)\vec{\mathbf{z}_k}, \\ \vec{g}_3 = \dfrac{\partial \varphi}{\partial \xi^3} = \vec{\mathbf{z}_3}. \end{cases} \tag{7}$$

Definition 3. (Covariant metric tensor). *The covariant metric tensor is the pairwise dot product of the covariant basis vectors:*

$$z_{ij} = \mathbf{z}_i . \mathbf{z}_j = \begin{bmatrix} \mathbf{z}_1.\mathbf{z}_1 & \mathbf{z}_1.\mathbf{z}_2 & \mathbf{z}_1.\mathbf{z}_3 \\ \mathbf{z}_2.\mathbf{z}_1 & \mathbf{z}_2.\mathbf{z}_2 & \mathbf{z}_2.\mathbf{z}_3 \\ \mathbf{z}_3.\mathbf{z}_1 & \mathbf{z}_3.\mathbf{z}_2 & \mathbf{z}_3.\mathbf{z}_3 \end{bmatrix}, \tag{8}$$

where $\mathbf{z}_i$ is in $\mathbb{R}^3$.

Suppose two vectors **A** and **B** are located at the same point and their components are A^i and B^i, then the dot product **A**.**B** is given by

$$\mathbf{A}.\mathbf{B} = A^i \mathbf{z}_i . B^j \mathbf{z}_j = \left(\mathbf{z}_i \cdot \mathbf{z}_j\right) A^i \cdot B^j = z_{ij} A^i B^j. \tag{9}$$

The length of a vector **B** can be expressed in terms of covariant metric tensor as

$$|\mathbf{B}| = \sqrt{z_{ij} B^i B^j} \tag{10}$$

Interestingly, covariant tensors are useful in modeling of higher-order initial lesions in human vascular system given by

1. $g_{ij} = \vec{g}_i \cdot \vec{g}_j = \mathbf{z}_{ij} - 2\xi^3 b_{ij} + \left(\xi^3\right)^2 c_{ij}$

2. $g_{i3} = \vec{g}_i \cdot \vec{g}_3 = 0$

3. $g_{33} = \vec{g}_3 \cdot \vec{g}_3 = 1$

Definition 4. (Contravariant metric tensor z^{ij}). *The contravariant metric tensor z^{ij} is the matrix inverse of the covariant metric tensor z_{ij}:*

$$z_{ij} \cdot z^{jk} = z_{ij} \cdot z^{kj} = \delta_k^i, \tag{11}$$

where δ_k^i is the Kronecker symbol.

Definition 5. (Contravariant basis $\mathbf{z}^i$). *The contravariant basis $\mathbf{z}^i$ is defined as*

$$\mathbf{z}^i = z^{ij} \mathbf{z}_j = z^{ji} \mathbf{z}_j \tag{12}$$

The bases $\mathbf{z}_i$ and $\mathbf{z}^i$ are mutually orthonormal:

$$\mathbf{z}_i \cdot \mathbf{z}^j = \delta_j^i. \tag{13}$$

Definition 6. (Christoffel symbol). *In affine and curvilinear coordinate systems, the covariant basis $\mathbf{z}_i$ is the same at all points and varies from one point to another,* *respectively. This variation can be described by the partial derivatives* $\partial z_i / \partial z^j$*. Using decomposition of partial derivatives* $\partial z_i / \partial z^j$ *with respect to the covariant basis* z_k*, the Christoffel symbol* Γ_{ij}^{k} *is given by*

$$\frac{\partial z_i}{\partial z^j} = \Gamma_{ij}^{k} z_k. \tag{14}$$

Note that the Christoffel symbol is symmetric in lower indices:

$$\Gamma_{ij}^{k} = \Gamma_{ji}^{k} = \mathbf{z}_k \cdot \frac{\partial \mathbf{z}_i}{\partial z^j}. \tag{15}$$

2.2 Fundamental forms

The *first fundamental form* of the surface is also known as the restriction of the metric tensor to the tangent plane. It is given by its components

$$\mathbf{z}_{ij} = \overrightarrow{\mathbf{z}_i} \cdot \overrightarrow{\mathbf{z}_j}.$$

Alternatively, its contravariant form is given by

$$\mathbf{z}^{ij} = \overrightarrow{\mathbf{z}^{i}} \cdot \overrightarrow{\mathbf{z}^{j}}.$$

Note that the first fundamental form can be used for the conversion of covariant components into contravariant components, such as

$$\mathbf{v}^i = \mathbf{z}^{ik} \mathbf{v}_k$$

The Euclidean norm of the two-dimensional tensors is denoted by $\|\cdot\|_\varepsilon$ and the corresponding inner product by $< \cdot\,, \cdot >_\varepsilon$. Note that the first fundamental form can be used for the evaluation of such norm quantities:

$$<\underline{u}, \underline{v}>_\varepsilon = u_i \mathbf{z}^{ij} v_j, \tag{16}$$

$$\|\underline{v}\|_\varepsilon^2 = v_i \mathbf{z}^{ij} v_j, \tag{17}$$

$$<\underline{\underline{T}}, \underline{\underline{U}}> = T_{ij} \mathbf{z}^{ik} \mathbf{z}^{jl} U_{kl}, \tag{18}$$

$$\left\|\underline{\underline{T}}\right\|_\varepsilon^2 = T_{ij} \mathbf{z}^{ik} \mathbf{z}^{jl} T_{kl}. \tag{19}$$

The second fundamental form

$$b_{ij} = \mathbf{z}_{\vec{3}} \cdot \mathbf{z}_{\overrightarrow{i,j}},$$

where

$$\mathbf{z}_{\overrightarrow{i,j}} = \frac{\partial^2 \varphi}{\partial \xi^i \partial \xi^j} = \mathbf{z}_{\overrightarrow{j,i}}$$

is the fundamental form of symmetry.

The second fundamental form is yet another important second-order tensor of the surface. It is also known as the *curvature tensor* since it provides information about the curvature of the surface. The values of these curvatures along the directions are called the *principal curvatures*. The product and the half-sum of the principal curvatures are classically known as the *Gaussian* curvature and *mean* curvature, respectively.

The third fundamental form

$$c_{ij} = b_i^k \, b_{kj}$$

It is a derivative along a curve lying on the surface. Note that the expressions of surface Christoffel symbols and surface covariant derivative are inferred from the third fundamental form.

Remark 1. $z_{\vec{3}} \cdot z_{\vec{3}} = 1 \Rightarrow z_{\overrightarrow{3,i}} \cdot z_{\vec{3}} = 0$ *that is* $z_{\overrightarrow{3,i}}$ *which lies in the tangent plane. Hence, we have*

$$z_{\overrightarrow{3,i}} = \left(z_{\overrightarrow{3,i}} \cdot z_k \right) \overrightarrow{z^k}, \tag{20}$$

and thus

$$z_{\overrightarrow{3,i}} = -b_{ik} \overrightarrow{z^k} = -b_i^k z_{\vec{k}}. \tag{21}$$

The initial tunica intimal lesion in AD is heterogenous in terms of various attributes such as shape, size, and conjugality among others. These notions of surface differential geometry are helpful to model these lesions as higher-order initial lesions. To illustrate, the surface of lesion modeled as initial lesion can be *elliptic*, *parabolic*, or *hyperbolic* according to whether its Gaussian curvature is positive, zero, or negative, respectively. Note that Gaussian curvature is derived from the second fundamental form. From now onwards, we simply use initial lesion model to describe initial lesion of aortic dissection.

3. Modeling of initial lesion

Normally, the aorta is composed of three layers, tunica adventitia, tunica media, and tunica intima (from outside to inside in cross section). Tunica adventitia is composed of linear palisades of collagen fibers as an envelope over tunica media that is a smooth muscle layer, capable of elastic recoil for propelling blood forward. Tunica intima is quite a thin innermost layer comprised of linear array of collagen fibers.

3.1 A simplistic view of initial lesions

To simplify, it is assumed that collagen fibers are straight and resist deformation caused by hemodynamic stresses. In addition, hemodynamic stress, normal to mid-surface of tunica media, is zero throughout the cardiac cycle. The modeling of initial lesion of AD based on the aforementioned kinematical assumptions can be interpreted as a truncation of the expansion of displacements across the thickness of the normal human aorta. The kinematical assumptions pertain to the displacements of points located on tunica intima layer of the aorta through the lesion thickness. Such points are orthogonal to mid-surface in the earlier pre-deformed configuration. Note that the kinematical assumptions connect the displacements of points located on the tunica intima layer that is orthogonal to the mid-surface of the tunica media layer in undeformed configuration. The displacement is expressed by the following equation:

$$\vec{\mathfrak{D}}(\xi^1,\xi^2,\xi^3) = \vec{d}(\xi^1,\xi^2) + \xi^3\vec{\theta}_k(\xi^1,\xi^2)\vec{\mathbf{z}}^k(\xi^1,\xi^2), \tag{22}$$

In Eq. (22), we consider the tunica intima layer in the direction of $\vec{\mathbf{z}}_3$ at the coordinate (ξ^1,ξ^2). The displacement $\vec{d}(\xi^1,\xi^2)$ represents a global infinitesimal displacement of the linearly arranged endothelial cells of the tunica intima on the line displacing by the similar amount. The displacement $\xi^3\vec{\theta}_k(\xi^1,\xi^2)\vec{\mathbf{z}}^k(\xi^1,\xi^2)$ is due to the rotation of the line measured by θ_1 and θ_2.

Hemodynamic flows can cause both linear and rotational strain. The linear strain is caused by laminar flow, while the rotational strain is caused by either turbulent flow and/or concomitant nonlinear geometry of the vessel. Thus, the measure of linear strain is not sufficient, rather inaccuracies emanate from the increments in rotation. We choose the principle of deformation gradient to calculate both the strains. The combined linear and nonlinear strains can be characterized by stretch tensor called Green-Lagrange strain tensor. The 3D-Lagrange-Green tensor, for which the components $e_{\alpha\beta}$ for general displacement $\vec{\mathfrak{D}}(\xi^1,\xi^2,\xi^3)$ are

$$e_{\alpha\beta} = \frac{1}{2}\left(\vec{g}_\alpha \cdot \vec{\mathfrak{D}}_{,\beta} + \vec{g}_\beta \cdot \vec{\mathfrak{D}}_{,\alpha}\right), \qquad \alpha,\beta = 1,2,3. \tag{23}$$

To calculate the components of Green-Lagrange strain tensor, we need to evaluate $\mathfrak{D}_{,\alpha} = \partial\mathfrak{D}/\partial\xi^\alpha$ (displacement of endothelial cells in a line on the tunica intima in ξ^α direction). For the specific displacement in (22), we compute the covariant components of the linearized strain tensor. We have

$$\frac{\partial d}{\partial \xi^i} = \frac{\partial}{\partial \xi^i}\left(d_k\mathbf{z}^k + d_3\mathbf{z}_3\right) \tag{24}$$

We peruse the fundamental forms to obtain

$$\frac{\partial}{\partial \xi^i}\left(d_k\mathbf{z}^k\right) = \mathbf{z}^k\frac{\partial d_k}{\partial \xi^i} + d_k\frac{\partial \mathbf{z}^k}{\partial \xi^i} = \mathbf{z}^k\frac{\partial d_k}{\partial \xi^i} + b_i^k d_k a_3. \tag{25}$$

Hence,

$$\begin{aligned}\frac{\partial d}{\partial \xi^i} &= d_{k|i}\mathbf{z}^k + b_i^k d_k\mathbf{z}_3 + d_{3,i}\mathbf{z}_3 + d_3\mathbf{z}_{3,i}\\ &= \left(d_{k|i} - b_{ki}d_3\right)\mathbf{z}^k + \left(d_{3,i} + b_i^k d_k\right)\mathbf{z}_3,\end{aligned} \tag{26}$$

where $d_{k|i} = \partial d_k/\partial\xi^i$. As we have calculated the derivative for linearized strain, we calculate the derivative for rotational strain. From (21)

$$\frac{\partial}{\partial \xi^i}\left(\theta_k\mathbf{z}_k\right) = \theta_{k|i}\mathbf{z}_k + b_i^k\theta_k\mathbf{z}_3. \tag{27}$$

The overall displacement in Eq. (22) is composed of linear displacement and rotational displacement. Therefore,

$$\frac{\partial\mathfrak{D}}{\partial \xi^i} = \frac{\partial d}{\partial \xi^i} + \frac{\partial}{\partial \xi^i}\left(\xi^3\theta_k\mathbf{z}_k\right) = \left(d_{k|i} - b_{ki}\mathbf{z}_3 + \xi^3\theta_{k|i}\right)\mathbf{z}_k + \left(d_{3,i} + b_i^k d_k + \xi^3 b_i^k\theta_k\right)\mathbf{z}_3. \tag{28}$$

Moreover,

$$\frac{\partial \mathfrak{D}}{\partial \xi^3} = \theta_k \mathbf{z}^k \tag{29}$$

Substituting Eqs. (28), (29), and (7) into (23)

$$\begin{cases} e_{ij} = \gamma_{ij}\left(\vec{d}\right) + \xi^3 \chi_{ij}\left(\vec{d}, \underline{\theta}\right) - \left(\xi^3\right)^2 \kappa_{ij}(\underline{\theta}), & i,j = 1,2 \\ e_{i3} = \zeta_i\left(\vec{d}, \underline{\theta}\right), & i = 1,2 \\ e_{33} = 0, \end{cases} \tag{30}$$

where

$$\begin{cases} \gamma_{ij}\left(\vec{d}\right) &= \dfrac{1}{2}\left(d_{i|j} + d_{j|i}\right) - b_{ij}\mathbf{z}_3 \\ \chi_{ij}\left(\vec{d}, \underline{\theta}\right) &= \dfrac{1}{2}\left(\theta_{i|j} + \theta_{j|i} - b_j^k d_{k|i} - b_i^k d_{k|j}\right) + c_{ij}\mathbf{z}_3 \\ \kappa_{ij}(\underline{\theta}) &= \dfrac{1}{2}\left(b_j^k \theta_{k|i} + b_i^k \theta_{k|j}\right) \\ \zeta_i\left(\vec{d}, \underline{\theta}\right) &= \dfrac{1}{2}\left(\theta_i + d_{3,i} + b_i^k d_k\right) \end{cases} \tag{31}$$

In the framework of the kinematical assumptions, the second-order tensors, $\underline{\underline{\gamma}}$ and $\underline{\underline{\chi}}$, and the first-order tensor $\underline{\zeta}$ are called the membrane strain, bending strain, and shear strain, respectively.

3.2 Higher-order model for initial lesion

In pathological conditions and even in physiological conditions strained to its limits, fluid-structure interaction in the aorta does not follow the kinematical assumptions because the arrangement of collagen fibers in the aortic wall is not straight. Tunica intima is comprised of a single layer of endothelial cells with a subendothelial layer of varying thickness. Tunica intimal surface is nonconformal depending upon the amount of subendothelial ground matrix, contrary to the conventional perspective of the conformal tunica intimal surface. The tunica media is a complex three-dimensional network of smooth muscle cells, elastin, and bundles of collagen fibrils. These well-defined concentrically oriented fibers are mutually reinforcing in radial direction. Tunica adventitia is comprised of fibroblasts, fibrocytes, collagen fibers (helically arranged), and ground matrix.

The constituents of the aortic wall including collagen fibers, elastin fibers, smooth muscle fibers, and ground matrix can stretch to deformation and recoil. Histologically and functionally, these constituents are viscoelastic; hence, aortic tissues resist deformation, albeit partially. Note that hemodynamic strain normal to mid-surface of tunica intima will not be zero.

The assumptions in the simplistic case (22) does not hold true in clinical settings. Thus, an initial lesion model is required to incorporate these attributes. An initial lesion model that is asymptotically consistent with three-dimensional solid mechanics without resorting to any independent kinematical assumptions on the strains requires correction for rotation inaccuracies, while only linearized strain tensor is perused for displacement equation (22). For initial lesion model, the displacement vector $\vec{\mathfrak{D}}(\xi^1, \xi^2, \xi^3)$ contains at least all terms up to degree two, namely,

$$\vec{\mathfrak{D}}(\xi^1, \xi^2, \xi^3) = \vec{d}(\xi^1, \xi^2) + \xi^3 \vec{\theta}(\xi^1, \xi^2) + (\xi^3)^2 \vec{\varrho}(\xi^1, \xi^2). \tag{32}$$

In the simplistic view, the strain normal to the tunica intima is zero since the vessel wall does not deform. In higher-order model, the vector $\vec{\theta}$ is arbitrary in the Euclidean space and not constrained to lie in the tangential plane. The modified expression for strain components is as follows:

$$\begin{cases} e_{ij}\left(\vec{\mathfrak{D}}\right) = \gamma_{ij}\left(\vec{d}\right) + \xi^3 \chi_{ij}\left(\vec{d}, \vec{\theta}\right) + \left(\xi^3\right)^2 \kappa_{ij}\left(\vec{\theta}, \vec{\varrho}\right) + \left(\xi^3\right)^3 l_{ij}\left(\vec{\varrho}\right) \\ e_{i3}\left(\vec{\mathfrak{D}}\right) = \zeta_i\left(\vec{d}, \vec{\theta}\right) + \xi^3 m_i\left(\vec{\theta}, \vec{\varrho}\right) + \left(\xi^3\right)^2 n_i\left(\vec{\varrho}\right) \\ e_{33}\left(\vec{\mathfrak{D}}\right) = \varpi\left(\vec{\theta}\right) + \xi^3 p\left(\vec{\varrho}\right) \end{cases} \tag{33}$$

where

$$\begin{cases} \gamma_{ij}\left(\vec{d}\right) & = \frac{1}{2}\left(d_{i|j} + d_{j|i}\right) - b_{ij} d_3 \\ \chi_{ij}\left(\vec{d}, \vec{\theta}\right) & = \frac{1}{2}\left(\theta_{i|j} + \theta_{j|i} - b_j^k d_{k|i} - b_i^k d_{k|j}\right) - b_{ij}\theta_3 + c_{ij} d_3 \\ \kappa_{ij}\left(\vec{\theta}, \vec{\varrho}\right) & = \frac{1}{2}\left(\varrho_{i|j} + \varrho_{j|i} - b_j^k \theta_{k|i} - b_i^k \theta_{k|j}\right) - b_{ij}\varrho_3 + c_{ij}\theta_3 \\ l_{ij}\left(\vec{\varrho}\right) & = -\frac{1}{2}\left(b_j^k \varrho_{k|i} + b_i^k \varrho_{k|j}\right) + c_{ij}\varrho_3 \\ \zeta_i\left(\vec{d}, \vec{\theta}\right) & = \frac{1}{2}\left(\theta_i + d_{3,i} + b_i^k d_k\right) \\ m_i\left(\vec{\theta}, \vec{\varrho}\right) & = \frac{1}{2}\left(2\varrho_i + \theta_{3,i}\right) \\ n_i\left(\vec{\varrho}\right) & = \frac{1}{2}\left(-b_i^k \varrho_k + \varrho_{3,i}\right) \\ \varpi\left(\vec{\theta}\right) & = \theta_3 \\ P\left(\vec{\varrho}\right) & = 2\varrho_3 \end{cases} \tag{34}$$

Here, the tensors, $\underline{\underline{\gamma}}$ and $\underline{\zeta}$, are called the membrane and shear strain tensors as defined in Eq. (31). The tensor $\underline{\underline{\chi}}$ is a generalization of the bending strain tensor, and $\underline{\underline{k}}$ is a generalization of $-\underline{\underline{\kappa}}$ in Eq. (31), since θ_3 appears in the expressions of $\underline{\underline{\chi}}$ and $\underline{\underline{k}}$ in Eq. (34). Because of different orders in ξ^3 in higher-order displacement vector, the newer tensors including $\underline{\underline{l}}$, $\underline{m}$, $\underline{n}$, $\underline{\varpi}$, and $\underline{p}$ are obtained. In initial lesion model, the different orders in ξ^3 introduces complex interplay of various tensors. The continuous interplay among tensors of different orders makes it difficult to calculate resultant displacement, comprised of linear and rotational displacements. It becomes necessary to peruse algebra for weak formulation of this complex interplay of tensors. The variational formulation using a test function on displacement which aids to evaluate displacement equation in higher-order is

$$\int_\Omega H^{\alpha\beta\lambda\mu} e_{\alpha\beta}\left(\vec{\mathfrak{D}}\right) e_{\lambda\mu}\left(\vec{\Delta}\right) dV = \int_\Omega \vec{F}.\vec{\Delta} dV, \tag{35}$$

where the function, $\vec{\Delta}(\xi^1, \xi^2, \xi^3)$, is called test function; for each $\vec{\Delta} \in \mathcal{V}$ (domain for initial lesion) there exists unique $\vec{\mathfrak{D}} \in \mathcal{V}$ such that Eq. (35) holds:

$$\vec{\Delta}(\xi^1, \xi^2, \xi^3) = \vec{\delta}(\xi^1, \xi^2) + \xi^3 \vec{\eta}(\xi^1, \xi^2) + (\xi^3)^2 \vec{\varsigma}(\xi^1, \xi^2) \tag{36}$$

It obviously comes to mind: what are kinematical assumptions in initial lesion model? Keeping the histological and functional perspective of the vessel wall from the biomechanical point of view, it is known that internal surface of the vessel wall is not smooth. It becomes obvious that it will not follow banal kinematical assumptions as mentioned earlier. Note that the initial lesion of AD might be evolving on the tunica intima due to medial degeneration. The lesion presence is spatially nonlinear. It seems plausible that it is governed by quadratic equation as higher-order tensor has quadratic components. The equation for kinematical assumption in higher-order displacement equation, setting $\tau = 2\xi^3/t$, is given by

$$\vec{\mathfrak{D}} = \frac{\tau(\tau-1)}{2} \vec{d}^{bot} + \left(1 - (\tau)^2\right) \vec{d}^{mid} + \frac{\tau(\tau+1)}{2} \vec{d}^{top} \tag{37}$$

Since the internal lining is not smooth, a gestalt view of affected tunica intima has initial lesions at differing heights. Lesions are on the tunica intima surface. To localize spatial dimension of lesions across tunica intima surface, correction terms are to be introduced to tunica intima surface levels, viz. top, mid, and bottom in form of $\tau(\tau-1)/2$, $1-(\tau)^2$, and $\tau(\tau+1)/2$, respectively.

4. Mathematical analysis of initial lesion

We did weak formulation to estimate strain tensors. Now, we assess net displacement. In order to do so, well-posedness of variational form (35) is the key. To understand the evolution of AD, the initial lesion from its inception to the advanced stage wherein the lesion contributes to nonlinear radial dilatation and marginal elongation of diseased aortic tissue needs to be evaluated. There are bounds to emergence of lesion, the lower bound is the status of primal lesion first noticed, and the upper bound is advanced stage of lesion that contributes to rupture of the aorta. Within these bounds, the blood flow acts on the lining of the aorta, adversely impacting the primal lesion that is susceptible to progression from lower bound to upper bound and contributing to the severity of disease.

The inherent nature of normal aortic tissue is to retain its earlier state despite varying interplay of tensors in higher-order. But this resistive tendency, called *coercivity*, weakens as primal lesion progresses towards upper bound. This transition from lower bound to upper bound depends on the complex interplay of various tensors in higher-order. Interestingly, the interplay between tensors in higher-order and coercivity is responsible for worsening of the disease. Intuitively, coercivity is inversely proportional to the progression towards upper bound. Thus, gaining information about the progression towards upper bound and concomitant decline in coercivity is vital to understand net displacement of initial lesion and progression of disease.

For a particular bound and coercivity for a test function $\vec{\Delta}\left(\vec{\delta}, \vec{\eta}, \vec{\varsigma}\right)$, there exists a unique $\vec{\mathfrak{D}}\left(\vec{d}, \vec{\theta}, \vec{\varrho}\right)$, exemplifying a particular state of disease. Furthermore, there are various states of displacement of initial lesion due to progression of the disease.

Such a compendium is used to characterize a particular displacement state. A higher-dimensional space, *Sobolev space* which is comprised of all such possible combinations, comes handy. It is constituted of functions with sufficiently many derivative including partial differential equations of fluid-structure interaction and equipped with the norm that measures size and regularity of these functions. The test function $\vec{\Delta}\left(\vec{\delta},\vec{\eta},\vec{\varsigma}\right)$ is a replica of $\vec{\mathfrak{D}}\left(\vec{d},\vec{\theta},\vec{\varrho}\right)$ in higher-dimensional metric space, $\mathcal{V}$. Since test function is an idealized version of net displacement vector in continuum mechanics, evaluating the interaction between test function $\vec{\Delta}\left(\vec{\delta},\vec{\eta},\vec{\varsigma}\right)$ and force $\vec{F}$ yields insights about $\vec{\mathfrak{D}}\left(\vec{d},\vec{\theta},\vec{\varrho}\right)$. $\vec{\Delta}\left(\vec{\delta},\vec{\eta},\vec{\varsigma}\right)$ is a test function present in Sobolev space. Gaining information about the characteristics of $\vec{\Delta}\left(\vec{\delta},\vec{\eta},\vec{\varsigma}\right)$, we can infer about $\vec{\mathfrak{D}}\left(\vec{d},\vec{\theta},\vec{\varrho}\right)$. It yields insight about the disease process wherein with progression the initial lesion is contributed by the blood flow. Interestingly, proving well-posedness of Eq. (35) gives insights about $\vec{\mathfrak{D}}\left(\vec{d},\vec{\theta},\vec{\varrho}\right)$ present in bilinear function $A\left(\vec{d},\vec{\theta},\vec{\varrho};\vec{\delta},\vec{\eta},\vec{\varsigma}\right)$, which is given by

$$A\left(\vec{d},\vec{\theta},\vec{\varrho};\vec{\delta},\vec{\eta},\vec{\varsigma}\right) = \int_{\Omega} H^{\alpha\beta\lambda\mu} e_{\alpha\beta}\left(\vec{d}+\xi^3\vec{\theta}+\left(\xi^3\right)^2\vec{\varrho}\right) e_{\lambda\mu}\left(\vec{\delta}+\xi^3\vec{\eta}+\left(\xi^3\right)^2\vec{\varsigma}\right) dV. \quad (38)$$

The linear function is given by

$$F\left(\vec{\delta},\vec{\eta},\vec{\varsigma}\right) = \int_{\Omega} \vec{F}.\left(\vec{\delta}+\xi^3\vec{\eta}+\left(\xi^3\right)^2\vec{\varsigma}\right) dV \quad (39)$$

The specification of the displacement space is given by

$$\mathcal{V} = \left\{\left(\vec{\delta},\vec{\eta},\vec{\varsigma}\right) \in H^1(\mathcal{S}) \times H^1(\mathcal{S}) \times H^1(\mathcal{S})\right\} \cap \mathscr{BC}, \quad (40)$$

where H^1 is the Sobolev space of order 1, $\mathscr{BC}$ is space for boundary conditions.
Lemma 1. *Let us consider* $\vec{\delta}$, $\underline{\eta} \in H^1(\mathcal{S})$ *and*

$$\left(\underline{\underline{\gamma}}\left(\vec{\delta}\right), \underline{\underline{\chi}}\left(\vec{\delta},\underline{\eta}\right), \underline{\zeta}\left(\vec{\delta},\underline{\eta}\right)\right) = \left(\underline{\underline{0}},\underline{\underline{0}},\underline{0}\right) \quad on \quad \mathcal{S}. \quad (41)$$

Then, the displacement (36) in $\mathscr{B}$ (higher-dimensional initial lesion body) corresponds to an infinitesimal rigid-body motion, i.e., there exists $\vec{T}$ and $\vec{R}$ a global translation vector and an infinitesimal rotation vector, respectively, such that

$$\vec{\delta}\left(\xi^1,\xi^2\right) = \vec{T} + \vec{R} \wedge \vec{\varphi}\left(\xi^1,\xi^2\right); \qquad \underline{\eta}\left(\xi^1,\xi^2\right) = \vec{R} \wedge z_3\left(\xi^1,\xi^2\right) \quad (42)$$

Lemma 2. *For any* $\left(\xi^1,\xi^2,\xi^3\right) \in \Omega$, *there exist two constants* $\mathfrak{c}, C>0$ *such that the following inequalities hold*

$$\mathfrak{c}\sqrt{\mathfrak{z}\left(\xi^1,\xi^2\right)} \leq \sqrt{g\left(\xi^1,\xi^2,\xi^3\right)} \leq C\sqrt{\mathfrak{z}\left(\xi^1,\xi^2\right)} \quad (43)$$

$$\mathfrak{c}\mathfrak{z}^{ij}\left(\xi^1,\xi^2\right)Y_iY_j \leq g^{ij}\left(\xi^1,\xi^2,\xi^3\right)Y_iY_j \leq C\mathfrak{z}^{ij}\left(\xi^1,\xi^2\right)Y_iY_j, \ \forall (Y_1,Y_2) \in \mathbb{R}^2. \quad (44)$$

$$\begin{aligned}&\mathfrak{c}\mathbf{z}^{ik}(\xi^1,\xi^2)\mathbf{z}^{jl}(\xi^1,\xi^2)Y_{ij}Y_{kl} \le g^{ik}(\xi^1,\xi^2,\xi^3)g^{jl}(\xi^1,\xi^2,\xi^3)Y_{ij}Y_{kl}\\ &\le \mathcal{C}\mathbf{z}^{ik}(\xi^1,\xi^2)\mathbf{z}^{jl}(\xi^1,\xi^2)Y_{ij}Y_{kl}, \quad \forall (Y_{11},Y_{12},Y_{21},Y_{22})\in\mathbb{R}^4.\end{aligned} \tag{45}$$

Lemma 3. *The gradient of a vector field is on average not distant from the space of skew-symmetric matrices, the gradient must not be a far from a particular skew-symmetric matrices. Thus, there exists a constant* $\delta_k > 0$ *such that for any first order surface tensor* $\underline{r} \in H^1(\mathcal{S})$,

$$|\underline{r}|_{H^1(\mathcal{S})} \le \delta_k\left(\left\|\underline{\underline{\epsilon}}(r)\right\|_{L^2(\mathcal{S})} + \|\underline{r}\|_{L^2(\mathcal{S})}\right), \; for \; \underline{\underline{\epsilon}}(r) = \frac{1}{2}\left(\underline{\underline{\nabla}}\;\underline{r} + \left(\underline{\underline{\nabla}}\;\underline{r}\right)^T\right), \tag{46}$$

where $\underline{\underline{\epsilon}}$ is symmetrized gradient tensor.

It is inferred from Lemma 2 that mapping of initial lesion is well-defined in curvilinear coordinate system wherein quantity g is volume measure. Also, Lemma 2 suggests that this function is well-defined and continuous. Because the initial lesion is defined over upper bound ($\mathcal{C}$) and lower bound ($\mathfrak{c}$), the set of bounds is a compact set. The mid-surface of initial lesion definitely lies within the bounds. Thus, the characterization of initial lesion is well-defined. In order to comment on net displacement of the initial lesion during the progression of disease, we prove the following theorem to establish well-posedness of weak formulation for displacement vector.

Theorem 1. *Assume* $\vec{F} \in L^2(\mathscr{B})$; *the essential boundary conditions enforced in* $\mathcal{V}$ *are such that no rigid-body motion is possible, i.e., the only element* $\left(\vec{\delta},\vec{\eta},\vec{\varsigma}\right)$ *in* $\mathcal{V}$ *satisfies Eq.* (42) *for some* $\left(\vec{T},\vec{R}\right)$ *is* $\left(\vec{0},\vec{0},\vec{0}\right)$.

Then there exists a unique $\left(\vec{d},\vec{\theta},\vec{\varrho}\right)$ in $\mathcal{V}$ that satisfies

$$A\left(\vec{d},\vec{\theta},\vec{\varrho};\vec{\delta},\vec{\eta},\vec{\varsigma}\right) = F\left(\vec{\delta},\vec{\eta},\vec{\varsigma}\right) \tag{47}$$

for any $\left(\vec{\delta},\vec{\eta},\vec{\varsigma}\right) \in \mathcal{V}$, and we have

$$\left\|\vec{d},\vec{\theta},\vec{\varrho}\right\|_1 \le C\left\|\vec{F}\right\|_{L^2(\mathscr{B})} \tag{48}$$

Proof. We prove coercivity of A and continuity of A and F. Coercivity argument is explained in three steps. We shall write f instead of function $f(,)$ to make equations more compact.

(i) First, we prove

$$A\left(\vec{d},\vec{\eta},\vec{\varsigma};\vec{\delta},\vec{\eta},\vec{\varsigma}\right) \ge \gamma\left(\left\|\underline{\underline{\gamma}}\right\|_0^2 + \left\|\underline{\underline{\chi}}\right\|_0^2 + \left\|\underline{\underline{k}}\right\|_0^2 + \left\|\underline{\underline{l}}\right\|_0^2 + \|\underline{\zeta}\|_0^2 + \|\underline{m}\|_0^2 + \|\underline{n}\|_0^2 + \|\varpi\|_0^2 + \|p\|_0^2\right). \tag{49}$$

From Eqs. (44) and (45), using $g^{\alpha\beta}g^{\lambda\mu}e_{\alpha\beta}e_{\lambda\mu} = \left(g^{\alpha\beta}e_{\alpha\beta}\right)^2 \ge 0$, we have

$$\begin{aligned} A\left(\vec{d},\vec{\eta},\vec{\varsigma};\vec{\delta},\vec{\eta},\vec{\varsigma}\right) &\ge \gamma\int_\Omega g^{\alpha\beta}g^{\lambda\mu}e_{\alpha\beta}e_{\lambda\mu}dV \\ &\ge \gamma\int_\Omega\left[g^{ik}g^{jl}e_{ij}e_{kl} + g^{ij}e_{i3}e_{j3} + (e_{33})^2\right]dV \\ &\ge \gamma\int_\Omega\left[\mathbf{z}^{ik}\mathbf{z}^{jl}e_{ij}e_{kl} + \mathbf{z}^{ij}e_{i3}e_{j3} + (e_{33})^2\right]dV \end{aligned} \tag{50}$$

Now using Eqs. (43) and (33) and integrating through the thickness, we obtain

$$A\left(\vec{d},\vec{\eta},\vec{\varsigma};\vec{\delta},\vec{\eta},\vec{\varsigma}\right)\geq\gamma\int_{\omega}t\{\mathbf{z}^{ik}\mathbf{z}^{jl}[\gamma_{ij}\gamma_{kl}+\frac{t^2}{12}\chi_{ij}\chi_{kl}+\frac{t^2}{6}\gamma_{ij}k_{kl}+\frac{t^4}{80}k_{ij}k_{kl}+\frac{t^4}{40}l_{ij}\chi_{kl}+\frac{t^6}{448}l_{ij}l_{kl}]+\left[\varpi^2+\frac{t^2}{12}p^2\right]+\mathbf{z}^{ij}\left[\zeta_i\zeta_j+\frac{t^2}{12}m_im_j+\frac{t^2}{6}\zeta_in_j+\frac{t^4}{80}n_in_j\right]\}dS \tag{51}$$

To simplify the above expression, we use the following inequality:

$$|ab|\leq\frac{1}{2}\left(\eta a^2+\frac{1}{\eta}b^2\right),\quad\forall\eta>0. \tag{52}$$

Now we have

$$\begin{aligned}|\frac{t^2}{6}\mathbf{z}^{ik}\mathbf{z}^{jl}\gamma_{ij}k_{kl}|&=\frac{1}{6}|\langle\gamma,t^2k\rangle|\\&\leq\frac{1}{12}\left(a_1\|\gamma\|_\varepsilon^2+\frac{t^4}{a_1}\|k\|_\varepsilon^2\right)\\&\leq\frac{1}{12}\mathbf{z}^{ik}\mathbf{z}^{jl}\left(a_1\gamma_{ij}k_{kl}+\frac{t^4}{a_1}k_{ij}k_{kl}\right),\end{aligned} \tag{53}$$

and similarly

$$|\frac{t^4}{40}\mathbf{z}^{ik}\mathbf{z}^{jl}l_{ij}\chi_{kl}|\leq\frac{1}{80}\mathbf{z}^{ik}\mathbf{z}^{jl}\left(a_2t^6l_{ij}l_{kl}+\frac{t^2}{a_2}\chi_{ij}\chi_{kl}\right). \tag{54}$$

$$|\frac{t^2}{6}\mathbf{z}^{ij}\zeta_in_j|\leq\frac{1}{12}\mathbf{z}^{ij}\left(a_3\zeta_i\zeta_j+\frac{t^4}{a_3}n_in_j\right), \tag{55}$$

where $a_1,a_2,a_3>0$. Using suitable values of the constants, $a_1=a_3=10$, $a_2=6/35$, and $t>0$, Eq. (51) becomes

$$\begin{aligned}A\left(\vec{d},\vec{\eta},\vec{\varsigma};\vec{\delta},\vec{\eta},\vec{\varsigma}\right)\ &\geq\ \gamma\int_{\omega}\{\mathbf{z}^{ik}\mathbf{z}^{jl}\left[\gamma_{ij}\gamma_{kl}+\chi_{ij}\chi_{kl}+k_{ij}k_{kl}+l_{ij}l_{kl}\right]\\&+\ \mathbf{z}^{ij}\left[\zeta_i\zeta_j+m_im_j+n_in_j\right]+\left[\varpi^2+p^2\right]\}dS\end{aligned} \tag{56}$$

Hence, Eq. (49) is proved. The bilinear function is bounded below by the sum of norm of strain tensors. This function for mid-surface is integrated through the thickness of the entire lesion giving semblance of the whole lesion.

(ii) Denoting

$$\|\eta_3,\vec{\varsigma}\|_*=\left(\left\|\underline{m}\left(\vec{\eta},\vec{\varsigma}\right)\right\|_0^2+\left\|\underline{n}\left(\vec{\varsigma}\right)\right\|_0^2+\left\|\underline{\underline{k}}\left(\left(\underline{0},\eta_3\right),\vec{\varsigma}\right)\right\|_0^2+\left\|\varpi\left(\vec{\eta}\right)\right\|_0^2+\left\|p\left(\vec{\varsigma}\right)\right\|_0^2\right)^{1/2} \tag{57}$$

We now show that $\|\cdot\|_*$ provides a norm equivalent to the H^1-norm over certain subspace of the Sobolev space. Note that $\varpi\left(\vec{\eta}\right)=p\left(\vec{\varsigma}\right)=0$ gives $\eta_3=\varsigma_3=0$ and $\underline{m}\left(\vec{\eta},\vec{\varsigma}\right)=\eta_3=0$ gives $\underline{\varsigma}=0$. Bounding the norm from above, we get

$$\left\| \eta_3, \vec{\varsigma} \right\|_* \leq C \left\| \eta_3, \vec{\varsigma} \right\|_1 \tag{58}$$

and we get

$$\left\| \eta_3, \vec{\varsigma} \right\|_* \leq \gamma \left\| \eta_3, \vec{\varsigma} \right\|_1 \tag{59}$$

Using Lemma 3 and Eq. (34), we have

$$\begin{aligned} |\underline{\varsigma}|_1^2 &\leq C\left(\left\| \underline{\underline{\varepsilon}}(\underline{\varsigma}) \right\|_0^2 + \|\underline{\varsigma}\|_0^2 \right) \\ &\leq C\left(\left\| \underline{\underline{k}}((\underline{0}, \eta_3), \vec{\varsigma}) \right\|_0^2 + \left\| \underline{\underline{b}} \varsigma_3 \right\|_0^2 + \left\| \underline{\underline{c}} \eta_3 \right\|_0^2 + \|\underline{\varsigma}\|_0^2 \right) \\ &\leq C\left(\left\| \underline{\underline{k}}\left((\underline{0}, \eta_3), \vec{\varsigma}\right) \right\|_0^2 + \|\varsigma_3\|_0^2 + \|\eta_3\|_0^2 + \|\underline{\varsigma}\|_0^2 \right) \end{aligned} \tag{60}$$

In addition, from the definition of $\underline{n}$ and $\underline{m}$ in Eq. (34), we have

$$|\varsigma_3|_1^2 \leq C\left(\left\| \underline{n}\left(\vec{\varsigma}\right) \right\|_0^2 + \|\underline{\varsigma}\|_0^2 \right) \tag{61}$$

$$|\eta_3|_1^2 \leq C\left(\left\| \underline{m}\left(\vec{\eta}, \vec{\varsigma}\right) \right\|_0^2 + \|\underline{\varsigma}\|_0^2 \right) \tag{62}$$

From Eqs. (60)–(62), we obtain

$$\begin{aligned} \left\| \eta_3, \underline{\varsigma} \right\|_1^2 &\leq C(\left\| \underline{m}\left(\vec{\eta}, \vec{\varsigma}\right) \right\|_0^2 + \left\| \underline{\underline{k}}(\underline{0}, \eta), \vec{\varsigma}) \right\|_0^2 \\ &\quad + \left\| \underline{n}\left(\vec{\varsigma}\right) \right\|_0^2 + \|\varsigma_3\|_0^2 + \|\eta_3\|_0^2 + \|\underline{\varsigma}\|_0^2) \\ &\leq C\left(\left\| \eta_3, \vec{\varsigma} \right\|_*^2 + \left\| \eta_3, \vec{\varsigma} \right\|_0^2 \right). \end{aligned} \tag{63}$$

Perusing the norm of the gradient of vector fields, the setting of lower and upper bounds is tantamount to estimating attributes of lesion at the initial and advanced stages, respectively. The sequence of all lower bounds corresponds to the initial stage of disease prevalent in affected population. Similarly, the sequence of all upper bounds corresponds to the advanced stage of disease prevalent in terminally ill patients. Note that each of these sequences is uniformly bounded in the H^1-norm. There exist a subsequence that converges to some limit for each of these sequences. The weak convergence in H^1 implies strong convergence in L^2 for the same norm to the same limit. Thus, the subsequence in L^2-norm converges strongly. This gives a stronger result about the sequences of upper and lower bounds. Clinically, it indicates various patients might report, at different stages of disease owing to different reasons, their disease initiation be an element, which is a limit point of the subsequence of lower bound sequence. Note that primal lesion presence in any patient whatsoever can be traced back by the convergence of subsequence of lower bound sequence. Its corollary equivalently applies to the advanced stage of the disease.

(iii) Coercivity bound: Coercivity is the measure of the ability of the initial lesion to withstand an external fluid-structure interaction without undergoing deformation. It is obviously dependent on the intensity of hemodynamic forces applied to the lesion. Thus, coercivity bound is the limit point of the ability of initial lesion to withstand deformation. Note that in due course of the progression of the disease, the evolution of the lesion at each stage is dependent on the increment of coercivity bound. In normal circumstances, it seems plausible that with ascension towards upper bound, coercivity reduces. Thus, the evaluation of coercivity bound is relevant.

The inequality (52) is valid for any norm. Hence, we infer

$$\left\|\vec{v}_1+\alpha\vec{v}_2\right\|^2+\left\|\vec{v}_2\right\|^2\geq\gamma\left(\left\|\vec{v}_1\right\|^2+\left\|\vec{v}_2\right\|^2\right), \tag{64}$$

where α is any real number. Using this inequality (64), we obtain

$$\begin{aligned}\left\|\underline{\underline{\chi}}\left(\vec{\delta},\vec{\eta}\right)\right\|_0^2+\left\|\varpi\left(\vec{\eta}\right)\right\|_0^2&=\left\|\underline{\underline{\chi}}(\vec{\delta},(\eta,0)-\underline{\underline{b}}\eta_3\right\|_0^2+\|\eta_3\|_0^2\\&\geq\gamma\left(\underline{\underline{\chi}}\left(\vec{\delta},(\eta,0)\right)\right)\Big\|_0^2+\|\eta_3\|_0^2,\end{aligned} \tag{65}$$

hence,

$$\begin{aligned}&\left\|\underline{\underline{\gamma}}\left(\vec{\delta}\right)\right\|_0^2+\left\|\underline{\underline{\chi}}(\vec{\delta},\vec{\eta})\right\|_0^2+\left\|\underline{\underline{\zeta}}(\vec{\delta},\vec{\eta})\right\|_0^2+\left\|\varpi\left(\vec{\eta}\right)\right\|_0^2\geq\gamma(\left\|\underline{\underline{\gamma}}\left(\vec{\delta}\right)\right\|_0^2+\left\|\underline{\underline{\zeta}}(\vec{\delta},\vec{\eta})\right\|_0^2\\&\quad+\left\|\underline{\underline{\chi}}(\vec{\delta},(\eta,0))\right\|_0^2+\|\eta_3\|_0^2\geq\gamma\left(\left\|\vec{\delta},\eta\right\|_1^2+\|\eta_3\|_0^2\right)\end{aligned} \tag{66}$$

Suppose $\vec{F}\in L^2(S)$ *and the essential boundary conditions enforced in* $\mathcal{V}$ *are such that no rigid-body motion is possible, i.e., the only element* $\left(\vec{\delta},\underline{\eta}\right)\in\mathcal{V}$ *satisfying* (42) *for some* $\left(\vec{T},\vec{R}\right)=\left(\vec{0},\underline{0}\right)$*. Then bilinear form* $\mathcal{A}$ *is coercive over* $\mathcal{V}$*.*

$$\left\|\underline{\underline{k}}\left(\vec{\eta},\vec{\varsigma}\right)\right\|_0^2+|\underline{\eta}|_1^2\geq\gamma\left(\left\|\underline{\underline{k}}\left(\left(\underline{0},\eta_3\right),\vec{\varsigma}\right)\right\|_0^2+\underline{\eta}\Big|_1^2\right) \tag{67}$$

hence,

$$\begin{aligned}&\left\|\underline{\underline{k}}\left(\vec{\eta},\vec{\varsigma}\right)\right\|_0^2+\left|\underline{\eta}\right|_1^2+\|\underline{m}(\vec{\eta},\vec{\varsigma})\|_0^2+\left\|\underline{n}\left(\vec{\varsigma}\right)\right\|_0^2+\left\|\varpi\left(\vec{\eta}\right)\right\|_0^2+\left\|p\left(\vec{\varsigma}\right)\right\|_0^2\\&\quad\geq\gamma\left(\left\|\underline{\underline{k}}((\underline{0},\eta_3),\vec{\varsigma})\right\|_0^2+\|\underline{m}(\vec{\eta},\vec{\varsigma})\|_0^2+\left\|\underline{n}\left(\vec{\varsigma}\right)\right\|_0^2+\left\|\varpi\left(\vec{\eta}\right)\right\|_0^2+\left\|p\left(\vec{\varsigma}\right)\right\|_0^2+\left|\underline{\eta}\right|_1^2\right)\\&\quad=\gamma\left(\|\eta_3,\vec{\varsigma}\|_*^2+\left|\underline{\eta}\right|_1^2\right)\\&\quad\geq\gamma\left(\|\eta_3,\vec{\varsigma}\|_1^2+\left|\underline{\eta}\right|_1^2\right).\end{aligned} \tag{68}$$

Therefore, from Eqs. (49), (66), and (68), we have

$$\begin{aligned}A\left(\vec{d},\vec{\eta},\vec{\varsigma};\vec{\delta},\vec{\eta},\vec{\varsigma}\right)&\geq\gamma\left(\|\gamma\|_0^2+\|\chi\|_0^2+\|k\|_0^2+\|l\|_0^2+\|\zeta\|_0^2+\|m\|_0^2+\|n\|_0^2+\|\varpi\|_0^2+\|p\|_0^2\right)\\&\geq\gamma\left(\left\|\vec{\delta},\underline{\eta}\right\|_1^2+\|\eta_3\|_0^2\left\|\underline{\underline{k}}\right\|_0^2+\|m\|_0^2+\|n\|_0^2++\|p\|_0^2\right)\\&\geq\gamma\left(\left\|\vec{\delta},\underline{\eta}\right\|_1^2+\|\eta_3,\vec{\varsigma}\|_1^2\right)=\gamma\left\|\vec{\delta},\vec{\eta},\vec{\varsigma}\right\|_1^2.\end{aligned} \tag{69}$$

The analysis about coercivity bound suggested that it is not membrane strain tensor alone which progressively alters the structural characteristics of the initial lesion. Rather it is the complex interplay of various tensors in Eq. (34) that breaks the coercivity bound at a particular stage of the disease.

(iv) Completion of the proof.

Since we have proved boundedness and coercivity of the initial lesion, the continuity of the variational formulation can be derived from the continuity of linear function (39) related to test function $\vec{\Delta}\left(\vec{\delta}, \vec{\eta}, \vec{\varsigma}\right)$. There is merit in proving the continuity of variational formulation because such continuous functions within closed interval are bounded:

$$\begin{aligned} |\int_{\Omega} \vec{F} \cdot \left(\vec{\delta} + \xi^3 \vec{\eta} + (\xi^3)^2 \vec{\varsigma}\right) dV| &\leq \left\|\vec{F}\right\|_{L^2(\mathscr{B})} \left\|\vec{\delta} + \xi^3 \vec{\eta} + (\xi^3)^2 \vec{\varsigma}\right\|_{L^2(\mathscr{B})} \\ &\leq \left\|\vec{F}\right\|_{L^2(\mathscr{B})} \left\|\vec{\delta}, \vec{\eta}, \vec{\varsigma}\right\|_0. \end{aligned} \tag{70}$$

Then, from the H^1-coercivity of A infer

$$\gamma \left\|\vec{d}, \vec{\theta}, \vec{\varrho}\right\|_1^2 \leq A\left(\vec{d}, \vec{\theta}, \vec{\varrho}; \vec{d}, \vec{\theta}, \vec{\varrho}\right) = F\left(\vec{d}, \vec{\theta}, \vec{\varrho}\right) \leq C \left\|\vec{F}\right\|_{L^2(\mathscr{B})} \left\|\vec{d}, \vec{\theta}, \vec{\varrho}\right\|_0. \tag{71}$$

Hence, we infer Eq. (48) directly follows. □

The primary result of this section is sufficient conditions for the variational formulation (35) which is proved well-posed. In initial lesion model, a fluid-structure interaction modeled by using bilinear and linear functions specified over displacement is well-posed. Any transverse point-wise loading in H^1 for any lesion implies transverse displacement in H^1. Clinically, progressive interaction of various tensors within lower and upper bounds implies changes in coercivity bounds. It is suggestive of progression of the disease. On the other hand, a fracture in internal lining of vessel wall around the lesion causing blood to flow between tunica intima and tunica media. This flow either remains static (if there is non-patent false lumen) or flow out (if there is a patent false lumen track). Thus, the merit of well-posedness of variational formulation (35) cannot be overemphasized.

5. Asymptotic analysis of the initial lesion

We aim to discuss the asymptotic behavior of initial lesion model. The initial lesion continues to temporally evolve under the influence of fluid-structure interaction; the asymptotic analysis is helpful in this regard. The nonlinearity of progression of the disease can be assessed by formulating bending strain cases because membrane and shear strain vanish with the strain $\varpi(\eta)$, wherein $\eta \equiv 0$ for the test function $\vec{\Delta}\left(\vec{\delta}, \vec{\eta}, \vec{\varsigma}\right)$ in space $\mathcal{V}$. Let us introduce the space of pure-bending displacements:

$$\mathcal{V}_0 = \left\{ \left(\vec{\delta}, \vec{\eta}, \vec{\varsigma}\right) \in \mathcal{V}, | \ \gamma_{ij}\left(\vec{\delta}\right) = 0, \ \zeta_i\left(\vec{\delta}, \vec{\eta}\right) = 0 \quad \forall \ \ i,j = 1,2 \right\}. \tag{72}$$

Based on peculiar geometry and associated bounds, the initial lesion may or may not have nonzero pure-bending displacements. Situation 1, when pure bending is inhibited

$$\mathcal{V}_0 \cap \left\{ \left(\vec{\delta}, \vec{\eta}, \vec{\varsigma} \right) \in \mathcal{V} \right\} = \{(0,0,0)\}, \tag{73}$$

and situation 2, when pure bending is non-inhibited

$$\mathcal{V}_0 \cap \left\{ \left(\vec{\delta}, \vec{\eta}, \vec{\varsigma} \right) \in \mathcal{V} \right\} \neq \{(0,0,0)\}, \tag{74}$$

Let us define higher-dimensional body force as

$$\vec{F} = \varepsilon^{(\rho-1)} \vec{G}, \tag{75}$$

where the exponent $(\rho - 1)$ is used for consistency when the external work involves an integration over the thickness which is relevant for general asymptotic analysis; $\vec{G}$ represents a force field:

$$\vec{G}(\xi^1, \xi^2, \xi^3) = \vec{G}_0(\xi^1, \xi^2) + \xi^3 \vec{B}(\xi^1, \xi^2, \xi^3), \tag{76}$$

where $\vec{G}_0$ is in $L^2(\mathcal{S})$ and $\vec{B}$ is a uniformly bounded function over $\mathscr{B}$ in t. Since it is improbable to obtain strong convergence result in context of asymptotic analysis, we make weaker assumption about $\vec{G}$. We also forgo regularity assumption in context of weak convergence to introduce abstract bilinear forms. Depending upon boundary conditions, nonzero pure-bending displacements of initial lesion are assessed. The displacement is in response to inhibited and non-inhibited pure-bending lesion as we have already argued that only bending strain matters in asymptotic analysis. In the current framework of asymptotic analysis for initial lesion of a given thickness, specific membrane-dominated bilinear form is given by

$$\begin{aligned} A_m\left(\vec{d}, \theta; \vec{\delta}, \vec{\eta}\right) = \int_\omega l[{}^0H^{ijkl} \gamma_{ij}\left(\vec{d}\right) \gamma_{kl}\left(\vec{\delta}\right) + {}^0H^{ij33}\left(\gamma_{ij}\left(\vec{d}\right) \varpi\left(\vec{\eta}\right) + \gamma_{ij}\left(\vec{\delta}\right) \varpi\left(\vec{\theta}\right)\right) \\ + 4\,{}^0H^{i3j3} \zeta_i\left(\vec{d}, \vec{\theta}\right) \zeta_j\left(\vec{d}, \vec{\eta}\right) + {}^0H^{3333} \varpi\left(\vec{\theta}\right) \varpi\left(\vec{\eta}\right)] dS, \end{aligned} \tag{77}$$

bending-dominated bilinear form is given by

$$\begin{aligned} A_b\left(\vec{d}, \vec{\theta}, \vec{\varrho}; \vec{\delta}, \vec{\eta}, \vec{\varsigma}\right) = \int_\omega \frac{l^3}{12} {}^0H^{ijkl} \chi_{ij}\left(\vec{d}, \vec{\theta}\right) \chi_{kl}\left(\vec{\delta}, \vec{\eta}\right) + {}^0H^{ij33}\left(\chi_{ij}\left(\vec{d}, \vec{\theta}\right) p\left(\vec{\varsigma}\right) + \chi_{ij}\left(\vec{\delta}, \vec{\eta}\right) p\left(\vec{\varrho}\right)\right) \\ + 4\,{}^0H^{i3j3} m_i\left(\vec{\theta}, \vec{\varrho}\right) m_j\left(\vec{\eta}, \vec{\varsigma}\right) + {}^0H^{3333} p\left(\vec{\varrho}\right) p\left(\vec{\varsigma}\right)] dS, \end{aligned} \tag{78}$$

where the tensor ${}^0H$ is defined by

$${}^0H^{\alpha\beta\lambda\mu} = H^{\alpha\beta\lambda\mu}\big|_{\xi^3=0},$$

and linear form is given by

$$G\left(\vec{\delta}\right) = \int_\omega l \vec{G}_0 \cdot \vec{\delta} dS.$$

We now discuss the cases of non-inhibited pure bending versus inhibited pure bending.

5.1 The impact of non-inhibited pure bending on the initial lesion

Assume that $\mathcal{V}$ displacement space for the initial lesion contains few nonzero elements. The terms of order zero in ξ^3 in the strain Eq. (33) vanishes by a penalization mechanism, and the appropriate scaling factor is then $\rho = 3$. We define the norm

$$\left\| \vec{\delta}, \vec{\eta}, \vec{\varsigma} \right\|_b = \left(\left\| \vec{\delta} \right\|_1^2 + \left\| \underline{\eta} \right\|_1^2 + \| \eta_3 \|_0^2 + \| \varsigma_3 \|_0^2 + \left\| \varsigma + \frac{1}{2} \underline{\nabla} \eta_3 \right\|_0^2 \right)^{\frac{1}{2}} \tag{79}$$

for which the convergence is anticipated. Since $\left(\vec{d}^{\varepsilon}, \vec{\theta}^{\varepsilon}, \vec{\rho}^{\varepsilon} \right)$ is uniformly bounded in the norm $\| \cdot \|_b$, we extract a subsequence weakly converging in $\mathcal{V}$ to a limit $\left(\vec{d}^{w}, \vec{\theta}^{w}, \vec{\varrho}^{w} \right)$. Since in the early stage of the disease, the internal lining of the vessel wall, tunica intima, is smooth, we can expand the constitutive tensor:

$$H^{\alpha\beta\lambda\mu}\left(\xi^1, \xi^2, \xi^3\right) = {}^0H^{\alpha\beta\lambda\mu}\left(\xi^1, \xi^2\right) + \xi^3 \overline{H}^{\alpha\beta\lambda\mu}\left(\xi^1, \xi^2, \xi^3\right), \tag{80}$$

where $\overline{H}^{\alpha\beta\lambda\mu}\left(\xi^1, \xi^2, \xi^3\right)$ is bounded over initial lesion body $\mathcal{B}$. Using the uniform boundedness of $\varepsilon \left\| \vec{d}^{\varepsilon}, \vec{\theta}^{\varepsilon}, \vec{\varrho}^{\varepsilon} \right\|_1$, we get

$$\lim_{\varepsilon \to 0} \frac{1}{\varepsilon} A\left(\vec{d}^{\varepsilon}, \vec{\theta}^{\varepsilon}; \vec{\rho}^{\varepsilon}, \vec{\delta}, \vec{\eta}, \vec{\varsigma} \right) = A_m\left(\vec{d}^{w}, \vec{\theta}^{w}; \vec{\delta}, \vec{\eta} \right), \tag{81}$$

where A_m is the bilinear form to assess net displacement caused by the membrane strain. This is equivalent to

$$\begin{aligned} \left| \frac{1}{\varepsilon} A\left(\vec{d}^{\varepsilon}, \vec{\theta}^{\varepsilon}, \vec{\varrho}^{\varepsilon}; \vec{\delta}, \vec{\eta}, \vec{\varsigma} \right) \right| &= \left| \frac{1}{\varepsilon} \int_{\Omega} \vec{F} . \left(\vec{\delta} + \xi^3 \vec{\eta} + \left(\xi^3\right)^2 \vec{\varsigma} \right) dV \right| \\ &\leq C\varepsilon^2 \left\| \vec{\delta}, \vec{\eta}, \vec{\varsigma} \right\|_b + C\varepsilon^2 \left\| \vec{\delta}, \vec{\eta}, \vec{\varsigma} \right\|_0. \end{aligned} \tag{82}$$

When $\left(\vec{\delta}, \vec{\eta}, \vec{\varsigma} \right)$ is fixed in $\mathcal{V}$, we get

$$A_m\left(\vec{d}^{w}, \vec{\theta}^{w}; \vec{\delta}, \vec{\eta} \right) = 0 \qquad \forall \left(\vec{\delta}, \vec{\eta}, \vec{\varsigma} \right) \in \mathcal{V}. \tag{83}$$

Using equivalence relations among norms and semi-norms, infer that $\left(\vec{d}^{w}, \vec{\theta}^{w}, \vec{\varrho}^{w} \right) \in \mathcal{V}$. This result (83) shows that bilinear form for the membrane strain tensor vanishes. In this case, non-inhibited pure bending, bending strain tensor predominates whose bilinear form is given by

$$A_b\left(\vec{d}^{w}, \vec{\theta}^{w}, \vec{\varrho}^{w}; \vec{\delta}, \vec{\eta}, \vec{\varsigma} \right) = G\left(\vec{\delta} \right), \qquad \forall \left(\vec{\delta}, \vec{\eta}, \vec{\varsigma} \right) \in \mathcal{V}_0. \tag{84}$$

Eq. (84) equivalently holds for any $\left(\vec{\delta}, \vec{\eta}, \vec{\varsigma}\right) \in \mathcal{V}_0$ (pure-bending subspace of initial lesion). The uniqueness of solution implies that $\left(\vec{d}^w, \vec{\theta}^w, \vec{\varrho}^w\right) = \left(\vec{d}^0, \vec{\theta}^0, \vec{\varrho}^0\right)$. If Eq. (83) equivalently holds for any weakly converging subsequence $\left(\vec{d}^w, \vec{\theta}^w, \vec{\varrho}^w\right)$, we affirmatively conclude that the whole sequence converges weakly to $\left(\vec{d}^0, \vec{\theta}^0, \vec{\varrho}^0\right)$.

5.2 The impact of inhibited pure bending on the initial lesion

We define pure-bending subspace $\mathcal{V}^\#$, of displacement space $\mathcal{V}$ for initial lesion such that

$$\mathcal{V}^\# = \left\{\left(\vec{\delta}, \vec{\eta}\right) \mid \left(\vec{\delta}, \vec{\eta}, \vec{0}\right) \in \mathcal{V}\right\}.$$

In this case, pure bending is inhibited; $\|\cdot\|_m$ gives a norm in pure-bending subspace $\mathcal{V}^\#$ such that

$$\left\|\vec{\delta}, \vec{\eta}\right\|_m = \left\|\underline{\underline{\gamma}}\left(\vec{\delta}\right)\right\|_0 + \left\|\underline{\underline{\zeta}}\left(\vec{\delta}, \vec{\eta}\right)\right\|_0 + \left\|\varpi\left(\vec{\eta}\right)\right\|_0.$$

Since $\left(\vec{d}^\varepsilon, \vec{\theta}^\varepsilon\right)$ is uniformly bounded in pure membrane subspace of displacement space for initial lesion, $\varepsilon^2\left(\vec{\delta}, \vec{\eta}, \vec{\varsigma}\right)$ is uniformly bounded in $H^1(\mathcal{S})$; we infer that the sequence $\left(\vec{d}^\varepsilon + \frac{t^2}{12}\vec{\varsigma}, \vec{\theta}^\varepsilon\right)$ is also uniformly bounded in $\mathcal{V}$. Due to the weak convergence in pure membrane subspace $\mathcal{V}_m$,

$$\left(\underline{\underline{\gamma}}\left(\vec{d}^\varepsilon + \frac{t^2}{12}\vec{\varrho}^\varepsilon\right), \underline{\underline{\zeta}}\left(\vec{d}^\varepsilon + \frac{t^2}{12}\vec{\varrho}^\varepsilon, \vec{\theta}^\varepsilon\right), \varpi\left(\vec{\theta}^\varepsilon\right)\right) \overset{\varepsilon \to 0}{\rightarrow} \left(\underline{\underline{\gamma}}\left(\vec{d}^w\right), \underline{\underline{\zeta}}\left(\vec{d}^w, \vec{\theta}^w\right), \varpi\left(\vec{\theta}^w\right)\right),$$

converges weakly in $L^2(\mathcal{S})$. Hence, for any fixed $\left(\vec{\delta}, \vec{\eta}, \vec{\varsigma}\right)$ in displacement space $\mathcal{V}$, we infer

$$\lim_{\varepsilon \to 0} \frac{1}{\varepsilon} A\left(\vec{d}^\varepsilon, \vec{\theta}^\varepsilon, \vec{\varrho}^\varepsilon; \vec{\delta}, \vec{\eta}, \vec{\varsigma}\right) = A_m\left(\vec{d}^w, \vec{\theta}^w; \vec{\delta}, \vec{\eta}\right). \tag{85}$$

We have

$$\frac{1}{\varepsilon} A\left(\vec{d}^\varepsilon, \vec{\theta}^\varepsilon, \vec{\varrho}^\varepsilon; \vec{\delta}, \vec{\eta}, \vec{\varsigma}\right) = G\left(\vec{\delta}\right) + \frac{R}{\varepsilon}. \tag{86}$$

Here, $\frac{R}{\varepsilon} \to 0$ when $\varepsilon \to 0$. As $\left(\vec{\delta}, \vec{\eta}, \vec{\varsigma}\right)$ is fixed, we infer

$$A_m\left(\vec{d}^w, \vec{\theta}^w; \vec{\delta}, \vec{\eta}\right) = G\left(\vec{\delta}\right) \qquad \forall\left(\vec{\delta}, \vec{\eta}\right) \in \mathcal{V}. \tag{87}$$

Eq. (87) equivalently holds for any $\left(\vec{\delta}, \vec{\eta}\right) \in \mathcal{V}_m$ (membrane subspace of initial lesion). From the uniqueness of the weak convergence result, it follows that $\left(\vec{d}^w, \vec{\theta}^w\right) = \left(\vec{d}^m, \vec{\theta}^m\right)$. If this equivalently holds for any weakly converging subsequence $\left(\vec{d}^{w}, \vec{\theta}^{w}, \vec{\varrho}^{w}\right)$, we affirmatively conclude that the whole sequence $\left(\vec{d}^{\varepsilon} + \frac{t^2}{12}\vec{\varsigma}, \vec{\theta}^{\varepsilon}\right)$ converges weakly to $\left(\vec{d}^{m}, \vec{\theta}^{m}\right)$ in $\mathcal{V}_m$.

Finally, asymptotic analysis, both types of initial lesion problems, including case of non-inhibited pure bending and case of inhibited pure bending, has weak convergence. Asymptotic analysis revealed that initial lesion is bending-dominated when pure bending is non-inhibited and that initial lesion is membrane-dominated when pure bending is inhibited. Clinically, the primal lesion undergoes transformations under the influence of membrane, bending, and shear tensors. In advanced stages, the transition towards upper bound occurs due to change in coercivity bounds. During the advanced stages of disease, the bending is responsible for introducing progressive disarray of collagen fibers, smooth muscle cells, and ground matrix and thus contributes to rapid progression. Asymptotic analysis suggests that bending strain is relevant for the progression of disease in advanced stages. Hence, asymptotic analysis is a valuable technique for theoretical supplementation to model building and provide insights into the behavior of initial lesion.

6. Concluding remarks

6.1 Conclusion

We constructed the model by using higher-order kinematical assumptions relevant to human cardiovascular system. We called this model the initial lesion model. The weak convergence of the solution to initial lesion model was mathematically substantiated. In the analysis of the initial lesion, we concentrated to seek biological and mathematical insights in order to understand early stages of AD. A general understanding of evolution of initial lesion in aortic dissection is presented. The results presented in this chapter are relevant for the assessment of shell-type lesion in biological systems including human physiology and pathology. At least two observations are to be noted. First, the mathematical analysis of the initial lesion model is distinct from classical shell models. Second, the asymptotic analysis of the initial lesion model is based on degenerating three-dimensional continuum to bending strains to initial lesion behavior. For very thin shells as seen in human vessels' internal lining, the analytical perspective to the initial lesion model given in this chapter can be used in the convergence studies.

6.2 Future scope

Clinically complex situations such as the formation of false lumen either blind or patent in advanced stage of AD merit mathematical analysis perusing coercivity bounds.

Author details

Vishakha Jadaun and Nitin Raja Singh*
Indian Institute of Technology, Delhi, New Delhi, India

*Address all correspondence to: nitinrsingh22@gmail.com

References

[1] Bathe KJ. Finite Element Procedures. 2006

[2] Argyris JH, Kelsey S. Energy Theorems and Structural Analysis. A Generalised Discourse with Applications on Energy Principles of Structural Analysis Including the Effects of Temperature and Non-linear Stress-Strain Relations. London: Butterworths; 1960

[3] Clough RW, Martin HC, Topp LJ, Turner MJ. Stiffness and deflection analysis of complex structures. Journal of the Aeronautical Sciences. 1956;**23**:9

[4] Ahmad S, Irons BM, Zienkiewicz OC. Analysis of thick and thin shell structures by curved finite elements. International Journal for Numerical Methods in Engineering. 1970;**2**(3): 419-451

[5] Reissner E. The effect of transverse shear deformation on the bending of elastic plates. Journal of Applied Mechanics. 1945;**12**:A-69-A-77

[6] Mindlin RD. Influence of rotatory inertia and shear on flexural motions of isotropic, elastic plates. Journal of Applied Mechanics. 1951;**18**:31-38

[7] Badger S et al. Endovascular treatment for ruptured abdominal aortic aneurysm. Cochrane Database of Systematic Reviews. 2017;**5**

[8] Zankl AR et al. Pathology, natural history and treatment of abdominal aortic aneurysms. Clinical Research in Cardiology. 2007;**96**(3):140-151

[9] Legarreta JH et al. Hybrid decision support system for endovascular aortic aneurysm repair follow-up. In: International Conference on Hybrid Artificial Intelligence Systems. Berlin, Heidelberg: Springer; 2010. pp. 500-507

[10] Sakalihasan N, Limet R, Defawe OD. Abdominal aortic aneurysm. The Lancet. 2005;**365**(9470): 1577-1589

Chapter 11

Problems of Control Motion of Solar Sail Spacecraft in the Photogravitational Fields

Vladimir Stepanovich Korolev, Elena Nikolaevna Polyakhova and Irina Yurievna Pototskaya

Abstract

The problems of spacecrafts with a solar sail-controlled motion lead to the study of mathematical models for translational orbital motion and for the spaceship rotation about the mass center in photogravitational fields. There are opportunities to choose the optimal maneuvering conditions to realize orbital motion or to move to a given orbit point. The realization of the given optimal sail orientation about the sunlight flow allows to obtain motions in the vicinity of a possible relative equilibrium or stationary state. This realization also takes into account the stability change according to the process models with perturbations. For the motion control, we can change the properties, dimensions, or location of sail system elements. The spacecraft flights using light pressure are already a reality. Such space sailing ships may soon be used to fly to the big and small planets, for asteroids and comets meeting, to form special motion conditions in the vicinity of the Sun or the Earth. New technologies will bring visible benefits for solving complex problems, based on the direct use of practically unlimited source of solar energy.

Keywords: spaceflight, solar sail, control, stability

1. Introduction

1.1 Solar sail history

The principle of movement in space by solar sail is based on the light pressure effect on all the bodies, which experimentally are detected and measured by Russian scientist P.N. Lebedev in 1899 [1]. The development of the first engineering project of a space flight under a solar sail belongs to a Russian scientist and engineer Friedrich Zander (1887–1933).

In 1920, F.A. Zander and K.E. Tsiolkovsky (1857–1935) suggested that a very thin flat sheet, illuminated by sunlight, is able to achieve high speeds in space. As for the question of whether this property of photons can be used for space motion, they answered positively. The idea to use this effect for space flight was advanced by scientist and inventor F.A. Zander in 1924 [2]. He proposed the construction of solar sails and developed the foundation of the spacecraft motion theory. He was the first person to realize the potential of large specularly reflecting surfaces for space flight, proposed to build solar sails and developed the basis of the theory of motion of spacecraft. It can be considered the founding father of the innovative concept of the solar sail, which was developed in two manuscripts but was not published until 1947. Flight spacecraft using light pressure energy is no longer a fantasy but the reality of the near future [3, 4]. The first attempt of the project implementation and deployment of a solar sail in space was made in 1993.

Over the years there has been appeared numerous studies of mathematical models of the motion and possible new versions of the form of solar sails [5, 6]. The technology of large-scale designs of sunlight reflectors is still in its initial state. The first practical development of flights with a solar sail began in the 1970s of the twentieth century. Of particular concern was the planned for 1986 flight of a spaceship on a solar sail to meet with Halley's comet. The first attempt to implement the project and deploy a solar sail in space was completed in 1993; a 20-m-diameter mirror sail was successfully deployed on the "Progress M-15" cargo ship as a result of the space experiment "Znamya-2."

If a real sail is at an angle to the flow, the force vector will be directed almost normal to the plane of the sail with a good reflection coefficient. The force of light pressure on the mirror at the same time would be almost twice as much on the black sail of equal area, which completely absorbs the radiation. If the solar sail is made of black material, its thrust is twice less than perfectly mirrored. In this case, the force is not directed along the normal to the surface and the direction of flow of sunlight. Light pressure forms a central photogravitational field, which operates when spacecraft sails moving in an interplanetary space. This will allow to select optimal control during maneuvering.

There are projects using the solar sail to put the spacecraft into geosynchronous orbit in the equatorial plane or to maintain motion in the orbit plane, which is parallel to the equatorial plane and has a nonzero latitude. These latitudinal orbits can create new systems for the deployment of satellite communication systems. One of the possible tasks is the creation of a cosmic solar screen located near the Lagrange point L1 of the Sun-Earth-spacecraft system, which can be useful for monitoring the global temperature of the Earth. Many promising projects for using the solar sail are published.

The solar radiation flux creates a force locally uniform pressure field on the surface of the spacecraft sails. If the surface has a symmetry, and the point of application of the resultant coincides with the center of mass of the spacecraft design, then any initial position relative to the light flux is a state of neutral equilibrium and peace. The action of other forces, even small in size, can produce disturbing moment, which can cause rotation about the center of mass. To damp or compensate the disturbance and to maintain the sail correct position with respect to the light flux it is necessary to use an additional control force. The special sails design allows to solve control and the spacecraft stability problems. Thrust vector and point spacecraft relative to the main body can be changed or if the value of the surface properties of the solar sail and arrangement of the elements may vary with respect to the device using the additional devices.

Note the basic problems [6, 7] of engineering and realization of flights of the spacecraft with the solar sail:

- Creation of an effective reflective polymer film for sails.
- Packing of sails in special containers for delivery to space.
- Take into account the restrictions on the total weight of the spacecraft with a sail at launch.
- Special tools for deploying sails of a large area in the working position.

- The formation of special frame elements for control and support of the sail.
- Providing required initial orientation of the elements of the sail.
- Motion control and stability of the given position in flight.

Only by resolving all problems can we talk about space travel and maneuvering in reality. It should provide a sufficiently sophisticated control sail itself, as desired by changing its size, shape, and position relative to the main body. We can use the sail elements that can change the reflection coefficient of the surface of a given program. Successful construction has been recognized as the slit-like sails helicopter rotor, each blade which is rolled out from the container and can be rotated radially relative to the axis of fixing at a predetermined angle. In some projects, the spacecraft with the sail offers spinning relative to the main axis for the stability of the sail shape under the action of light pressure and disturbing forces.

The most successful may be the design of the sail system, which provides the installation of the desired orientation and control over its preservation. After creating and placing such mirror elements with certain proportions in orbit, we obtain orientation stability relative to the Sun for coplanar trajectories of the transition to a new orbit or preservation of a given final orbit. More complex options and models make it possible to program sequential control of the orbital or rotational motions of a spacecraft with a solar sail.

The efficiency of using solar sails is primarily associated with the angle of their orientation relative to the beam of rays. In a complex system of mirror surfaces, the beam path can be adjusted in such a way that the direction of the incident and reflected light beam is independent, creating new opportunities for a given direction of the thrust vector.

Only having solved all the problems, we can talk about space travel. This can be ensured by a difficult choice of controlling the elements of the sailing system, which will allow you to change the size or shape and position relative to the main body and the flow of sunlight. We can use sail elements that can change the reflection coefficient of a part of the surface according to a given program.

Of interest is the unique possibility of functioning in special zones in the vicinity of the Sun, even near the solar corona, where the sail can simultaneously play the role of a reliable heat-resistant screen that protects the main instrument compartment from overheating. This design will be indispensable for studying solar space and observing sunspots from close range. The disclosure of the sail creates a force in the direction that compensates for the force of gravity and, therefore, will change the parameters of a possible orbit.

Many options relate to near-Earth space maneuvers. The use of the solar sail is possible to put the spacecraft into a given orbit and to support further stable operation of satellite systems without additional fuel consumption.

1.2 Forces and their moments

The resulting light flux pressure force $\overrightarrow{F_i}$ is proportional to the surface S_i area of the sail elements with a parameter (1), which is determined by a reflection coefficient q_i and inversely proportional to the square of the distance r from the Sun (**Figure 1a**).

It also depends on the direction force (**Figure 1b**) of the vector normal $\vec{n}$ to the surface of the respective element $b(\vartheta_i)$ relative to the radial direction with angle ϑ.

Stability may be ensured by the torques of pressure forces about the center of mass, which can modify the value at change the settings. The main vector of the

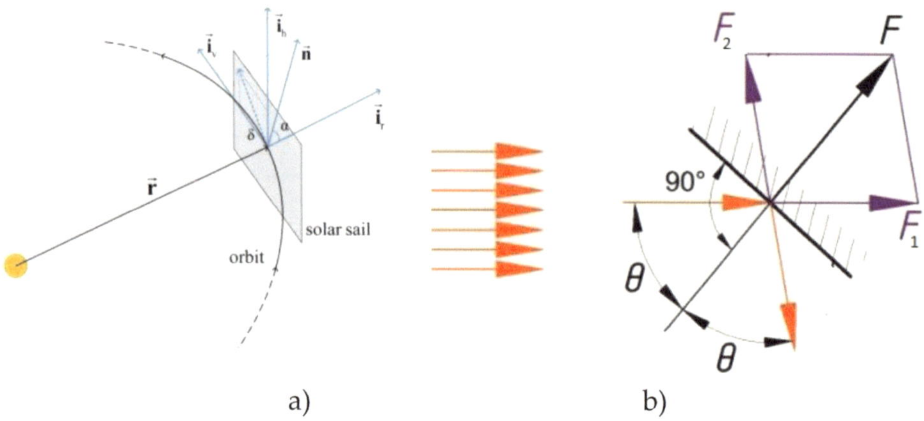

Figure 1.
The formation of the resultant force of the light flux acting on the surface of the sail.

forces and the sum of the moments of all the forces $\vec{F_i}$ acting on the sail with a relative position $\vec{\rho}_i$

$$\vec{F} = \sum_i \vec{F}_i = \sum_i q_i S_i \frac{b(\vartheta_i)}{r^2} \vec{n}\,(\vartheta_i), \;\; \vec{M} = \sum_i \vec{\rho}_i \times \vec{F}_i(\vartheta_i). \tag{1}$$

These values determine the motion of the spacecraft center of mass and rotation relative to the orbital system that accompanies motion in a central field.

The main vector of the moment of acting forces relative to the center of mass may differ from zero for light pressure forces if the elements have different areas or angles of mutual arrangement. This determines the ability to return in the right direction in the event of a change in orientation by random interference or the use of new elements of the basic layout of the spacecraft for orientation in the right direction [8, 9].

2. Equations of motion

2.1 Different types of equations

The equations of motion with allowance for disturbances can be represented in different forms, based on models of the problem of two or three bodies using convenient coordinate systems and basic parameters. Heliocentric flight to planets, asteroids, or to the Sun can be considered, to a first approximation, the motion in a photogravitational field as a two-body problem under the action of the additional light pressure of the rays on the sail surface for a fixed angle of the normal position, taking into account the influence of additional disturbances.

Motion in a photogravitational field can be considered as a two-body problem or a central force field without taking into account the influence of other forces, when the action of additional light pressure from the rays reduces the influence of gravitational interaction. This change is especially noticeable for the case of a large sail surface.

When creating orbits near the Earth or to place a spacecraft at the libration points of the Sun-Earth-spacecraft system, it is necessary to use a more general model of the photogravitational restricted three-body problem, which takes into account the movement of two main bodies, as well as the direction of the propagation of light radiation and direction from the spacecraft to the gravitational center: in this case it may not be the same.

Control using solar sails leads to complex problems and solving equations of mathematical models. There are basic versions of the equations of motion of the spacecraft in the central gravitational field, taking into account perturbations depending on the choice of the reference frame, absolute Cartesian, spherical and cylindrical coordinates, or Kepler's elements [5, 6, 10] for the orbit.

Changes in Cartesian coordinates x_i of the spacecraft center of mass in the absolute coordinate system based on the main operating force of gravity and the center of the field of light pressure at movement in three-dimensional case can be described by a second-order equation:

$$\frac{d^2x_i}{dt^2} + \frac{\mu}{r^3}x_i = \frac{\partial U}{\partial x_i} + F_i, \quad i = 1, 2, 3, \tag{2}$$

where we have used the notations x, the Cartesian coordinates; r, the module of the radius vector; μ, the gravitational parameter; U, the force function of potential forces of the considered disturbances; and F_i, the nonpotential acceleration and control, including of the light pressure forces in projections on the axis coordinate system or jet forces on the active phases of orbit.

You can use the polar coordinates (r, φ) in the study of movement in the orbital plane:

$$\frac{d^2r}{dt^2} + \frac{\mu}{r^2} - r(\dot{\varphi})^2 = P_1, \quad \frac{d}{dt}\left(r^2\dot{\varphi}\right) = P_2, \tag{3}$$

where P_i $(i = 1, 2)$ is the radial and transversal components of the perturbing acceleration, which depend on the installation angle of the sail elements to implement the control law. The position relative to the flow of sunlight is taken into account at the corresponding pressure value far from the Sun.

The control algorithms $u(t)$ are numerous and are determined through the parameters of the initial and final orbits or by the tasks of maneuvering the spacecraft in the process of movement. A fixed constant angle will determine the change in the parameters of the orbit. Without taking into account all perturbations and controls, we can use the classical solution of the two-body problem.

The movement in the central gravitational field has a solution, which in the absence of disturbing forces is determined by the initial values of the radius vector, velocity vector, and the gravitational parameter of the central body. They determine the constant Kepler elements $k = (a, e, i, \Omega, \omega, M_0)$ which allow us to calculate the Cartesian coordinates $x_i(t)$ and components $v_i(t)$ of the velocity vector for the unperturbed motion at any time using the following formulas:

$$\begin{aligned}
x_1 &= r(\cos u \cos \Omega - \sin u \sin \Omega \cos i),\\
x_2 &= r(\cos u \sin \Omega + \sin u \cos \Omega \cos i),\\
x_3 &= r \sin u \sin i,\\
v_1 &= \alpha(\cos u \cos \Omega - \sin u \sin \Omega \cos i)\\
&\quad -\beta(\sin u \cos \Omega + \cos u \sin \Omega \cos i),\\
v_2 &= \alpha(\cos u \sin \Omega + \sin u \cos \Omega \cos i)\\
&\quad -\beta(\sin u \sin \Omega + \cos u \cos \Omega \cos i),\\
v_3 &= \alpha \sin u \sin i + \beta \cos u \sin i.
\end{aligned} \tag{4}$$

Notation used here

$$r = a(1 - e\cos E), \quad p = a(1 - e^2),$$

$$\alpha = \sqrt{\frac{\mu}{p}}\, er^{-1} \sin\vartheta, \quad \beta = \sqrt{\mu p}\, r^{-1},$$

and Kepler's equation

$$E - e\sin E = M_0 + n(t - t_0) = M. \tag{5}$$

Moving time between two points of the orbit can be determined from the above equation which is called the equation of Kepler (5).

The equations of motion while taking into account the perturbations can be represented as osculating elements $k(t) = (a, e, i, \Omega, \omega, M_0)$. The orbit elements for the perturbed motion of the spacecraft are functions of time $k(t)$. You can use Euler differential equations in which the functions on the right-hand sides of the equations are determined by the current values of the elements and the projections of the disturbing acceleration on the axis of the orbital coordinate system.

The contribution of radiation pressure is determined by the angle ϕ of deviation, the normal vector $\vec{n}$ from direction $\vec{r}_0$ of flow. If we turn the flat mirror sail at an angle to the rays, the momentum transferred to the solar sail will be directed almost perpendicular to the reflective surface. Part of the momentum directed parallel to the sail, the photons will remain at home, so that the sail will get less than in the full disclosure of the rays. Turning the sail, we are able to control the direction of the thrust vector. However, for it to pay its value. If the vector of normal for flat sail is perpendicular to the flow of rays, the sail does not give any traction. The projections of the vector on the radial and transverse directions will be influenced by a change in the parameters of the orbit motion. The projection on the normal to the plane of the orbit will allow to change its inclination with respect to the initial position. Acceleration, which tells the stream of rays, also depends on the ratio of the area of the sail to the weight of the entire structure and the surface properties.

Equations Hill-Clohessy-Wiltshire [11–15] which managed orbital motion of the moving coordinate system in the spatial case are

$$\begin{aligned} &\ddot{x} + 2\omega\dot{y} = u_x(t), \\ &\ddot{y} - 2\omega\dot{x} - 3\omega^2 y = u_y(t), \\ &\ddot{z} + \omega^2 z = u_z(t), \end{aligned} \tag{6}$$

The solution nonlinear equations (6) can be presented in the form of changes or deviations from the given movement of the reference point and then add a particular solution with the selected control function. The solution of a homogeneous system can be represented as the following (7) system of six equations:

$$x(t) = \left(x_0 - 2\frac{\dot{y}_0}{\omega}\right) - 3\omega t\left(2y_0 + \frac{\dot{x}_0}{\omega}\right) + 2\left(3y_0 + \frac{2\dot{x}_0}{\omega}\right)\sin\omega t + 2\frac{\dot{y}_0}{\omega}\cos\omega t,$$

$$y(t) = 2\left(2y_0 + \frac{\dot{x}_0}{\omega}\right) - \left(3y_0 + \frac{2\dot{x}_0}{\omega}\right)\cos\omega t + \frac{\dot{y}_0}{\omega}\sin\omega t,$$

$$z(t) = z_0\cos\omega t + \frac{\dot{z}_0}{\omega}\sin\omega t, \tag{7}$$

$$\dot{x}(t) = -3\omega\left(2y_0 + \frac{\dot{x_0}}{\omega}\right) + 2\omega\left(3y_0 + \frac{2\dot{x_0}}{\omega}\right)\cos\omega t - 2\dot{y}_0 \sin\omega t,$$

$$\dot{y}(t) = \omega\left(3y_0 + \frac{2\dot{x_0}}{\omega}\right)\sin\omega t + \dot{y}_0 \cos\omega t,$$

$$\dot{z}(t) = \dot{z_0}\cos\omega t - z_0\omega\sin\omega t,$$

The unfolding of the sails on a circular heliocentric orbit will lead to the fact that the light pressure partially compensates the Sun's gravity. This is a reason to use a mathematical model of photogravitational force field [3, 7, 10].

The orbital elements for perturbed motion of the spacecraft are functions of time. We can use the differential equations of Euler, where the right-hand sides are determined by the current values of the elements $k(t) = (a, e, i, \Omega, \omega, M_0)$ and projections of the perturbing acceleration P_i on the axes of the orbital coordinate system:

$$\begin{gathered}
\frac{da}{dt} = 2a^2\left(esin\vartheta P_1 + pr^{-1}P_2\right), \\
\frac{de}{dt} = p(\,sin\;\vartheta P_1 + \;cos\;\vartheta P_2 + \;cos\;EP_2), \\
\frac{di}{dt} = rcos\vartheta P_3, \quad \frac{d\Omega}{dt} = rsinu\,sin^{-1}i\;P_3, \\
\frac{d\omega}{dt} = e^{-1}[(r+p)\,sin\;\vartheta P_2 - pcos\vartheta P_1] - \;cos\,i\frac{d\Omega}{dt}, \\
\frac{dM_0}{dt} = \sqrt{e^{-2} - 1}[(pcos\vartheta - 2er)P_1 - (r+p)\,sin\;\vartheta P_2].
\end{gathered} \tag{8}$$

Here P_i $(i = 1, 2, 3)$ are the components of the disturbing acceleration in the projection on the axis of the orbital coordinate system. They depend on the installation angle of the sail elements for the implementation of the control law and determine a further change in the parameters of the orbit. The position relative to the stream of sunlight is taken into account with the corresponding value of light pressure far from the Sun.

The third and fourth equation of system shows that the plane orbit state is maintained if there is no projection of the disturbing forces in the normal to the plane. The behavior and properties of the solutions are analyzed in a dynamic system, which are simulating the controlled processes. The research methods of nonlinear continuous or discrete systems' quality of the movements, absolute or asymptotic stability, can be obtained [6, 12, 16].

2.2 Control of motion

The influence of solar pressure on the sail is determined by the angle of deviation of the normal vector to the surface from the direction of flow. If the plane of an ideal specular sail is at an angle to the rays, then the momentum transmitted to the solar sail will be directed almost perpendicular to the reflecting surface. By turning the elements of the sailing system, you can control the direction of the thrust vector. The arrangement of elements relative to the housing can be changed with the help of electric motors, supporting their work on the basis of solar batteries.

If the spacecraft with the folded solar sail is already delivered into orbit around the Earth or move around the Sun, the container of the sails will provide disclosure for the spacecraft new thrusters, providing a virtually unlimited supply of energy.

However, the sail has one major drawback: unlike jet engines, we cannot use its thrust in any direction with the same efficiency. It is necessary to orient the sail in a special way, to achieve the desired changes in the orbital parameters of outer space.

When you turn the sail so that the photons are reflected back relative to the direction of orbital motion, we get an additional force that gradually accelerates the spacecraft, which will move in a spinning spiral. When you turn the sail in a different direction, you get a decrease in speed or braking on the way to the sun.

To change the inclination of the orbital plane of the spacecraft using sails, it is necessary to direct the reflected flow perpendicular to the initial plane. In addition, the elongated elliptical orbit can rotate sequentially, changing the longitude of the pericenter relative to the central body, so that over time, it approaches the orbit of an asteroid or comet for a possible encounter [17].

Even more interesting maneuvers can be performed using a solar sail in the near-Earth space, as the propagation direction of light emission in this case coincides with the direction of the center of gravity. In addition, throughout the year the Sun makes a complete revolution in the celestial sphere relative to the Earth, so be patient, you can wait for the right time of the year for optimum flexibility and translate using the spacecraft sails in the desired orbit.

We get the opportunity to control the movement by changing the direction of the thrust vector of the current light pressure when the sail plane is rotated. Vector projections of the force on the radial and transverse directions will affect the change in the basic parameters of the orbit (size, shape) in the process of movement.

The projection of the force to the normal to the orbit plane changes the inclination relative to the initial position of the orbit plane. Acceleration also depends on the ratio of the sail area to the mass of the entire structure and surface properties.

Control algorithms $u(t)$ are numerous. They are determined by the parameters of the orbit or the conditions and purpose of the maneuver. Motion in a gravitational field can be precisely determined if disturbing forces are absent. Then the orbit is determined by the initial values of the radius vector, velocity vector, and gravitational parameter of the central body. For a fixed position of the sail, we can consider a photogravitational field with a small perturbation, which determines the corrections to the parameters of the orbit, reducing the size of the orbit (**Figure 2a**) or increase the size of the orbit (**Figure 2b**).

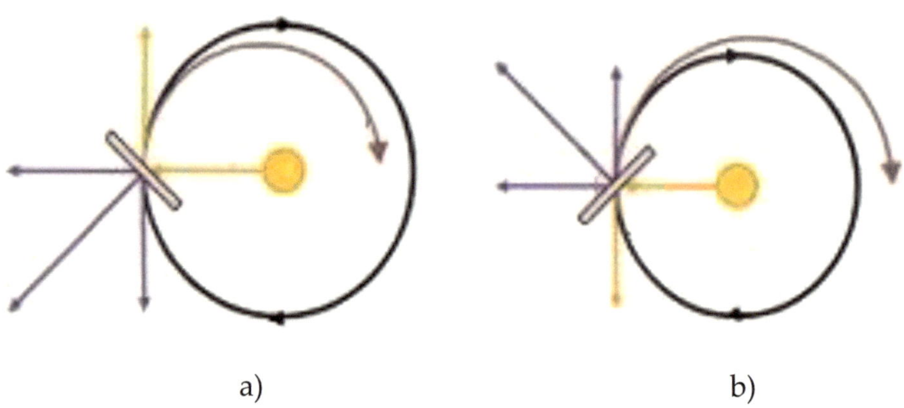

Figure 2.
Motion control is carried out by the position of the solar sail, when the vector of light pressure forces determines braking (a) or acceleration (b). In case (a) there will be a decrease in the size of the orbit; case (b) increases the size of the orbit.

The presence of perturbations of a periodic or random nature can change the nature of the solutions of such systems. The properties of solutions of dynamic systems are determined by the selected feedback control function. The orbital stability of trajectories or the stability with respect to a part of variables [11–13] is also considered.

2.3 The stability of the orbital motion

The system is called stable if it returns to the equilibrium or rest state after the termination of external influences that moved it out of this position. If after the termination of the external impact of the system does not return to a state of equilibrium, it is unsustainable. Stable equilibrium position becomes asymptotically stable with the addition of dissipative forces with complete dissipation.

Determining the motion of any mechanical system is often required to assess the stability and control of motion states. The strict definition of a stable equilibrium position and other solutions of dynamical systems were given in 1892 by the Russian scientist Lyapunov [6–9, 14, 18].

The movement or behavior of the solution of the dynamical system is called Lyapunov stable if small variations in the initial data from the reference phase variables selected for the study, solving a system of differential equations, lead to small deviations in the future. If the deviation over time tends to be zero, the reference solution is called asymptotically stable.

The system is called instable in case when even very small perturbation influence leads to large deviations or change the motion character, including the equilibrium position displacement, which is not stable if the velocity initial value is different from zero.

We also consider the stability of the orbital trajectory or stability of some of the variables [6–12]. In this case, it appears that the phase trajectory and its projection onto the corresponding subspace are close enough to the base path, although the representative points can be arbitrarily disperse, away from each other over time. Periodic solution of the system is not asymptotically stable. But if in such system all multipliers' modules but one are less than one, then according to Andronov-Witt theorem [14], the trajectory of the system periodic solution is asymptotically orbital stable.

Stability with respect to part of variables for partial differential equations involved Rumyantsev [19, 20], who published an article on the analogue of the theory of the second Lyapunov method for the stability problems for some of the variables [21]. He and his students developed methods for research on some of the variables of stability problems.

If the dynamic equations are written in the canonical form, and there are n first integrals, then the Arnold's theorem [12–14], all phase trajectories lie on the n-dimensional torus, and the motion are conditionally periodic system. This set is called the equilibrium or stationary state of motion of the system.

In the field of action of the geopotential excluding other perturbing forces exists stable equilibrium for body position while maintaining the orientation of the major axis of the ellipsoid of inertia in the direction of the center of gravity.

2.4 Control of orientation motion

In the Cartesian coordinate system for the main body, the known parameters of the axial moments of inertia are taken into account. In the field of the geopotential force without other perturbing forces, there are stable equilibrium positions for the body while maintaining the orientation of the main axes (x, y, z) of the inertia ellipsoid with the main moments towards the center of attraction, taking into account the angular velocity of the orbit.

If we consider only the first linear approximation, then the equation of oscillations of the spacecraft with a small perturbation or deviation from the equilibrium position has the form, which at the next step turns into the equation of harmonic oscillations with additional control functions.

Therefore, small oscillations of the mathematical pendulum will be isochronous. We get the orbital stability of motion or stability with respect to part of the variables. For small vibrations in the vicinity of the equilibrium position, the damping effect of additional gyroscopic devices or motors can be used. The action of disturbances can be compensated by a change in the size or reflective properties of the elements of the sailing ship, as well as their relative position. This creates additional moments that can be used for control and stability.

In the case of possible oscillations of the satellite in the orbit plane while maintaining the orientation of the other major axis orthogonal to this plane, the law of change in kinetic momentum takes into account the effect of the Earth's gravitational field [9, 12, 16, 22]. This leads to the equation.

$$I_z\ddot{\vartheta} = M_z = 3\omega_0^2 (I_y - I_x) \sin\vartheta \cos\vartheta. \tag{9}$$

Let is denote $\omega^2 = 3\omega_0^2 (I_y - I_x)(2I_z)^{-1}$, where I_x, I_y, I_z are moments of inertia and ω_0 is the angular velocity motion on orbit and a new variable $\varphi = 2\vartheta$. Then we come to the ordinary equations of oscillations with perturbation. If we consider the first linear approximation only, the equation of oscillations of a spacecraft under a small perturbation or deviation from the equilibrium position has a form, which on the next step turns into the equation of harmonic oscillations with additional control function:

$$\ddot{\varphi} = -\omega^2 \sin\varphi + u(t, \varphi), \tag{10}$$

The period in linear approximation depends on the initial data. Therefore, small oscillations of mathematical pendulum

$$\ddot{\varphi} + \omega^2 \varphi = u(t, \varphi) \tag{11}$$

will be isochronous. We get the orbital stability of the movement or the stability with respect to the part of the variables. For the small oscillations in the neighborhood of the equilibrium position, it is possible to use the damping action of additional gyroscopic devices or engines [15, 17]. To damp the oscillations, a control is proposed in the form of piecewise constant functions

$$u(t, \varphi) = -u_{max} sign(\varphi) \tag{12}$$

of a relay type with a switching period, which is determined by the frequency ω.

Euler's equation (8) of rotation of a rigid solid about a center of mass show that there are three options for steady motions in the form of stationary rotations about the three principal axes of the ellipsoid of inertia when the two components of the angular velocity are equal to zero and the third is a constant [7, 8].

The Euler equation in the general case [8, 22] determines the rotation of a body under the action of moments of force relative to the center of mass. In the case of the body rotation around the instantaneous axis we need the forces moment about the axis to turn the body or to stop rotation.

3. Conclusions

When a rigid body moves in the Earth's gravitational field, there are location options for stable orientation relative to the orbital system. Consideration of the effect of light pressure on a sailing spacecraft leads to the appearance of other possible provisions or stability conditions that can be used in the process of motion control.

The perturbations action can be compensated by the varying of size or reflective properties of spacecraft sail's elements, as well as their mutual arrangement. It creates additional torques, which can be used for a control and stability.

In the case where the main forces can be considered the gravitational interaction with the Sun and its light pressure, you can use the model photogravitational central field to interplanetary space flights to asteroids or other planets.

In the case of movement in orbits near the Earth, the directions of the main acting forces (gravity and light pressure) do not coincide. However, as a first approximation, it can be assumed that the luminous flux determines the force of a constant value, which is directed collinearly to a straight line passing through the two main bodies of the Sun-Earth-spacecraft system in a restricted circular three-body problem.

Then amendments to control the orientation by the changing in the angular position or shape of the sail can be taken into account.

The particular interest is the case of placing the spacecraft in the vicinity of the Euler–Lagrange libration points where a small disturbing force determine the motion character and stability. Optimal control theory leads to complicated formulations of the problem for the solving of additional equations of mathematical models that can use the Pontryagin maximum principle or Bellman equation [15, 18] for different cases and tasks. There are analytical and numerical methods of research and analysis of the basic properties of the equations that allow to obtain exact or approximate solution of set of the necessary conditions of the extremum of the quality functional [16, 17].

Solar sailing in space is a matter of the future. This will require sophisticated design solutions and space technology [7–12, 23–27]. A special spacecraft's control uses solar sail as a motioned forcement of the thruster units.

Then we take into account corrections for attitude control due to changes of the angular or rotation the sail geometry.

Thus, the nonlinear equations of motion will include a permanent disturbance, which can easily be taken into account. Of particular interest is the case of location the spacecraft in the vicinity of the libration points, where small perturbation forces will be determined by stability conditions.

Author details

Vladimir Stepanovich Korolev*, Elena Nikolaevna Polyakhova and
Irina Yurievna Pototskaya
Saint-Petersburg State University, Russia

*Address all correspondence to: vokorol@bk.ru

References

[1] Lebedev PN. Collected Works. Moscow: Nauka; 1963 (in Russian)

[2] Zander FA. Problems of Spaceflights with Jet Propulsion Engines. Interplanetary Travels. Moscow: Oborongiz; 1961 (in Russian)

[3] Polyakhova EN. Space Flight with a Solar Sail: Problems and Perspectives. Moscow: Nauka; 1986 (in Russian)

[4] Polyakhova EN, Koblik VV. Solar Sail – Science Fiction or Space Sailing Reality. Moscow: URSS; 2016 (in Russian)

[5] Polyakhova EN, Korolev VS. Control of the solar sail space vehicle. In: International Conference "Stability and oscillations of nonlinear control systems". Moscow: IPU RAN; 2016. pp. 294-297 (in Russian)

[6] Polyakhova EN, Korolev VS. Problems of spacecraft control by solar sail. In: International Conference "Stability and Oscillations of Nonlinear Control Systems" (Pyatnitskiy's Conference 2016); 2016. DOI: 10.1109/STAB.2016.7541214

[7] Polyakhova EN, Vjuga AA, Titov VB. Orbital Space Flight in Problems with Detailed Solutions and in Numbers: A Tutorial. Moscow: URSS; 2016 (in Russian)

[8] Kirpichnikov SN, Kirpichnikova ES, Polyakhova EN, Shmyrov AS. Planar heliocentric roto-translatory motion of a spacecraft with a solar sail of complex shape. Celestial Mechanics and Dynamical Astronomy. 1995;**63**(3–4): 255-269

[9] Koblik VV, Polyakhova EN, Sokolov LL. Controlled solar sail transfers into near-Sun regions combined with planetary gravity-assist flybys. Celestial Mechanics and Dynamical Astronomy. 2003;**86**(1): 59-80

[10] Koblik VV, Polyakhova EN, Sokolov LL. Solar sail near the Sun: Point-like and extended models of radiation sources. Advances in Space Research. 2011;**48**(11):1717-1739

[11] Clohessy WH, Wiltshire RS. Terminal guidance system for satellite rendezvous. Journal of the Aerospace Sciences. 1960;**27**(9):653-674

[12] Korolev VS, Pototskaya IY. Integration of dynamical systems and stability of solution on a part of the variables. Applied Mathematical Sciences. 2015;**9**(15):721-728. DOI: 10.12988/ams.2015.4121004

[13] Korolev VS, Pototskaya IYu. Problems of stability with respect to a part of variables. In: International Conference on Mechanics - Seventh Polyakhov's Reading; 2015. DOI: 10.1109/POLYAKHOV.2015.7106739

[14] Korolev VS. Determining movement of navigation satellites in view of disturbances. Bulletin of the Saint Petersburg State Institute of Technology. 2004;**10**(3):39-46 (in Russian)

[15] Korolev VS. Problems of spacecraft multi-impulse trajectories modeling, International Conference on Stability and Control Processes in Memory of V. I. Zubov. In: SCP-2015—Proceedings - 7342072; 2015. pp. 91-94

[16] Martyusheva A, Oskina K, Petrov N, Polyakhova E. Solar radiation pressure influence in motion of asteroids, including near-earth objects. In: International Conference on Mechanics - Seventh Polyakhov's Reading. 2015; DOI: 10.1109/POLYAKHOV.2015.7106756

[17] Forward RL. Light-levitated geostationary cylindrical orbits. Journal

of the Astronautical Sciences. 1981; **29**(1):73-80

[18] Novoselov VS, Korolev VS. Control of a Hamiltonian system subject to disturbances. Innovations in Science. 2015;**51**:23-29 (in Russian)

[19] Rumyantsev VV. On the Stability of Stationary Motion of Satellites. Moscow: Computing Center of the USSR Academy of Sciences; 1967. 141p

[20] Rumyantsev VV. On the optimal stabilization of controlled systems. Journal of Applied Mathematics and Mechanics. 1970;**34**(3):440-456

[21] Egorov VA, Pomazanov MV. Solar Sail: The Principles of Design. Driving-Set and Flights to Asteroids. Preprints IPM Keldysh. Moscow: IPM; 1997 (in Russian)

[22] Beletsky VV. Motion of an Artificial Satellite about its Center of Mass. Moscow: Nauka; 1965 (in Russian)

[23] Forward RL. Future Magic: How Today's Science Fiction Will Become Tomorrow's Reality. New York: Avon Books; 1988

[24] Friedman L. Starsailing: Solar Sailing and Interstellar Travel. New York: Wiley Science Editions; 1988

[25] Kulakov F, Alferov G, Efimova P. Methods of remote control over space robots. International Conference on Mechanics - Seventh Polyakhov's Reading; 2015. C. 7106742. DOI: 10.1109/POLYAKHOV.2015.7106742

[26] Mcinnes CR. Solar Sailing: Technology, Dynamics and Mission Applications. Berlin, Germany: Springer-Praxis; 1999

[27] Polyakhova EN, Korolev VS. The solar sail: Current state of the problem. In: AIP Conference Proceedings (8th Polyakhov's Reading: Proceedings of the International Scientific Conference on Mechanic); 2018. C. 040014

Chapter 12

Nonlinear Friction Model for Passive Suspension System Identification and Effectiveness

Ali I. H. Al-Zughaibi

Abstract

To achieve a high level of performance, frictional effects have to be addressed by considering an accurate friction model, such that the resulting model faithfully simulates all observed types of friction behaviour. A nonlinear friction model is developed based on observed measurement results and dynamic system analysis. The model includes a stiction effect, a linear term (viscous friction), a nonlinear term (Coulomb friction) and an extra component at low velocities (Stribeck effect). During acceleration, the magnitude of the frictional force at just beyond zero velocity decreases due to the Stribeck effect, which means the influence of friction reduces from direct contact with bearings and body into the mixed lubrication mode at low velocity. This could be due to lubricant film behaviour. In respect of acceleration and deceleration when the direction changes for the mass body, friction almost depends on this direction, while the static frictional force exhibits springlike characteristics. However, friction is not determined by current velocity alone, it also depends on the history of the relative wheel and body velocities and movements, which are responsible for friction hysteresis behaviour.

Keywords: nonlinear friction model, stiction region, Stribeck effect, viscous friction, passive suspension system model

1. Introduction

Friction occurs almost everywhere. Many things, including human acts, depend on it. It is usually present in machines. Usually, friction is not required, so a great deal is done to reduce it by design or by control. Friction is often quantified by a coefficient of friction (μ), expressing the ratio of the friction force to the applied load [1].

The spearheading work of Amontons, Coulomb and Euler, who attempted to clarify the friction phenomenon regarding the mechanics of relative movement of rough surfaces in contact with one another, is mentioned by [2]. From that point forward, only sporadic consideration has been paid to the vital question of friction as a dynamic process that changes on contact. Instead, the most significant proportion of the investigation has concentrated on describing and evaluating complex mechanisms, such as adhesion and deformation that contribute to development of frictional resistance, while frequently ignoring the dynamic aspects of the issue. Consequently, despite those mechanisms being relatively well researched, characterised and understood, no efficient and comprehensive model has emerged for the evolution of the friction force as a function of the states of the system, namely, time, displacement and velocity. The requirement for such a model is now becoming more urgent, since the consideration of the friction force dynamics proves essential to understanding and control of systems, including rubbing elements, from machines to earthquakes. Therefore, if it were possible to qualify and quantify this friction force dynamic, it would be a relatively simple step to treat the dynamics of a whole system comprising friction; thus, our results are consistent with their findings.

Friction is a very complicated phenomenon arising from the contact of surfaces. Experiments indicate a functional dependence upon a large variety of parameters, including sliding speed, acceleration, critical sliding distance, normal load, surface preparation and, of course, material combination. In many engineering applications, the success of models in predicting experimental results remains strongly sensitive to the friction model.

A fundamental, unresolved question in system simulation remains: what is the most appropriate way to include friction in an analytical or numerical model and what are the implications of the chosen friction model?

From a control point of view, control strategies that attempt to compensate for the effects of friction, without resorting to high gain control loops, inherently require a suitable friction model to predict and compensate for the friction. Even though no exact formula for the friction force is available, friction is commonly described in an empirical model. Nevertheless, for precision/accuracy requirement, a good friction model is also necessary to analyse stability, predict limit cycles, find controller gains, perform simulations, etc. Most existing model-based friction compensation schemes use classical friction models, such as Coulomb and viscous friction. In applications with high-precision positioning, the results are not always satisfactory. Friction is a natural phenomenon that is quite difficult to model and is not yet completely understood [3].

From a friction-type point of view, in fluid- or grease-lubricated mechanisms, friction decreases as the velocity increases away from zero. In general terms, this effect is understood. It is due to the transition from boundary lubrication to fluid lubrication. In boundary lubrication, extremely thin, perhaps monomolecular, layers of boundary lubricants that adhere to the metal surfaces separate metal parts. These lubricant additives are chosen to have low shear strength, so as to reduce friction, proper bonding and a variety of other properties such as stability, corrosion resistance or solubility in the bulk lubricant. Boundary lubricants are standard in greases and oils specified for precision machine applications. With the exception of when lubricants and the friction properties of boundary lubricants are a secondary consideration [4], therefore, this study considers transition friction.

This study found that friction helps to remove a vibration, or oscillation, from mass body displacement as the damping contributes in the test rig. That was unexpected because it always caused the system to deteriorate and friction to be incorporated with the primary target of suspension system performance. Therefore, it is vital to consider friction in this study, and this novel contribution takes into account the friction with the test rig and implements a ¼-car suspension model [5]. In addition, the author hopes to contribute towards a reconsideration of friction with conventional car suspension models.

2. Why considering friction within this study?

In the test rig, a ¼-car, to achieve the primary target of this test rig and the design requirements, the designer had to force the mass body to move in vertical lines. Therefore, a 240 kg mass plate, used to represent a ¼-car body, is organised to move vertically via two linear bearings. Two rails, THK type HSR 35CA, 1000 mm long and parallel to each other, are used with each linear bearing. A double wishbone suspension linkage was chosen because it preserves the geometry of a wheel in an upright position independent of the suspension type used. The wheel hub is connected to the chassis, which is attached to the car body. The test rig passive suspension photograph is shown in **Figure 1**, while the schematic diagram is shown here in **Figure 2**.

Surawattanawan [6] conducted a simulation and experimental study for the same test rig without consideration of the real position for the spring and viscous damper (S and VD); as a result, the friction effects were ignored. However, in the author's opinion, the real inclined position of S and VD should be considered. Accordingly, the test rig design and the input type help to generate a normal force at the body bearings and a vertical force relative to body movement, as will be demonstrated by the free body diagram of test rig later. This force is responsible for generating Coulomb friction at body lubricant bearings. In addition, the mass body has been slipped on lubricant bearings; this will undoubtedly generate viscous friction. Therefore, it is essential to consider these frictions in the current study, qualified by the critical effects of friction in any system, as well as their effects on results.

Figure 1.
Photograph of the passive test rig.

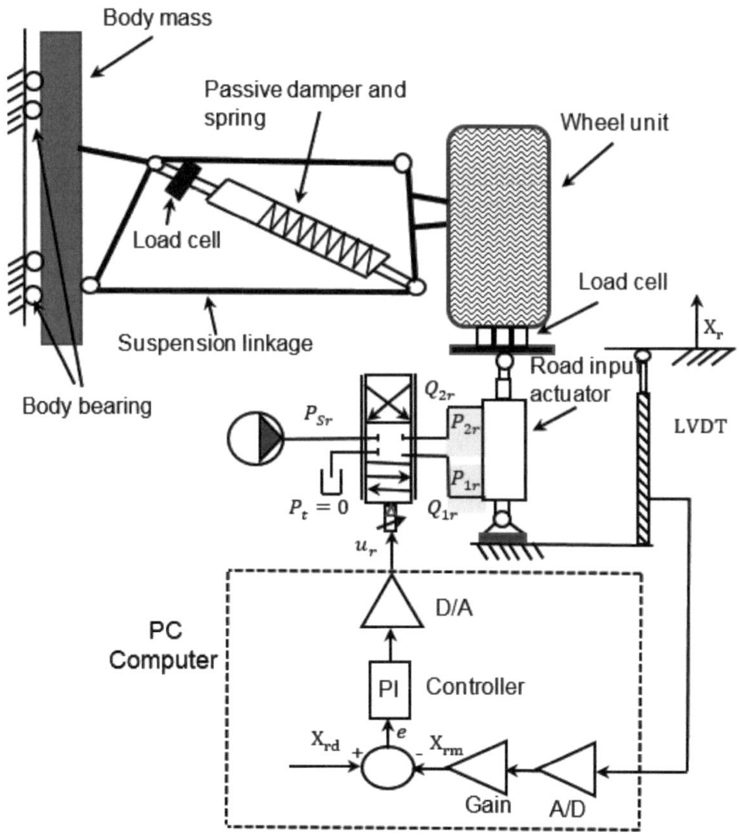

Figure 2.
Schematic diagram of the test rig.

3. Mathematical model of passive suspension system

Vehicle suspensions are designed to minimise car body acceleration $\ddot{X}_b$, within the limitation of the suspension displacement $X_w - X_b$ and tyre deflection $X_r - X_w$. Hence, the vehicle response variables that need to be examined are:

1. Car body acceleration, $\ddot{X}_b$

2. Suspension displacement, $X_w - X_b$

3. Tyre deflection, $X_r - X_w$

Using Newton's second law, the equation of motion for the mass body passive system of ¼-car model is:

$$M_b.\ddot{X}_b = \left[k_s(X_w - X_b) + b_d\left(\dot{X}_w - \dot{X}_b\right)\right] \quad (1)$$

while the dynamic equation of motion for the mass wheel is:

$$M_w.\ddot{X}_w = -\left[k_s(X_w - X_b) + b_d\left(\dot{X}_w - \dot{X}_b\right)\right] + k_t(X_r - X_w) + b_t\left(\dot{X}_r - \dot{X}_w\right) \quad (2)$$

The constant parameters taken from the test rig are as follows:

Car body mass, $M_b = 240$ kg.
Wheel unit mass, $M_w = 40$ kg.
Tyre stiffness, $k_t = 920000$ N/m.
Tyre damping rate, $b_t = 3886$ N/ms^{-1}
Suspension stiffness, $k_s = 28900$ N/m
Suspension damping, $b_d = 260$ N/ms^{-1}

Following completion of the passive suspension experimental work, an attempt to model these experimental tests is by developing a passive suspension model. The simulation was achieved through developing code in C++. An issue arose in that a considerable difference was found between the body displacement observed in experiments and in the simulation results. From this aspect, the idea of considering friction force emerged. There were two clear indicators from observation measurements, which helped to quantify the friction effects; these are discussed in the following sections.

4. The dynamic indicator

From the simulation results, it was found there are definite fluctuations in body displacement, as generally expected from a quarter-car suspension model, regarding our experience and references. Watton [7] mentioned in his book *Modelling, Monitoring and Diagnostic Techniques for Fluid Power Systems* on pages 182–186, regarding the same test rig, there was an oscillation at the car body in both experimental and simulation results, as shown in **Figure 3**.

It is clearly seen that the body displacement oscillates in the current simulation results, without implementing friction forces, as shown in **Figure 4**. There were differences in the periods of oscillation between **Figures 3** and **4**. This is relative to the different models and parameters used.

There were no such fluctuations in the experimental results, as shown, for example, in **Figure 5**.

Figures 4 and **5** display the desired input, without filter, for the road and the responses of the wheel and body for the present simulation and experimental results, respectively. From these figures, it is clearly seen that the wheel displacement follows the road displacement in both experiment and simulation results, while the body travel follows the wheel with a pure delay, which will be shown in

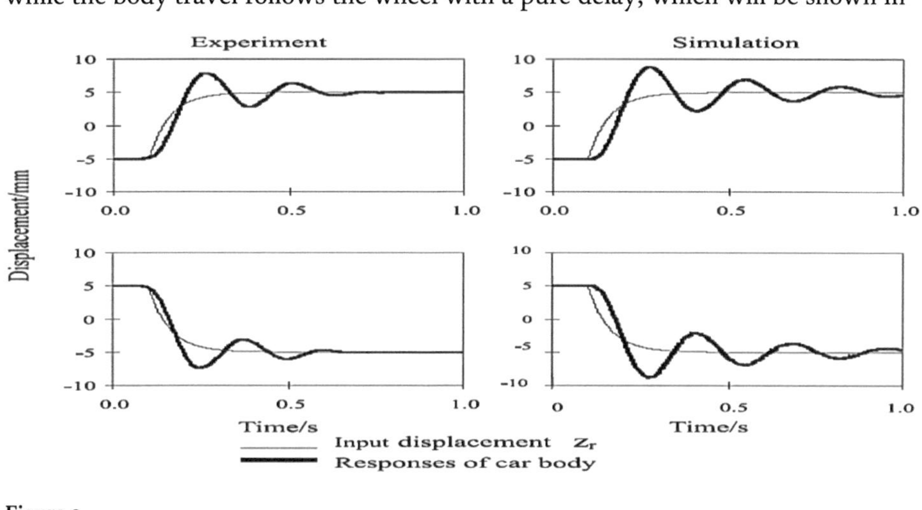

Figure 3.
Typical 1 DOF test result [7].

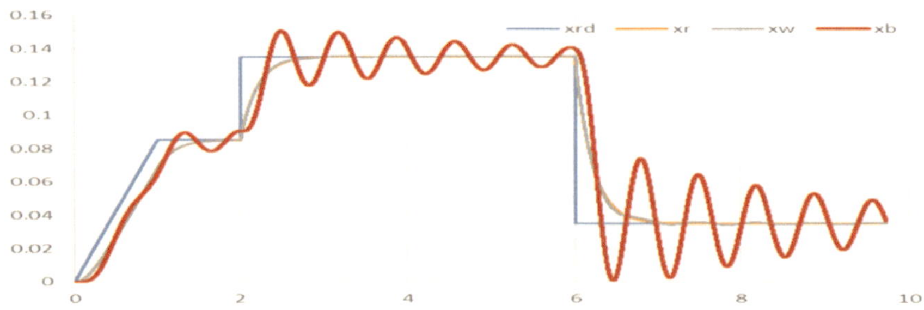

Figure 4.
Simulation results for X_r, X_w and X_b *(m).*

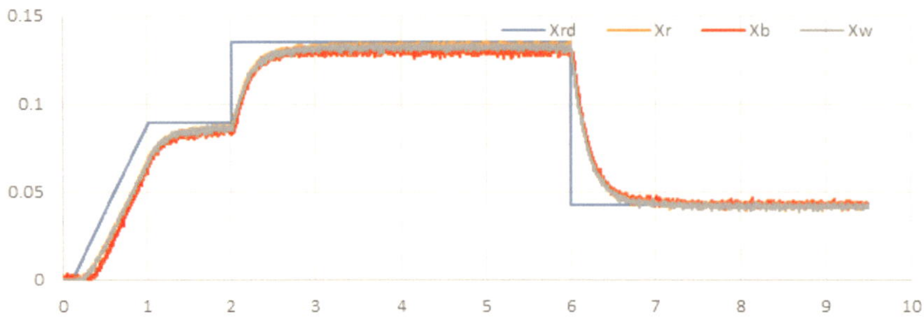

Figure 5.
Experimental results for X_r, X_w and X_b *(m).*

more detail in Section 5, and fluctuates in the simulation results. The author named this disparity 'a dynamic friction indicator'. This name was not unique, having been used by several authors before. From this point of view, it could be said that for the experimental work, the friction forces at body lubricant bearings are responsible for eliminating the oscillation from the body travels.

5. The static indicator

In demonstrating the measured body and wheel movements, a delay is illustrated between them when the wheel rises up or falls; the body similarly travels after pure delay. The early and later stages of the wheel rise and fall, respectively; the results to system input and the body delay are shown in **Figure 6**.

For more convenience, the experimental data of the relative travel between the wheel and body ($X_w - X_b$) was used. These are available from test rig from linear variable differential transformer (LVDT) sensors, and the result is shown in **Figure 7**. The evident noise is attributed to sensor and experimental characteristics. From this figure, it is clearly seen that there is zero difference between X_w and X_b at the start of the test or for a short period, approximately 0.3 s. This is believed to be due to data acquisition delays. The differences gradually increase; while the wheel starts to move up, the differences between X_w and X_b steadily increase until reaching the maximum. During this period the body sticks without movement ($X_b = 0.0$); when the resulting force overcomes the static friction, the body will start to move. The relative travel difference between them slowly reduces, approximately 0.5–1.5 s, until reaching zero or near zero at steady state (SS), after 1.5 s.

This observation, which the author named 'static friction indicator', leads to an investigation of the body stiction. It was found that this could be regarded as the effect of static friction force.

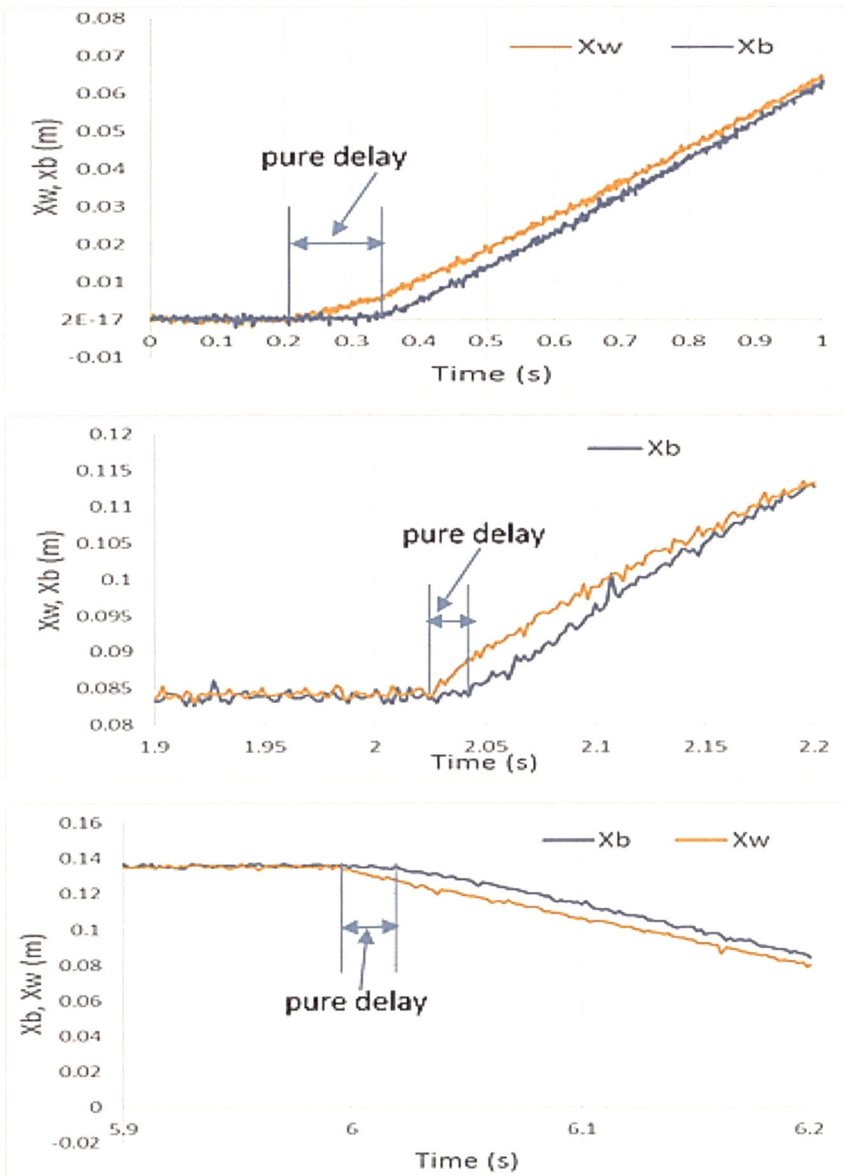

Figure 6.
Measurements of pure delay of X_b *from* X_w *at three positions.*

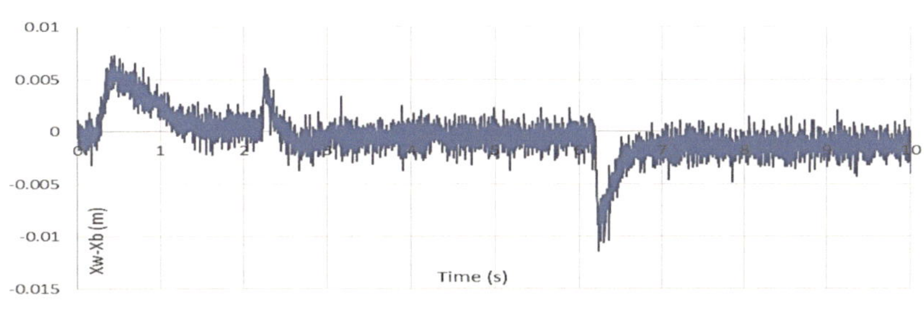

Figure 7.
Experimental results of difference displacements between X_w and X_b.

To include knowledge about friction in the simulation model, consideration of conventional friction was pursued, drawing upon published information. The following section reviews the approach.

6. Conventional friction model

The traditional friction model considered the construction of a more comprehensive friction prediction model that accounts for the various aspects of this particular phenomenon so that mechanical systems with friction can be more accurately identified and, consequently, better controlled. Most of the existing model-based friction compensation schemes use classical friction models, such as Coulomb and viscous friction. In applications with high-precision positioning and with little velocity tracking, the results are not always satisfactory. A better description of the friction phenomena for small speeds, especially when crossing zero velocity, is necessary. Friction is a natural phenomenon that is quite difficult to model and is usually modelled as a discontinuous static map between velocity and friction torque that depends on the velocity's sign. Typical examples are different combinations of Coulomb friction, viscous friction and Stribeck effect, as mentioned in [3, 8, 9]. However, there are several exciting properties observed in systems with friction that cannot be explained by static models. This is necessarily due to the fact that friction does not have an instantaneous response to a change in velocity, i.e. it has internal dynamics. Examples of these dynamic properties [3, 10] are:

- Stick-slip motion, which consists of limit cycle oscillation at low velocities, caused by the fact that friction is more significant at rest than during motion

- Presiding displacement which shows that friction behaves like a spring when the applied force is less than the static friction breakaway force

- Frictional lag which means that there is some hysteresis in the relationship between friction and velocity

The general description of friction is a kind of relation between velocity and friction force, depending on the velocity situations, described in several types of research. For example, Tustin's model consists of Coulomb and viscous friction [11]. The inclusion of the Stribeck effect, with one or more breakpoints, gives a better approximation at low velocities, as shown in **Figure 8**.

Now, in order to start establishing the real bearing friction model, it should involve the dynamic analysis of the test rig as follows:

6.1 How to account for the vertical force

The following explains in detail the main features of the friction model and will begin with how to account for the vertical force that is responsible for generating Coulomb friction by drawing a free body diagram of the test rig.

6.1.1 Free body diagram of the test rig

Figure 9 shows the free body diagram of the test rig; the friction force acts as an internal force in the tangential direction of the contacting surfaces. This force obeys a constitutive equation, such as Coulomb's law, and acts in a direction opposite to the relative velocity. Therefore, the inclination position of S and VD and the system

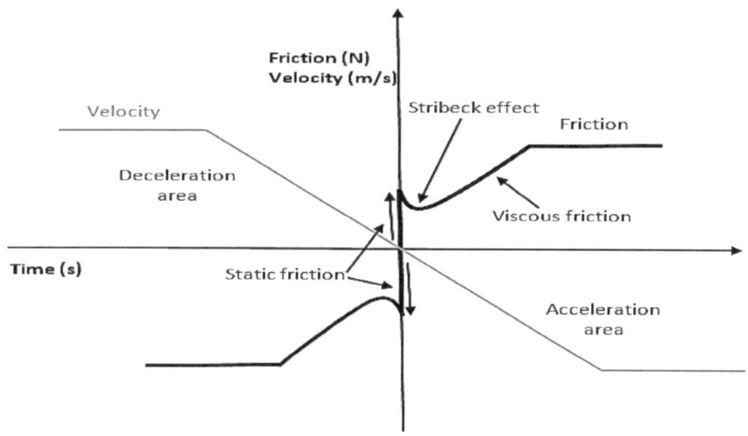

Figure 8.
Conventional friction model [11].

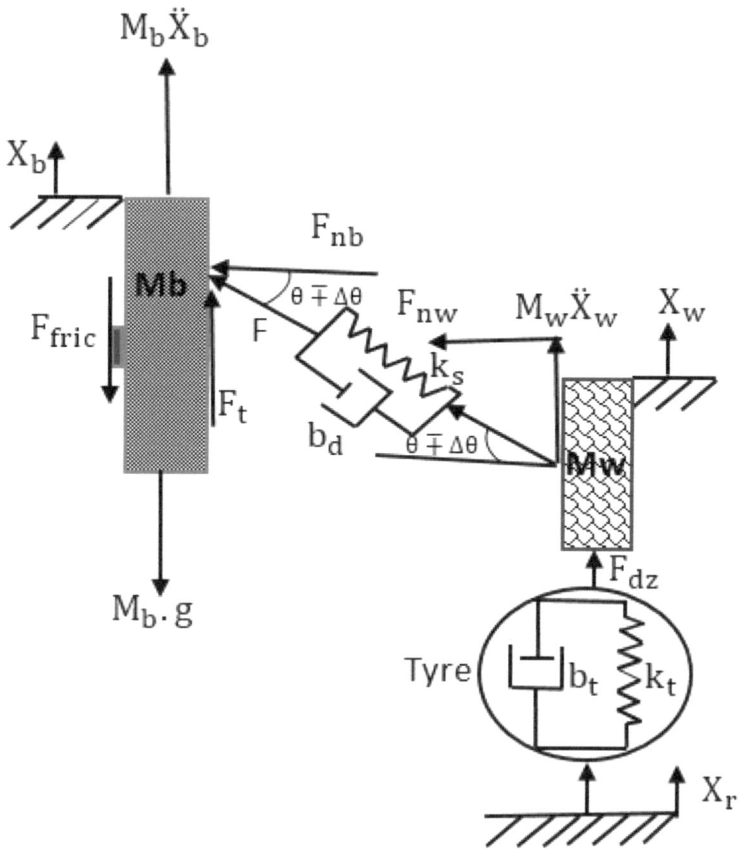

Figure 9.
Free body diagram of the test rig.

type inputs help to generate the kinematic bearings body friction relative to this normal force component. From **Figure 9**, the following analysis should be conducted to account for this friction force:

$$F = k_s(X_w - X_b) + b_d(\dot{X}_w - \dot{X}_b)/\sin(\theta \mp \Delta\theta) \tag{3}$$

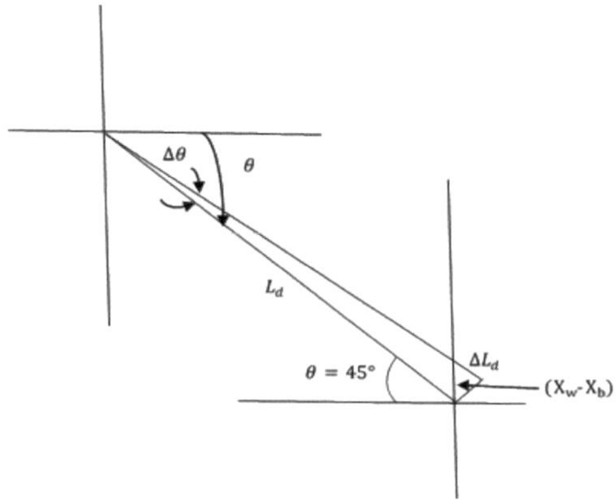

Figure 10.
Engineering geometry of passive units.

$$F_{nb} = F\cos(\theta \mp \Delta\theta) \tag{4}$$

$$F_{nb} = k_s(X_w - X_b) + b_d(\dot{X}_w - \dot{X}_b)/\tan(\theta \mp \Delta\theta) \tag{5}$$

$$F_{fricC} = \mu F_{nb} \tag{6}$$

where F_{fricC} is Coulomb friction, μ is the friction coefficient, F_{nb} is the body normal force component and F is the spring and damping forces.

6.1.2 Dynamic linkage angle expression

The construction linkage angle is dynamically changed by $\mp\Delta\theta$.

From engineering geometry of passive units, as shown in **Figure 10**, it can be found that

$$\frac{L_d - \Delta L_d}{\sin(90 - \theta)} = \frac{X_w - X_b}{\sin(\Delta\theta)}, \theta = 45^\circ \tag{7}$$

$$\sin(\theta) = \frac{\Delta L_d}{X_w - X_b} \tag{8}$$

$\Delta L_d = (X_w - X_b)\sin(\theta)$, where ΔL_d is the dynamic change in S and VD length. Then,

$$\frac{L_d - (X_w - X_b)\sin(\theta)}{\sin(\theta)} = \frac{X_w - X_b}{\sin(\Delta\theta)} \tag{9}$$

$$\sin\Delta\theta = \frac{(X_w - X_b)\sin(\theta)}{L_d - (X_w - X_b)\sin(\theta)} \tag{10}$$

$$\Delta\theta = \sin^{-1}\left[\frac{(X_w - X_b)\sin(\theta)}{L_d - (X_w - X_b)\sin(\theta)}\right] \tag{11}$$

7. Nonlinear friction model

To achieve a high level of performance, frictional effects have to be addressed by considering an accurate friction model, such that the resulting model faithfully simulates all observed types of friction behaviour.

A nonlinear friction model is developed based on observed measurement results and dynamic system analysis. The model includes a stiction effect, a linear term (viscous friction), a nonlinear term (Coulomb friction) and an extra component at low velocities (Stribeck effect). During acceleration, the magnitude of the frictional force at just beyond zero velocity decreases due to the Stribeck effect, which means the influence of friction reduces from direct contact with bearings and body into the mixed lubrication mode at low velocity. This could be due to lubricant film behaviour.

In respect of acceleration and deceleration when the direction changes for the mass body, friction almost depends on this direction, while the static frictional force exhibits springlike characteristics. However, friction is not determined by current velocity alone, it also depends on the history of the relative wheel and body velocities and movements, which are responsible for friction hysteresis behaviour.

This model, which has now become well established, has provided a more satisfactory explanation of observed dynamic fluctuations of body mass. It will be attempted to heuristically 'fit' a dynamic model to experimentally observed results. The resulting model is not only reasonably valid for the ¼-car test rig behaviour but is also reasonably suitable for most general friction lubricant cases.

The model simulates the symmetric hysteresis loops observed in the bearings' body undergoing small amplitude ramp and step forcing inputs. As might be expected, they are capable of reproducing the more sophisticated pre-sliding behaviour in particular hysteresis. The influence of hysteresis phenomena on the dynamic response of machine elements with moving parts is not yet thoroughly examined in the literature. In other fields of engineering, where hysteretic phenomena manifest themselves, more research has been conducted. In Ref. [12], for example, adaptive modelling techniques were proposed for dynamic systems with hysteretic elements. The methods are general, but no insight into the influence of the hysteresis on the dynamics is given. Furthermore, no experimental verification is provided. Altpeter [13] made a simplified analysis of the dynamic behaviour of the moving parts of a machine tool where hysteretic friction was present.

In this study, the friction model, despite its simplicity, can simulate all experimentally observed properties and facets of low-velocity friction force dynamics. Because of the test rig schematic and the system input signal, with historic travel, there are three circumstances depending on whether the body velocity is accelerating or decelerating. Firstly, the velocity values start from zero and just after velocity reversals, reach the highest and are restored to zero, or close to zero at SS. Secondly, the velocity starts from SS with a sharper increase than in the first stage and will extend to peak before returning to zero or near to zero at second SS. Thirdly, it starts from the second SS and after velocity reversals will touch a maximum value, twice rather than at case two, and go back down at a third SS. In the all these velocity cases, the velocity behaviours will make friction hysteretic loops, possibly because of increases in body velocity differing from decreases. The historical action of relative travels of wheel and body contributes to friction hysteresis.

In general, this friction model considers the static, stiction region and dynamic friction, which consists of the Stribeck effect, viscous friction and Coulomb friction. The mathematical model and summary for each part will be demonstrated in the next step.

7.1 Mathematical friction model

The mathematical expression for establishing the friction model gave the constituent terms described in order to accurately represent the observed phenomena, as shown in Eq. (12).

$$F_{fric} = \begin{cases} k_s(X_w - X_b) + b_d(\dot{X}_w - \dot{X}_b) & \dot{X}_b = 0.0 \\ C_e e^{(|\dot{X}_b|/e1)} + \left[\dfrac{\mu(k_s(X_w - X_b) + b_d(\dot{X}_w - \dot{X}_b))}{\tan(\theta \mp \Delta\theta)}\right] + \sigma_v \dot{X}_b & \dot{X}_b > 0.0 \\ -C_e e^{(|\dot{X}_b|/e1)} + \left[\dfrac{\mu(k_s(X_w - X_b) + b_d(\dot{X}_w - \dot{X}_b))}{\tan(\theta \mp \Delta\theta)}\right] + \sigma_v \dot{X}_b & \dot{X}_b < 0.0 \end{cases} \tag{12}$$

Eq. (12) shows the friction model, which includes the two main parts of friction: static when, $\dot{X}_b = 0.0$, and dynamic, when $\dot{X}_b > 0.0$. The latter is presented by two expressions, depending on the velocity direction, and is discussed in detail later. In static friction, the stiction area is solely dependent on the velocity because the body velocity should be close to zero velocity or frequently just beyond zero velocity. The static model is accounted by the force balance of the test rig when the body was motionless, while the wheel was moved and describes the static friction sufficiently accurately. However, a dynamic model is necessary which introduces an extra state which can be regarded as transition and Coulomb and viscous friction. In addition to these friction models, steady physics state is also briefly discussed in this study.

7.2 Static friction model

After a test starts, the wheel begins to move respective to the road inputs, and initially the body remains motionless. This results from the static bearing friction and is undoubtedly a stick region body, $X_b = 0.0$. This friction component can be considered via the test rig vertical force balance $\sum F_v = 0.0$.

For the test rig, the following conventional model represents a ¼-car without considering body friction as aforementioned by Eq. (1), the first reported implementation of friction forces within Newton's second law for a ¼-car model [14], which leads to a new dynamic equation of motion for the mass body:

$$M_b.\ddot{X}_b = \left[k_s(X_w - X_b) + b_d(\dot{X}_w - \dot{X}_b)\right] - F_{fric} \tag{13}$$

As described in the short period where the body remains motionless $X_b = 0.0$ and $\ddot{X}_b = 0.0$, Eq. (13) becomes

$$0.0 = \left[k_s(X_w - X_b) + b_d(\dot{X}_w - \dot{X}_b)\right] - F_{fricS} \tag{14}$$

then

$$F_{fricS} = \left[k_s(X_w - X_b) + b_d(\dot{X}_w - \dot{X}_b)\right] \tag{15}$$

where F_{fricS} is the static friction, which is a function of the relative displacements and relative velocities between the wheel and body multiplied by spring stiffness and viscous damper coefficients, with direction totally dependent on the next stage $\dot{X}_b$ direction. This is considered as pre-sliding displacement, which exhibits how friction characteristics behave like a spring when the applied force is less than the static friction breakaway force. From the experimental work, amplitude input = 50 mm, it was found that the maximum stick friction force occasionally occurs at $(X_w - X_b) \leq 0.0069$ and $X_b \cong 0.0$.

7.3 Dynamic friction model

Earlier studies (see, e.g. [8, 10, 15]) have shown that a friction model involving dynamics is necessary to describe the friction phenomena accurately. A dynamic model describing the springlike behaviour during stiction was proposed by [16]. The Dahl model is essentially Coulomb friction with a lag in the change of friction force when the direction of motion is changed. The model has many commendable features and is theoretically well understood. Questions, such as the existence and uniqueness of solutions and hysteresis effects, were studied in an interesting paper by [17]. The Dahl model does not include the Stribeck effect. An attempt to incorporate this into the Dahl model was made by [18] where the authors introduced a second-order Dahl model using linear space-invariant descriptions. The Stribeck effect in this model is only transient; however, following a velocity reversal, it is not present in the steady-state friction characteristics. The Dahl model has been used for adaptive friction compensation [19, 20], with improved performance as a result. There are also other models for dynamic friction; Armstrong-Helouvry [8] proposed a seven-parameter model. This model does not combine the different friction phenomena but is, in fact, one model for stiction and another for sliding friction. Another dynamic model suggested by [21] had been used in connection with control by [15]. This model is not defined at zero velocity.

In this study, it was proposed that a nonlinear dynamic friction model combines the transition behaviour from stiction to the slide regime including the Stribeck effect, the Coulomb friction with consideration of the normal dynamic force at body bearings with suitable friction coefficient and the viscous friction dependent on the body velocity and appropriate viscous coefficient. This model involves arbitrary steady-state friction characteristics. The most crucial results of this model are to highlight precisely the hysteresis behaviours of friction relative to body velocity behaviour.

Referring to Eq. (12), there are two forms of dynamic friction, depending on the body velocity direction; it will be shown in detail as follows:

For $\dot{X}_b > 0.0$ the dynamic friction form is

$$F_{fricD} = \left\{ C_e e^{\left(|\dot{X}_b|/e1\right)} + \left[\frac{\mu\left(k_s(X_w - X_b) + b_d\left(\dot{X}_w - \dot{X}_b\right)\right)}{\tan\left(\theta \mp \Delta\theta\right)} \right] + \sigma_v \dot{X}_b \right\} \tag{16}$$

From Eq. (16), it is clearly seen that dynamic friction consists of three parts. A summary is given for each: part one form is

$$F_{fricT} = C_e e^{\left(|\dot{X}_b|/e1\right)} \tag{17}$$

where F_{fricT} is transition friction, C_e is attracting parameter, e1 is the curvature degree and the absolute body velocity value meaning the direction of velocity is not affected. The transition friction has exponential behaviour with degrees identified experimentally and completely agrees with the literature review of most research studies regarding lubricant friction, which begins from the maximum value at the sticky region and quickly dips when the body just begins to move, or the body velocity is increased.

Secondly, F_{fricC} represents Coulomb friction, which is equal to the normal bearing force times the friction coefficient (μ), as follows:

$$F_{fricC} = \left\{ \frac{\mu\left(k_s(X_w - X_b) + b_d\left(\dot{X}_w - \dot{X}_b\right)\right)}{\tan\left(\theta \mp \Delta\theta\right)} \right\} \tag{18}$$

where F_{fricC} is Coulomb friction with the opposite sign to velocity direction.

Finally, F_{fricV} represents viscous friction, which, because there is a lubricant contact between bearing and body, is counted by multiplying the body velocity with an appropriate viscous coefficient (σ_v).

$$F_{fricV} = \sigma_v \dot{X}_b \tag{19}$$

when $\dot{X}_b < 0.0$, the overall dynamic friction expression becomes

$$F_{fricD} = \left\{ -C_e e^{\left(|\dot{X}_b|/e1\right)} + \left[\frac{\mu\left(k_s(X_w - X_b) + b_d\left(\dot{X}_w - \dot{X}_b\right)\right)}{\tan\left(\theta \mp \Delta\theta\right)} \right] + \sigma_v \dot{X}_b \right\} \tag{20}$$

Eq. (20) is similar to Eq. (16) as they have the same three terms but with a negative sign added in just for the transition friction term. This is because these values will describe the development friction in the opposite direction in the negative friction region.

The underlying motivation is that when the dynamic behaviour of the ¼-car model is thoroughly understood, the knowledge can be used to design appropriate feedback controllers for active suspension systems with compensation for the friction forces.

7.4 Steady-state friction

It is vital to consider the friction behaviour within SS period. From **Figure 11** of body displacement as function of time, it is clear that the historical movement demeanour, which starts to move from the stiction region, $X_b = 0.0$ and $\dot{X}_b \cong 0.0$, is the first SS, stage (A), and then reaches the second SS, stage (B), at the mid-point of the road hydraulic actuator $X_b = 0.085$ m and $\dot{X}_b \cong 0.0$. Secondly, the body starts moving from the second SS and will reach the highest with a total amplitude $X_b = 0.135$ m and $\dot{X}_b \cong 0.0$ at the third SS, stage (C). Finally, it will start to move from the third SS stage and reach the lowest value of amplitude $X_b = 0.035$ m, travelling twice the distance compared with the second stage. Thus, it will finally achieve the four SS (D) at $X_b = 0.035$ m and $\dot{X}_b \cong 0.0$.

At body stiction and SS station, $\ddot{X}_b$ is equal to zero. Therefore, the friction at steady state should be similar to static friction as mentioned in Section 7.2.

7.5 Simple friction model

Eq. (12) gives a general form for nonlinear friction occurring at the body supported lubricant bearings. This model could be studied from a different point of

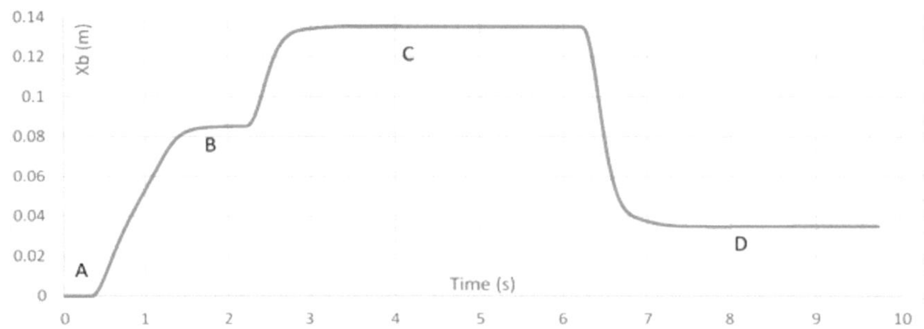

Figure 11.
Body displacement (X_b) with time.

view, whereby it can be returned to two dominant parameters, the body velocity and the normal body force, that could be termed damping friction relative to the body velocity and Coulomb friction qualified to normal body force.

For simplicity, even though the friction model, Eq. (12), reflected most of the observations measured using the system dynamics analysis and was used with the passive suspension model, it can still be employed in simple form through overlooking Coulomb friction. Therefore, the simple expression of friction without Coulomb friction is

$$F_{fric} = \begin{cases} k_s(X_w - X_b) + b_d(\dot{X}_w - \dot{X}_b)\dot{X}_b = 0.0 \\ C_e e^{(|\dot{X}_b|/e1)} + \sigma_v \dot{X}_b \dot{X}_b > 0.0 \\ -C_e e^{(|\dot{X}_b|/e1)} + \sigma_v \dot{X}_b \dot{X}_b < 0.0 \end{cases} \quad (21)$$

In Eq. (21), this model has the same three various forms dependent on $\dot{X}_b$, value and direction. Part one is the static friction, which has precisely the same shape for general friction, while the dynamic formula, damping friction, depending only on the body velocity in a different form by ignoring the Coulomb term. The interesting point is that, by implementing these simple friction forms, the simulation results also acquire a good agreement in comparison with the experimental results regarding system response parameters, which encouraged its use with the active suspension system. The question arises as to which one is more suitable for our case. Although the general friction model system, Eq. (12), gives more detail, depending on the system dynamics, and has the ability to highlight the hysteresis phenomenon that should occur with this system type, the simple friction model has lost this hysteresis.

However, the simple form also provides a real accord between the experimental and simulation results for system response, with little variation relative to that gained from considering general friction. From this point of view, a mathematical analysis is used, by using the residual mean square (RMS).

The RMS is defined as 'a measure of the difference between data and a model of that data'. Therefore, two measured signals, X_b and $X_w - X_b$, will be used to show the accuracy of considering the general or simple friction forms.

RMS accounts for the measurement and simulation with and without Coulomb friction for relative movements between the wheel and body, as illustrated:

$$(RMS)c = \sqrt{\frac{1}{N}\sum((X_w - X_b)_m - (X_w - X_b)_{Sc})^2} \quad (22)$$

and

$$(RMS) = \sqrt{\frac{1}{N}\sum((X_w - X_b)_m - (X_w - X_b)_S)^2} \quad (23)$$

where (RMS)c and (RMS) are the RMS between the measured and simulation values with and without considering Coulomb friction, respectively, $(X_w - X_b)_m$ is the measured relative displacement, $(X_w - X_b)_{Sc}$ and $(X_w - X_b)_S$ are the simulation data with and without implementing Coulomb friction and N is the total number of sample. The RMS results are shown in **Table 1**.

From **Table 1**, the RMS results show that using the friction model considering Coulomb friction is more accurate.

Signal	$(RMS)c$	(RMS)
$(X_w - X_b)$	0.006362	0.006366
X_b	0.096267	0.096386

Table 1.
RMS results.

8. Results

8.1 Friction results for general form (considering Coulomb friction)

Figure 12 shows friction force as a function of body velocity for the input force when amplitude = 50 mm, while the other cases when amplitude is = 30 or 70 mm. Accordingly, with the same friction behaviour, the same friction model can be used. It is apparent that the friction behaves as a hysteresis loop. Therefore, both sets of curves form a circle, enclosing a nonzero area, which is typical of dynamic friction besides the starting static friction. The loops enclosed three areas relating to velocity increases, decreases and directions. This is similar to expectations from the results of a dynamic friction model discussed in Sections 8.1 and 8.3. The upper portion of the curve shows the behaviour for increasing velocity when $\dot{X}_b > 0.0$ in two circumstances, while the lower portion shows the behaviour for decreasing velocity when $\dot{X}_b < 0.0$. This phenomenon may be a consequence of the dynamics of the process rather than of the nonlinearity; this phenomenon is often referred to as hysteresis. The hysteretic friction is, moreover, not a unique function of the velocity, but depends on the previous hysteresis of the movements.

In fact, there are two urgent situations that should be highlighted: the first is when the velocity equals zero, the body is motionless, and the friction values are similar to static friction values, as discussed in Section 7.2, while the second important situation is when the values of friction are within the SS situation, which has already been specified in the previous analysis in Section 7.4.

However, **Figure 12** shows the behaviour of friction relative to the body velocity. It is evident that the reaction in the stick region, or static friction at $X_b=0.0$, friction values start from zero and reach a maximum at the breakaway threshold force. From the experimental test, the breakaway force at the maximum relative displacement between X_w and X_b and the corresponding values for wheel and body velocity, accounted by Eq. (15), could be estimated. As a result, it was found to be equal to 193.8 N. Therefore, after the first positive position of static friction, because the direction of displacement moves up, whenever the body starts to move,

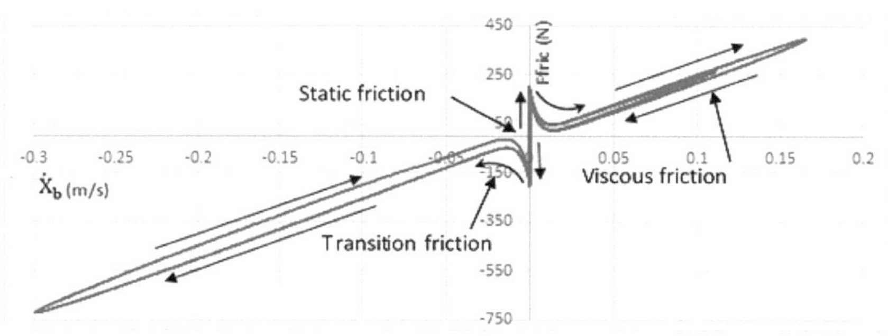

Figure 12.
Friction as function of the body velocity.

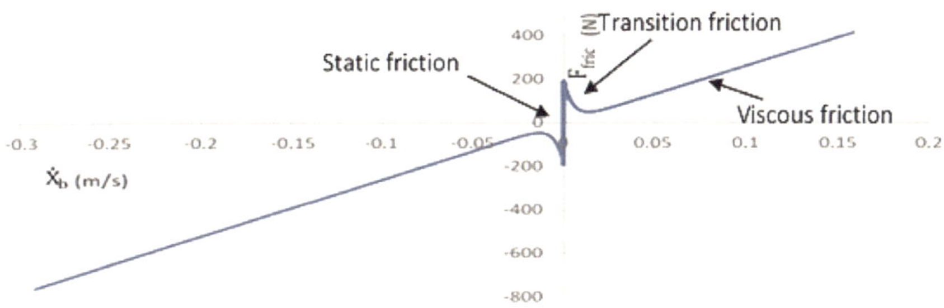

Figure 13.
Damping friction as a function of the body velocity.

when $X_b > 0.0$, the friction hardly dips relative to the transition area from direct contact between body and bearings to mixed hydraulic contact. This clearly shows the Stribeck effects relative to hydraulic layer behaviour: a squeeze-film effect. Following the system inputs and velocity value when $\dot{X}_b > 0.0$, the friction firstly draws a small, enclosed, positive cycle. After that, the body velocity returns to the second SS and increases to reach a maximum value before returning to the third SS with friction drawing a larger enclosed cycle in a positive direction. When $\dot{X}_b < 0.0$, the static values are equal to those for $\dot{X}_b > 0.0$ in the opposite direction, while the friction draws the most massive enclosed nonzero cycle with a value twice that of the larger enclosed cycle in the positive direction. This is because of the friction value and guidance following the road input and velocity values.

8.2 Friction results for simple form (without Coulomb friction)

In considering friction, while disregarding the Coulomb effects relative to the vertical force from the force inputs and the construction of the test rig, the inclination of the spring and damper from one side and the distance between the wheel unit and body mass from another side allows a promotion friction formula, damping friction, to be obtained. Although some features of friction characteristics, the hysteresis behaviour, will have been lost in considering this friction model with the passive suspension system design, success also has been achieved close to the experimental data. **Figure 13** shows the damping friction as a function of the body velocity when amplitude = 50 mm. It is approved that there is no hysteresis performance.

Meanwhile, **Figure 14** illustrates the association between damping and Coulomb friction. Although the damping friction is dominant, it remains vital to reflect the Coulomb friction in the general friction model, because it is responsible for bringing hysteresis performance to the model, and, as mentioned, this is quite essential to our system type.

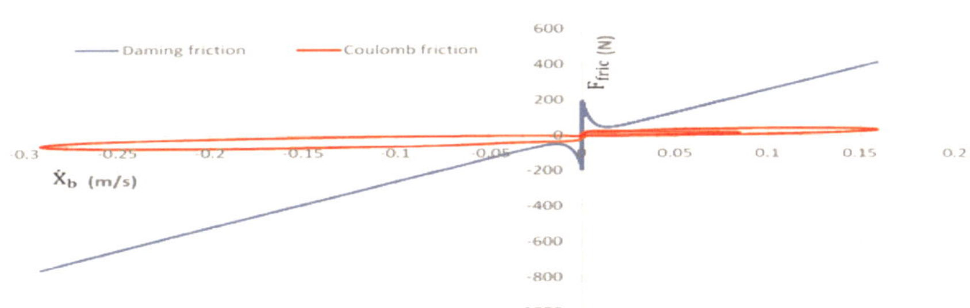

Figure 14.
Damping and Coulomb friction as a function of the body velocity.

9. Discussion

This chapter was set up to question the aspects of friction that merit inclusion with the ¼ car model. After a brief stating of the general frictional considerations, this discussion will review and summarise the findings.

Friction is a highly complex phenomenon, evolving at the contact of surfaces. Experiments demonstrate a functional addiction upon a significant change in parameters, including sliding speed, acceleration, critical sliding distance, normal load, surface preparation and material combination. In many engineering applications, the success of models in predicting experimental results remains strongly sensitive to the friction model. Friction is a natural phenomenon that is quite difficult to model and is not yet completely understood.

The investigation of the principal questions to inform the simulation framework were tested as follow: what is the most suitable technique for including friction in an analytical or numerical model, and what are the inferences of friction model superiority? The constituent elements are discussed in turn as follows:

9.1 The main reasons for considering friction

In this study, as shown in Section 3, considering and implementing the friction model within the equation of motion for the mass body is qualified for the following reasons:

1. Friction itself is crucial to find in any mechanical system. Friction exists everywhere, since degradation, precision, monitoring and control system are strongly affected by friction.

2. From the experiment test, it is clearly seen that there is no oscillation of mass body travels, while that was found with simulation model results. Therefore, a new term should be considered to overcome the issue, that is to say, a friction term.

3. In addition, from experimental measurements in Section 4.4, it is apparent that at the start of the test, while the wheel began to move in relation to road inputs, the body remained motionless for a period.

10. Conventional friction model

The majority of current model-based friction compensation schemes utilise classical friction models, such as Coulomb and viscous friction. In applications with high-precision positioning and with low-velocity following, the outcomes are not generally acceptable. Typical types are different combinations of Coulomb friction, viscous friction and the Stribeck effect, as has been mentioned in several researchers' works as shown in Section 5.

In this review, the established friction model, irrespective of its extreme effortlessness, can recreate all, that we are aware of, conditionally watched properties and features of low-velocity friction force dynamics. Considering the test rig schematic and the force information, there are three conditions, depending upon whether the body speed is speeding up or decelerating. Firstly, the velocity qualities start from zero, and soon after, velocity reversals reach the most elevated level and are maintained at zero, or close to zero, at SS. Secondly, the velocity begins from SS with a sharper increment than in the first stage and will be stretched to the ultimate before it returns to zero, or near to zero, at SS. Thirdly, it will begin from SS and, after velocity reversals, will reach the highest estimate, twice the time as for case two, and spine to SS. In every one of these velocity cases, the velocity behaviour will make friction hysteretic loops that could account for the increments of body speed in a variety of paths from reductions.

In general, this friction model deliberates the static, stiction region and dynamic friction, which consists of the Stribeck effect, viscous friction and Coulomb friction, which rely on the dynamic tangential force, which evolves in the test rig contact bearings. Therefore, there are general and simple friction forms as follows:

10.1 Mathematical friction model

$$F_{fric} = \begin{cases} k_s(X_w - X_b) + b_d(\dot{X}_w - \dot{X}_b) & \dot{X}_b = 0.0 \\ C_e e^{(|\dot{X}_b|/e1)} + \left[\dfrac{\mu\left(k_s(X_w - X_b) + b_d(\dot{X}_w - \dot{X}_b)\right)}{\tan(\theta \mp \Delta\theta)}\right] + \sigma_v \dot{X}_b & \dot{X}_b > 0.0 \\ -C_e e^{(|\dot{X}_b|/e1)} + \left[\dfrac{\mu\left(k_s(X_w - X_b) + b_d(\dot{X}_w - \dot{X}_b)\right)}{\tan(\theta \mp \Delta\theta)}\right] + \sigma_v \dot{X}_b & \dot{X}_b < 0.0 \end{cases} \tag{24}$$

This mathematical eq. (24) incorporates two primary parts of friction: static and dynamic friction. The latter as two expressions and is influenced by the velocity track. In static friction, the stiction area is exclusively not subject to the velocity because the body velocity should be close to zero velocity or just beyond zero velocity. Frequently, the static models are numbered by the strength adjustment of the test rig when the body sticks, while the wheel is moved and depicts the static friction sufficiently precisely. A dynamic model is vital to present an additional state, which can be viewed as the transition, Coulomb and viscous friction. In addition to these friction models, steady physics state is also briefly discussed in this study.

10.2 Simple friction model

When be ignored the Coulomb friction, the previous nonlinear friction model shown in Eq. (12) becomes a simple model, as illustrated in Eq. (21), despite losing some features of friction characteristics with this model, in comparison with experimental data that also obtained close results. From this aspect, another approach should be found to discover which approach obtains more accurate results by comparing with measured results. By using RMS mathematical analysis, the results shown in **Table 1** prove, as an outcome, that the friction model is more accurate with a consideration of Coulomb friction.

11. Conclusion

An accurate nonlinear dynamic model for friction has been presented. The model is simple yet captures most friction phenomena that are of interest for simulated test results. The low-velocity friction characteristics are particularly important for high-performance pointing and tracking. The model can describe arbitrary steady-state friction characteristics. It supports hysteretic behaviour due to frictional lag and springlike behaviour in stiction and gives a different breakaway force depending on the rate of change of the applied force. All these phenomena are unified into static, steady-state and dynamic friction equations. The model can be readily used in simulations of systems with friction.

It is essential to consider friction in this study, in the hope that the study creates an opening and contributes towards a reconsideration of the role of friction using the current quarter in half- and full-car suspension models.

Simulation leads to the same conclusion as proven by the experimental results obtained from the test rig test. Comparison between experimental and simulation results show that the proposed general friction model is more accurate than the conventional models (simple model).

Acknowledgements

I highly appreciate my original university, Kerbala University, and Cardiff University for giving me a chance to do this work. At the same time, I thank the technician people in the laboratory for helping.

Author details

Ali I. H. Al-Zughaibi
Engineering College, Kerbala University, Karbala, Iraq

*Address all correspondence to: ali.i@uokerbala.edu.iq

References

[1] Nichols S. MANE 6960 Friction & Wear of Materials; 2007

[2] Al-Bender F et al. A novel generic model at asperity level for dry friction force dynamics. Tribology Letters. 2004;**16**(1):81-93. DOI: 10.1023/B:TRIL.0000009718.60501.74

[3] De Wit CC et al. A new model for control of systems with friction. IEEE Transactions on Automatic Control. 1995;**40**(3):419-425. DOI: 10.1109/9.376053

[4] Rabinowicz E. Friction and wear of materials. Journal of Applied Mechanics. 1965;**33**(1966):479. DOI: 10.1115/1.3625110

[5] Al-Zughaibi AIH. Experimental and analytical investigations of friction at lubricant bearings in passive suspension systems. An International Journal of Nonlinear Dynamics and Chaos in Engineering Systems. 2018;**94**(2):1227-1242. DOI: 10.1007/s11071-018-4420-x (Open Access)

[6] Surawattanawan P. The influence of hydraulic system dynamics on the behaviour of a vehicle active suspension [thesis]. Cardiff, UK: Cardiff University; 2000

[7] Watton J. Chapter 3: Modelling, Monitoring and Diagnostic Techniques for Fluid Power Systems. London: Springer Science & Business Media; 2005. pp. 182-186. ISBN-13: 9781846283734

[8] Armstrong-Helouvry B. Control of Machines with Friction. Vol. 128. New York: Springer Science & Business Media; 2012

[9] Lischinsky P et al. Friction compensation for an industrial hydraulic robot. IEEE Control Systems Magazine. 1999;**19**(1):25-32. DOI: 10.1109/37.745763

[10] Armstrong-Hélouvry B et al. A survey of models, analysis tools and compensation methods for the control of machines with friction. Automatica. 1994;**30**(7):1083-1138. DOI: 10.1016/0005-1098(94)90209-7

[11] Tsurata K et al., editors. Genetic algorithm (GA) based modelling of nonlinear behaviour of friction of a rolling ball guide way. In: Proceedings 6th International Workshop on Advanced Motion Control; 30 March-1 April 2000. Nagoya, Japan: IEEE; 2002

[12] Smyth AW et al. Development of adaptive modelling techniques for non-linear hysteretic systems. International Journal of Non-Linear Mechanics. 2002;**37**(8):1435-1451. DOI: 10.1016/S0020-7462(02)00031-8

[13] Altpeter F. Friction modelling, identification and compensation. École Polytechnique FÉdÉrale de Lausanne; 1999. DOI: 10.5075/epfl-thesis-1988

[14] Al-Zughaibi A et al. A new insight into modelling passive suspension real test rig system, quarter race car, with considering nonlinear friction forces; ImechE, part D. Journal of Automobile Engineering. 2018;**233**(8):2257-2266. DOI: 10.1177/0954407018764942

[15] Dupont PE. Avoiding stick-slip through PD control. IEEE Transactions on Automatic Control. 1994;**39**(5):1094-1097. DOI: 10.1109/9.284901

[16] Dahl PR. A solid friction Model. No. TOR-0158 (3107-18)-1. Segundo, CA: Aerospace Corp El; 1968

[17] Bliman P-A. Mathematical study of the Dahl's friction model. European Journal of Mechanics. A, Solids. 1992;**11**(6):835-848 ISSNs: 0997-7538

[18] Bliman P, Sorine M. Friction modelling by hysteresis operators. Application to Dahl, stiction and Stribeck effects. In: Proc. Conf. on Models of Hysteresis; Trento. 1991. p. 10

[19] Walrath CD. Adaptive bearing friction compensation based on recent knowledge of dynamic friction. Automatica. 1984;**20**(6):717-727. DOI: 10.1016/0005-1098(84)90081-5

[20] Leonard NE, Krishnaprasad PS, editors. Adaptive friction compensation for bi-directional low-velocity position tracking. In: Proceedings of the 31st IEEE Conference on Decision and Control; 16-18 December 1992. Tucson, AZ, USA: IEEE; 2002

[21] Ruina A, Rice J. Stability of steady frictional slipping. Journal of Applied Mechanics. 1983;**50**(2):343-349. DOI: 10.1115/1.3167042

Chapter 13

Electrostatically Driven MEMS Resonator: Pull-in Behavior and Non-linear Phenomena

Barun Pratiher

Abstract

This chapter deals with the investigation on stability and bifurcation analysis of a highly non-linear electrically driven micro-electro-mechanical resonator has been established. A non-linear model of this system will briefly be described considering both transverse and longitudinal displacement of the resonator. A short description to explore the need of incorporating higher-order correction of electrostatic pressure has been highlighted. The pull-in results and consequences of higher-order correction on the pull-in stability will be reported. In addition, consequences of air-gap, electrostatic forcing parameter, and effective damping on non-linear phenomena have been studied to highlight the possible undesirable catastrophic failure at the unstable critical points. Basins of attractions that postulate a unique response in multi-region state for a specific initial condition will also be studied. This chapter can enable a significant adaptation to identify the locus of instability in micro-cantilever-based resonator when subjected to AC voltage polarization with the understanding of theoretical ideas for controlling the systems and optimizing their operation.

Keywords: micro-beam, electrostatic actuation, pull-in analysis, higher-order-electrostatic distribution, non-linear phenomena, stability

1. Introduction and state-of-art research

The development of electrostatically actuated micro-system has been extensively carried out by the research community in order to develop low cost and high durability, and further improve the performance of sensors and actuators for wide applications. The use of electrostatic actuation offers a simplicity in design with low-cost fabrication, fast response, the ability to achieve rotary motion, and low power consumption. However, this actuation often leads into a complex non-linear phenomenon. As a result, structural movability becomes suspicious due to pull-in occurrence for any finite air-gap thickness. Furthermore, stable deflection range due to active electrostatic actuation is always being restricted since the movable substrate or electrode gets collapsed onto the stationary plate. Thus, computing pull-in voltage is inevitable and plays a decisive factor indicating a critical voltage under which stable operations and structural reliability may be asserted.

Generally, micro-electro mechanical system mathematically model by considering either a thin beam or a thin plate having cross-section in the order of microns and length in the order of hundreds of microns with an efficient electrostatic actuation. When the range of air-gap between stationary electrode and movable electrode is relatively large, typically of the order of 10^{-2}–10^{-1} or even higher for designing small-size of electro-statically actuated devices, parallel approximation theory of capacitor becomes ill-suited and in-valid. For a high air gap, it is unavoidable to develop a large deflection model for an electrostatically actuated micro-beam considering both higher order distribution of electrostatic pressure and mid-plane stretching exist. In the following section, a brief current research on modeling and dynamics of micro-electro-mechanical systems (MEMS) structures has been cited.

A number of researchers have attempted to develop numerous models over the times to improve the design characteristics and investigating related dynamics. Luo and Wang [1] investigated analytically and numerically the chaotic motion in the certain frequency band of a MEMS with capacitor non-linearity. Pamidighantam et al. [2] derived a closed-form expression for the pull-in voltage of fixed-fixed micro-beams and fixed-free micro-beams by considering axial stress, non-linear stiffening, charge re-distribution, and fringing fields. They carried out an extensive analysis of the non-linearities in a micro-mechanical clamped-lamped beam resonator. Abdel-Rahman et al. [3] presented a non-linear model of electrically actuated micro-beams with consideration of electrostatic forcing of the air-gap capacitor, restoring force of the micro-beam, and axial load applied to the micro-beam. The response of a resonant micro-beam subjected to an electrical actuation has been investigated by Younis and Nayfeh [4]. Xie et al. [5] performed the dynamic analysis of a micro-switch using invariant manifold method. They considered micro-switch as a clamped-clamped micro-beam subjected to a transverse electrostatic force. An analytical approach and resultant reduced-order model to investigate the dynamic behavior of electrically actuated micro-beam-based MEMS devices have been demonstrated by Younis et al. [6]. The natural frequency and responses of electrostatically actuated MEMS with time-varying capacitors have been investigated by Luo and Wang [7]. Authors have demonstrated that the numerically and analytically obtained predictions were in good agreement with the findings obtained experimentally. A simplified discrete spring-mass mechanical model has been considered for the dynamic analysis of MEMS device. In Teva et al. [8], a mathematical model for an electrically excited electromechanical system based on lateral resonating cantilever has been developed. The authors obtained static deflection and the frequency response of the oscillation amplitude for different voltage-polarization conditions. Kuang and Chen [9] and Najar et al. [10] studied the dynamic characteristics of non-linear electrostatic pull-in behavior for shaped actuators in micro-electro-mechanical systems (MEMS) using the differential quadrature method (DQM). Zhang and Meng [11] analyzed the resonant responses and non-linear dynamics of idealized electrostatically actuated micro-cantilever-based devices in micro-electromechanical systems (MEMS) by using the harmonic balance (HB) method. Rhoads et al. [12] proposed a micro-beam device, which couples the inherent benefits of a resonator with purely parametric excitation with the simple geometry of a micro-beam. Krylov and Seretensky [13] developed higher-order correction to the parallel capacitor approximation of the electrostatic pressure acting on micro-structures taking into account the influence of the curvature and slope of the beam on the electrostatic pressure. The higher-order approximation has validated through a comparison with analytical solutions for simple geometries as well as numerical results. Decuzzi et al. [14] investigated the dynamic response of a micro-cantilever beam used as a transducer in a biomechanical sensor. Here, Euler-Bernoulli beam theory was introduced to model the cantilever motion of the transducer. They also considered Reynolds equation of lubrication for the analysis of hydrodynamic interactions. A number of review papers [15–18] provided an overview of the fundamental research on modeling and dynamics of electrostatically actuated MEMS devices under working different conditions. Nayfeh et al. [19] studied that the characteristics of the pull-in phenomenon in the presence of AC loads differ from those under purely DC loads. Zhang et al. [20] furnished a survey and analysis of the electrostatic force of importance in MEMS, its physical model, scaling effect, stability, non-linearity, and reliability in details. Chao et al. [21] predicted the DC dynamics pull-in voltages of a clamped-clamped micro-beam based on a continuous model. They derived the equation of motion of the dynamics model by considering beam flexibility, inertia, residual stress, squeeze film, distributed electrostatic forces, and its electrical field fringing effects. Shao et al. [22] demonstrated the non-linear vibration behavior of a micro-mechanical clamped-lamped beam resonator under different driving conditions. They developed a non-linear model for the resonator by considering both mechanical and electrostatic non-linear effects, and the numerical simulation was verified by experimental findings. Moghimi et al. [23] investigated the non-linear oscillations of micro-beams actuated by suddenly applied electrostatic force, including the effects of electrostatic actuation, residual stress, mid-plane stretching, and fringing fields in modelling. Chatterjee and Pohit [24] introduced a non-linear model of an electrostatically actuated micro-cantilever beam considering the non-linearities of the system arising out of electric forces, geometry of the deflected beam and the inertial terms. Furthermore, one may use the review articles [15–18] as a source of information to the overall images about the electromechanical model of MEMS devices actuated by electrostatically and related dynamics. A detailed review of perturbation techniques to obtain the non-linear solution of such systems/structures can be found in [25]. A detailed description of the forced and parametrically excited systems has been highlighted in [26–28].

Several researchers have studied the pull-in behavior of micro-mechanical system under various driving conditions about its static beam positions. In addition, it has been learnt that researchers are still considering simple geometry ignoring the non-linear effect or components in their mathematical model to investigate the theoretical and experimental aspects of dynamic performance of MEMS devices. Moreover, in order to highlight a proper insight and a better understanding into the MEMS devices, the accurate simulations of mechanical behaviors with a faithful mathematical model is fairly inevitable that can exhibit a more realistic shape of the bending deflection of the micro-beam and the development of resulting electrostatic pressure distribution. Here, author has been attempted to investigate the dynamic stability and bifurcation analysis of electrostatically actuated MEMS cantilever along with pull-in behavior, both statically and dynamically accounting for the effect of mid-plane stretching and non-linear distribution of electrostatic pressure. The main focus here is to investigate the assessment of the system stability and subsequent bifurcations, which usually demonstrate the locus of instability. Pull-in voltage and its response under the non-linear effects have been computed. The method of multiple scales has been used to analyze the stability and bifurcation of the steady state solutions via frequency-response characteristics, time responses, and basin of attractions.

2. Pull-in

Though the major focus of this study is to explore the non-linear behavior of an electrostatically actuated MEMS device, it is of vital importance to study the pull-in behavior of electrostatically driven MEMS device as well. Both static and dynamic pull-in have been albeit briefly discussed in the coming sub-section followed by the system's non-linear behaviors. Before proceeding with an understanding of the MEMS dynamics, especially non-linear dynamics, it is prudent to briefly explore the range of operating applied voltage under which the system model and attendant analysis are considered to be sufficiently accurate for the predictive design.

2.1 Problem description

Differential equation of motion of a continuous micro-cantilever beam subjected to AC potential difference by stationary electrode has been shown in **Figures 1** and 2, while the associated boundary conditions are being expressed in [25, 29]. However, the electrostatic force is considered to be uniform across the width, while transverse $v(x,t)$ and axial $u(x,t)$ displacement component holds a constraint equation known as in-extensibility condition

$$w'^2 + (1+u')^2 = 1.$$

$$\ddot{w} + 2\varsigma\dot{w} + w'''' + (d/l)^2\left[(w'')^3 + 4w'w''w''' + (w')^2 w''''\right] + (d/l)^2$$

$$\left[w'\int_0^{\xi}\left\{\ddot{w}'w' + (\dot{w}')^2\right\}d\bar{\xi}\right] - (d/l)^2\left[w''\int_{\xi}^{1}\int_0^{\bar{\eta}}\left\{\ddot{w}'w' + (\dot{w}')^2\right\}d\bar{\xi}\,d\bar{\eta}\right]$$

$$= \frac{6\varepsilon_0 l^4 V^2}{Eh^3 d^3}\left(1 + 2w + 3w^2 + HOD - \left\{\frac{(d/l)}{3}\left(2w'' + 2ww'' + w'^2 + HOD\right)\right\}\right). \tag{1}$$

2.2 Static analysis

It has been practically observed that the most common failure mode considered in the design of electrostatically driven MEMS devices is static pull-in condition beyond which it leads to desterilize the system for any further applied voltage. This failure is majorly occurred in the excess of electrostatic load in comparison to the static load-bearing capacity. As a result, system undergoes a negative stiffness in the

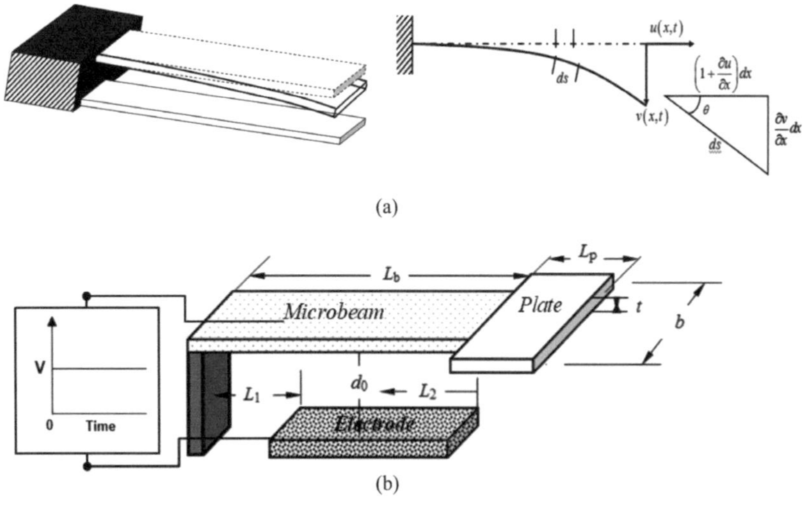

Figure 1.
(a) A pictorial diagram of micro-cantilever beam separated from a stationary electrode at a distance of d [25]. (b) Graphical representation of cantilever-based micro-structure coupled with rigid plate [29].

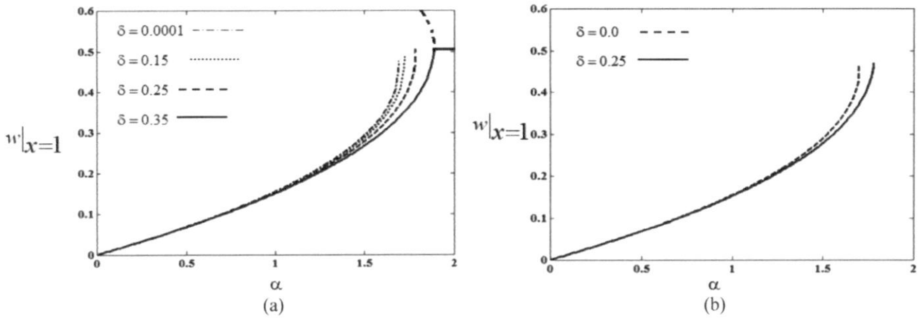

Figure 2.
(a) Variations of the non-dimensional tip deflection $w|_{x=1}$ *with variable* α *for various values of* δ*. (b): Effect of higher-order correction of electrostatic pressure on the non-dimensional tip deflection* $w|_{x=1}$ *with variable* α *[25].*

system's equation of motion. Pull-in instability as to calculate the pull-in voltage is inevitable to understand its limit to perform the desire task under a static voltage beyond which the movable electrode collapses onto the stationary electrode. As a result, the system is statically unstable as the electrostatic force overshadows the internal resistance as restoring force.

In most of the communication and power-circuits systems, electrostatically actuated micro-switch works only to alternate ON and OFF conditions by tuning a bias voltage across the pull-in back and forth. Therefore, it is advisable to have a micro-system, which can operate at low actuation voltage for performing the task mostly suited in power communications. The pull-in voltage can be obtained by demonstrating the static deflection of the tip of the micro-cantilever beam directly solving the boundary value problem by setting all time derivatives in Eq. (1) equal to zero. **Figure 1** shows the tip-deflection with the voltages ranging from zero to forcing level, where the pull-in instability takes place as explained details in [25, 29]. Recalling the fact that system leads to a pull-in condition when the system's net stiffness becomes negative. Here, the pull-in condition starts at α equal to 1.69 that indicates 66.83 compared to the pull-in voltage 66.78 obtained in Ref. [24] and 68.5 obtained in Ref. [23] considering the same design parameters.

However, the obtained static pull-in voltage may increase with increase in gap-length ratio ($\delta = {}^{d_0}/_{l}$). It can be noted that the pull-in voltage may occur at a lower when the effect of non-linear curvature is considered while calculating the electrostatic pressures. Further, the higher-order correction factor may lead to lower value of pull-in voltage, which provides a most suitable for the design of a micro-system having significant gap-length.

2.3 Dynamic analysis

The loss of stability in dynamic responses occurs when the deformable electrode comes into contact with the fixed electrodes under an instantaneous electrostatic actuation that is lower than the static pull-in voltages known as dynamic pull-in phenomenon. Analysis of dynamic pull-in of an electrostatically actuated is complex due to its non-linear nature of electrostatic forces along with time integration of the momentum equations. Along with time-dependent terms, the transient part of the applied voltage is being neglected while calculating static pull-in voltage. However, calculating the actual pull-in voltage in dynamic condition when a bias alternative voltage source exists is obligatory and different as that of static pull-in voltage. Hence, there is a need to calculate the pull-in voltage called as dynamic pull-in voltage, considering the dynamics of the micro-beam instead of static state only.

In calculating the dynamic pull-in voltage, inertia and dissipative elements along with the components storing the strain energy for elastic deformation play an influential role in a dynamic condition. Here, the dynamic pull-in behavior has been depicted and investigated by directly simulating the Eq. (1) using the well-known R-K method. A qualitative phase-plane analysis has been illustrated to capture the global behavior of the response trajectories. Hence, the dynamic pull-in voltage has been illustrated in the phase portrait, i.e., in the plane of velocity ∝ displacement for the every applied voltage. The voltage turns out to be the critical voltage, i.e., pull-in voltage until the trajectories lead to an intersection of the orbits with the origin. The voltage at which monoclinic orbits are passing through the saddle node or degenerate singularity point is known as dynamic pull-in voltage. It is however advised to go through the article [29] for the detail explanation of obtaining the dynamic pull-in voltage.

Beyond the critical voltage, the system is found to be dynamically unstable. It has been found that the dynamic pull-in voltage is well below the static pull-in voltage nearly 80–95% of static pull-in voltage depending upon geometric configurations and physical properties **Figure 3.**

For an applied voltage less than critical one, the trajectories exhibit closed periodic orbits with steady response amplitude lower than the dynamic pull-in deflection. Hence, a periodic solution initiated always from an initial guess for a

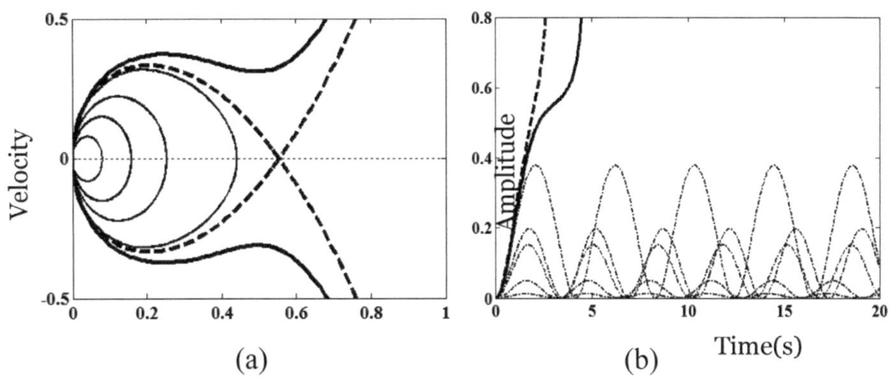

Figure 3.
(a) Dynamic pull-in voltage of undamped system for electrode length 45 μm [29]. (b) Time responses at pull-in voltage in [29].

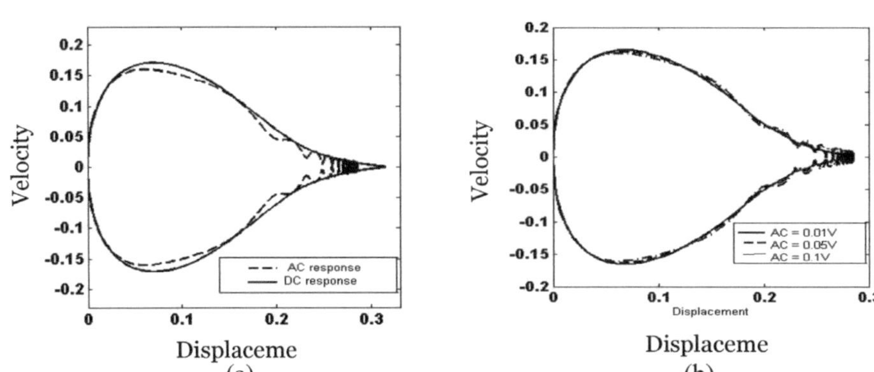

Figure 4.
(a) Dynamic pull-in voltage under DC and combined actuation. (b) Dynamic response at pull-in voltage under combined actuation at various voltages of AC actuation.

certain voltage leads to a closed trajectory as time approaches infinity. Hence, for every applied voltage V < VDPI (voltage in dynamic pull-in), the system shows an isolated stable closed trajectory. For any voltages greater than the dynamic pull-in voltage, the closed loop curves merge into a single curve, thus leading to an unstable domain. An effective DC contribution considering both DC and AC components has been depicted in the phase plane as shown in **Figure 4**, while the presence of AC component leads to a distortion in the solution trajectories. The presence of electrostatic actuation combining both DC and AC voltages system undergoes pseudo-dynamic pull-in, which is expected to all voltage-input combinations exceeding the predicted margin. However, this phenomenon generally holds true for small AC component, while inconsistency observes for a larger AC voltages.

3. Non-linear analysis

A comprehensive knowledge of non-linear dynamics in MEMS resonator is of great importance for the optimum design and operational stability. Thus, an understanding of the conditions to explore the non-linear phenomena arise; e.g., multiple-solutions; bifurcation can be implemented for further predicting chaotic responses in micro/nano-resonators. In this section, system's non-linear response at three distinct resonant conditions in the parametric space has been discussed. The non-linear phenomena have been reported here with the understanding of its instability via bifurcation. The characteristics form of the system non-linearity and electrostatic pressures and their effects on the system stability along with the effect of input voltages offer great flexibility toward designing the resonant sensors and filters. In order to obtain a detail understanding about the non-linear phenomena in MEMS systems, one may go through the articles [1, 5, 7, 11, 12, 19, 25, 30].

3.1 Problem description

Adopting the Galerkin's techniques and replacing $w = \Phi(x)X(\tau)$, where $\Phi(x)$ is admissible function obtained by satisfying the boundary conditions only and with similar procedures used in [25], the partial governing equation is then discretized into non-autonomous, time-dependent equation of motion with considering viscous damping effect. Expanding the non-linear electrostatic force developed due to applied voltage by Taylor series, one may obtain the following non-autonomous equation of motion.

$$\begin{aligned} &\ddot{X} + X + a\dot{X} + bX^3 + c\dot{X}^2X + d\ddot{X}X^2 \\ &= F\cos\Omega\tau + GX\cos\Omega\tau + KX^2\cos\Omega\tau \end{aligned} \tag{2}$$

Here, X is the non-dimensional displacement function or time modulation, while τ and Ω are the non-dimensional time and frequency, respectively. The expression for the co-efficient of non-autonomous $(a - d, F - G, K)$ is expressed in [25]. The equation of motion is further reduced to another form of micro-system neglecting effect of higher-order electrostatic distribution pressure, and mid-plane stretching effect.

$$\ddot{X} + X + a\dot{X} + bX^3 + c\dot{X}^2X + d\ddot{X}X^2 = F\cos\Omega\tau. \tag{3}$$

A huge number of researchers still consider either simple lumped-spring-mass model or Euler-Bernoulli beam theory with small air-gap assumption to carry out the theoretical and experimental investigation of dynamic performance of MEMS devices considering the mid-plane stretching effect.

$$\ddot{X} + X + a\dot{X} + bX^3 = F\cos\Omega\tau. \tag{4}$$

3.2 Bifurcation and stability

Equation of motions [Eqs. (2)–(4)] for various MEMS devices comprise linear and non-linear terms, direct forced, parametric term, and non-linear parametric terms due to non-linear electrostatic actuation. Since, the temporal equation of motion holds non-linear terms; it is difficult to find closed form solution. Hence, one may go for approximate solution by using the perturbation method. Here, method of multiple scales as explained in [25, 29–33] is used to obtain the set of algebraic equations turning into non-autonomous equations of motion for three resonance conditions, viz. primary resonance, parametric resonance condition, and third-order sub-harmonic conditions are being expressed under steady state conditions. The procedures used to derive the reduced order equation are similar to those explained in [25, 29–33]. Based on numerical values of the coefficients of the damping, forcing, and non-linear terms, they are one order less than the coefficients of the linear terms, which have a value of unity in this case and as result, in the following technique, co-efficient are expressed as $a = 2\varepsilon\varsigma,\ b = \varepsilon b,\ c = \varepsilon c,\ d = \varepsilon d,$ $F = \varepsilon F,\ G = \varepsilon G,$ and $K = \varepsilon K$ for sake of simplicity. By using method of multiple scales with the procedure as explained in [25, 30, 32, 33], substituting $T_n = \varepsilon^n\tau,\ n = 0, 1, 2, 3\cdots$ and displacement $X(\tau;\varepsilon) = X_0(T_0, T_1) + \varepsilon X_1(T_0, T_1) + O(\varepsilon^2)$ in Eq. (2) and equating the coefficients of like powers of ε, one may obtain the following expressions:

Order

$$\varepsilon^0 : {D_0}^2X_0 + X_0 = 0, \tag{5}$$

Order

$$\begin{aligned}\varepsilon^1 &: {D_0}^2X_1 + X_1 \\ &= -2D_0D_1X_0 - 2i\zeta X_0 - bX_0^3 - c\left(D_0^2X_0\right)X_0 - d(D_0X_0)^2X_0^2 + F\cos\Omega T_0 \\ &\quad + GX_0\cos\Omega T_0 + KX_0^2\cos\Omega T_0.\end{aligned} \tag{6}$$

General solutions of Eq. (5) can be written as

$$X_0 = A(T_1)\exp(iT_0) + \overline{A}(T_1)\exp(-iT_0). \tag{7}$$

After substituting Eq. (7) into Eq. (8), we have

$$\begin{aligned}&{D_0}^2X_1 + X_1 = -2iD_1A\exp(iT_0) - 2i\zeta A\exp(iT_0) - 3(b-c+d)A^2\overline{A}\exp(iT_0) - \\ &(b-c+d)A^3\exp(3iT_0) + \frac{F}{2}\exp(i\Omega T_0) + \frac{G}{2}A\exp i(\Omega+1)T_0 + \frac{G}{2}\overline{A}\exp i(\Omega-1)T_0 + \\ &\frac{K}{2}A^2\exp i(\Omega+2)T_0 + \frac{K}{2}A^2\exp i(\Omega-2)T_0 + \frac{K}{2}A\overline{A}\exp(i\Omega T_0) + cc.\end{aligned} \tag{8}$$

Any solution from the above equation may lead to an unbounded solution due to the existence of small divisor and secular terms in the equation. The terms associated e^{iT_0} or $\approx e^{i\Omega T_0}$, $\approx e^{i(\Omega-1)T_0}$, $\approx e^{i(\Omega-2)T_0}$ are known as small divisor and secular terms. These terms are required to be eliminated to obtain any bounded solution.

It may be observed that these terms exist when $\Omega \approx 1$, $\Omega \approx 3$, or $\Omega \approx 2$. In the following sub-sections, three resonance conditions, i.e., primary resonance, parametric resonance condition, and third-order sub-harmonic conditions have been briefly discussed. A details derivation and explanation is being carried out in [25].

3.2.1 Primary resonance condition

3.2.1.1 Reduced order model

Here, the resonance condition occurs when the frequency of applied voltage becomes equal to that of one of the natural frequencies, i.e., fundamental natural frequency. Following reduced equations are obtained as given below replacing $X = a(T_0)e^{i\phi(T_0)}$. A detail explanation about to obtain these reduced equations is being carried out in [25].

$$8a' = -8\zeta a + 4F\sin\phi + Ka^2\sin\phi, \tag{9}$$

$$8a\,\phi' = 8a\sigma - (3b - 3c + d)(\simeq\mu)a^3 + 4F\cos\phi + 3Ka^2\cos\phi. \tag{10}$$

Here, system only exhibits only non-trivial responses, i.e., $a \neq 0$ obtained from (Eqs. 9–10). Dynamic responses are being determined by solving the set of algebraic equations obtained by converting differential equations into set of algebraic equations under steady state conditions, i.e., $a' = 0$, and $\phi' = 0$. Here, stability of the steady state responses has been analyzed by investigating eigenvalues of the Jacobian matrix, which has been obtained by perturbing the algebraic equations with $a = a_o + a_1$ and $\gamma = \gamma_0 + \gamma_1$, where a_0, γ_0 are the singular points.

3.2.1.2 Results and discussions

Here, the condition at which the resonator has been excited with a frequency of the applied alternative voltage nearly equal to the fundamental frequency of the resonator is being discussed. In this vibrating state, the amplitude of vibration is always found to be a non-zero solution, while both stable and unstable non-trivial solutions are being observed. The study of bifurcation is being carried out on to see the losses the stability of system when a parameter passes through a critical value called a bifurcation point. The sudden change in amplitude undergoes catastrophic failure of the system. The graphical illustration of the vibration amplitude with varying the system control parameter has been constructed.

Figure 5a represents a typical frequency response curves for a specific air-gap $(\simeq\mu)$ between resonator and stationary. It is noteworthy that amplitude of vibration becomes increasing with increase in forcing frequency. The non-linear mode of operation possesses most likely hardening when the effective structural non-linearity becomes $\mu = +2.0$. In this condition, restoring forces due to geometric non-linearity overshadows the inertia effect of the device that leads to hardening spring effect. When the effective structural non-linearity becomes $\mu = -2.0$, vice versa effect is observed. For sweeping up the frequency moving toward E from point D, system undergoes a sudden jump down when the frequency reaches to its critical value regarded as saddle-node fixed bifurcation point. Similarly, a sudden upward jump in the amplitude from lower to higher amplitude undergoes for a sweeping down frequency. With further increase in frequency, the amplitude of vibratory motion decreases and follows the path DA. Hence, with decrease in frequency leads to the lower amplitude of responses from a higher value continuously.

With the experimental investigation, it has been noted that these jump phenomena may lead to mechanical crack across the width of the beam. The growth of the crack may further propagate with the experiences of subsequent jump up and down in response amplitude (**Figures 6** and 7).

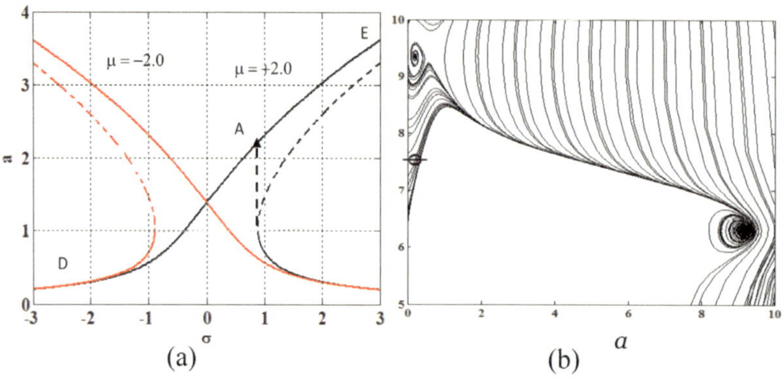

Figure 5.
(a) Frequency response curves for $\varsigma = 0.1, F = 1.2, K = 0.12$. (b) Basins of attraction at bi-stable point [25].

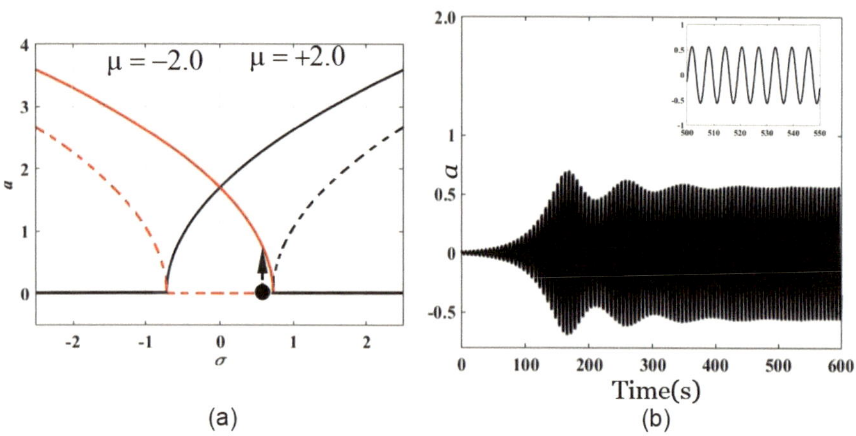

Figure 6.
(a) Frequency characteristics curves for $\varsigma = 0.1, H = 1.5$. (b) Time histories at unstable point [25].

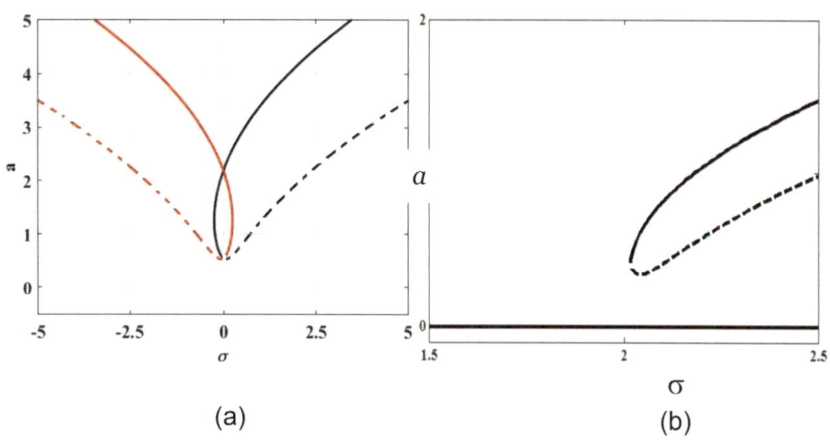

Figure 7.
(a) Frequency characteristics curve for $\varsigma = 0.1$, and $K = 1.5$. (b) Frequency characteristics curve for $d/l = 0.2, \varsigma = 0.2$ and $K = 1.0$.

It is observed that multiple solutions exist at some frequencies in the entire frequency range from 0.5 to 1.5. Being existence of multiple solutions, it is desirable to check whether all solutions are found to be stable or unstable or mixed solution. For a specific initial condition, trajectories have been drawn in the plane of amplitude and phase as time goes infinity. It is being observed that the system possesses the condition of bi-stability at some regions. Thus, in this region, wrong selection of initial conditions mostly results the wrong output response. It is thus keyed to opt out an appropriate condition for a specific solution that can prevail physically by the system.

3.3 Principal parametric resonance ($\Theta \approx 2$)

3.3.1 Reduced order model

Here, the resonance condition occurs when the frequency of applied voltage becomes twice the natural frequencies, i.e., fundamental natural frequency. Following reduced equations are obtained as given below:

$$4a' = -4\zeta a + Ha \sin\phi, \tag{11}$$

$$8a\phi' = 8a\sigma - (3b - 3c + d)(\simeq\mu)a^3 + 4H\cos\phi. \tag{12}$$

Here, system possesses both trivial $a = 0$ and non-trivial $a \neq 0$ responses determined by solving the reduced non-linear algebraic equations at steady state condition by using Newton's method, simultaneously. The stability of the steady state responses has also here obtained by replacing a with $a_o + a_1$, γ with $\gamma_0 + \gamma_1$, respectively, and then investigating the eigenvalues of the resulting Jacobian matrix (J).

3.3.1.1 Results and discussions

The electrostatically actuated micro- beam is vibrating with a frequency of the applied voltage nearly equal to the twice the fundamental frequency of the resonator. Unlike, primary resonance case, here, the system possesses both trivial and non-trivial solutions. Here, the vibration amplitude may vary from zero to non-zero value and vice-versa depending upon the state of vibration being considered. Depending upon the selected values of control parameters, the trivial and non-trivial solutions are noticed as stable and unstable for a specific frequency of the AC voltages. Sub-critical pitchfork bifurcation leads to sudden change in amplitude. This discontinuity in amplitude results catastrophic failure of the system.

Approximate solutions obtained by using the method of multiple scales have been compared with those found by numerically solving the temporal Eq. (2). Time response clearly shows that the trajectory initiated from the unstable trivial response finally moves to stable non-trivial fixed point response. Responses mostly obtained by solving the temporal equation of motion are being in good agreement with those findings by perturbation technique.

3.4 Sub-harmonic resonance case ($\Theta \approx 3$)

Similarly, one may have following reduced equations when the frequency of applied voltage is nearly equal to that of three times of natural frequency

$$8a' = -8\zeta a + Ka^2 \sin\phi, \tag{13}$$

$$8a\phi' = 8a\sigma - (3b - 3c + d)(\simeq\mu)a^3 + 3Ka^2\cos\phi. \tag{14}$$

Similar to the previous resonance case, here also system possesses both trivial $a = 0$ and non-trivial $a \neq 0$ responses obtained by solving the equations obtained after setting a' and γ' equal to zero using similar Newton's method for different system parameters. Similar procedure one may follow to find out the stability of the steady state response of this case by investigating the nature of the equilibrium points.

This resonance takes place when the frequency of applied voltage is nearly equal to thrice the fundamental frequency of the resonator. Here, the amplitude of vibration may shift from non-zero to zero value depending on the initial operating point. In this resonance case, trivial solutions are found to be stable for any frequency and control parameters. The loss of stability of the system depends on the position of critical point and selection of control parameters, while the system can bring down to stable condition by simply choosing the frequency and other system parameters, appropriately. The bifurcation present here is known as addle-node bifurcation point. The jump length is found to be increased with increase in control parameters. Similarly, it has been observed that jump length will increase with increase in both forcing parameter and damping.

4. Conclusions

The investigation on stability and bifurcation analysis of a highly non-linear electrically driven MEMS resonator along with pull-in behavior has been established. A non-linear mathematical model has been briefly described accounting off mid-plane stretching and non-linear electrostatic pressure under both DC and AC actuation. A short description of perturbation method to study the steady state responses has been highlighted. The pull-in results and consequences of non-linear effects on dynamics responses have been reported. Non-linear phenomenon has been studied to highlight the possible undesirable catastrophic failure at the unstable critical points, i.e., bifurcation. Basins of attractions that postulate a unique response in multi-region state for a specific initial condition has been demonstrated. This chapter enables a significant understanding about the locus of instability in micro-cantilever-based resonator when subjected to DC and AC potentials. A theoretical understanding for controlling the systems and optimizing their operation is being reported here.

Author details

Barun Pratiher
Department of Mechanical Engineering, Indian Institute of Technology Jodhpur, India

*Address all correspondence to: barun@iitj.ac.in

References

[1] Luo ACJ, Wang FY. Chaotic motion in a Micor-electro-mechanical system with non-linearity from capacitors. Communications in Nonlinear Science and Numerical Simulation. 2002;7:31-49

[2] Pamidighantam S, Puers R, Baert K, Tilmans H. Pull-in voltage analysis of electrostatically actuated beam structures with fixed–fixed and fixed–free end conditions. Journal of Micromechanics and Microengineering. 2002;**12**:458-464

[3] Abdel-Rahman EM, Younis MI, Nayfeh AH. Characterization of the mechanical behavior of an electrically actuated microbeam. Journal of Micromechanics and Microengineering. 2002;**12**:759-766

[4] Younis M, Nayfeh A. Study of the nonlinear response of a resonant microbeam to an electric actuation. Nonlinear Dynamics. 2003;**31**:91-117

[5] Xie W, Lee H, Lim S. Nonlinear dynamic analysis of MEMS switches by nonlinear modal analysis. Nonlinear Dynamics. 2003;**3**:243-256

[6] Younis MI, Abdel-Rahman EM, Nayfeh AH. A reduced-order model for electrically actuated microbeam-based MEMS. Journal of Microelectromechanical Systems. 2003;**12**:672-680

[7] Luo ACJ, Wang F-E. Nonlinear dynamics of a micro-electro-mechanical system with time-varying capacitors. Journal of Vibration and Acoustic. 2000;**126**:77-83

[8] Teva J, Abadal G, Davis ZJ, Verd J, Borrise X, Boisen A, et al. On the electromechanical modeling of a resonating nano-cantilever-based transducer. Ultramicroscopy. 2004;**100**:225-232

[9] Kuang JH, Chen CJ. Dynamic characteristics of shaped micro-actuators solved using the differential quadrature method. Journal of Micromechanics and Microengineering. 2004;**14**:647-655

[10] Najar F, Houra S, El-Borgi S, Abdel-Rahman E, Nayfeh A. Modeling and design of variable-geometry electrostatic microactuators. Journal of Micromechanics and Microengineering. 2005;**15**:419-429

[11] Zhang W, Meng G. Nonlinear dynamical system of micro-cantilever under combined parametric and forcing excitations in MEMS. Sensors and Actuators A. 2005;**119**:291-299

[12] Rhoads JF, Shaw SW, Turner KL. The nonlinear response of resonant microbeam systems with purely-parametric electrostatic actuation. Journal of Micromechanics and Microengineering. 2006;**16**:890-899

[13] Krylov S, Seretensky S. Higher order correction of electrostatic pressure and its influence on the pull-in behaviour of microstructures. Journal of Micromechanics and Microengineering. 2006;**16**:1382-1396

[14] Decuzzi P, Granaldi A, Pascazio G. Dynamic response of microcantilever-based sensors in a fluidic chamber. Journal of Applied Physics. 2007;**101**:024303

[15] Batra R, Porfiri M, Spinello D. Review of modelling electrostatically actuated micro-electromechanical systems. Smart Materials and Structures. 2007;**16**:23-31

[16] Fargas MA, Costa CR, Shakel AM. Modeling the Electrostatic Actuation of MEMS: State of the Art 2005. Barcelona, Spain: Institute of Industrial and Control Engineering; 2005

[17] Lin RM, Wang WJ. Structural dynamics of microsystems—Current state of research and future directions. Mechanical Systems and Signal Processing. 2006;**20**:1015-1043

[18] Rhoads J, Shaw SW, Turner KL. Nonlinear dynamics and its applications in micro- and Nano-resonators. In: Proceedings of DSCC 2008, ASME Dynamic Systems and Control Conference. Ann Arbor, Michigan, USA; October 20-22 2008

[19] Nayfeh A, Younis MI, Abdel-Rahman EM. Dynamic pull-in phenomenon in MEMS resonators. Nonlinear Dynamics. 2007;**48**:153-163

[20] Zhang WM, Meng G, Chen D. Stability, nonlinearity and reliability of electrostatically actuated MEMS devices. Sensors. 2007;7:760-796

[21] Chao PCP, Chiu C, Liu TH. DC dynamics pull-in predictions for a generalized clamped–clamped micro-beam based on a continuous model and bifurcation analysis. Journal of Micromechanics and Microengineering. 2008;**18**:115008

[22] Shao L, Palaniapan M, Tan W. The nonlinearity cancellation phenomenon in micromechanical resonators. Journal of Micromechanics and Microengineering. 2008;**18**:065014

[23] Moghimi ZM, Ahmadian M, Rashidian B. Semi-analytic solutions to nonlinear vibrations of microbeams under suddenly applied voltages. Journal of Sound and Vibration. 2009; **325**:382-396

[24] Chatterjee S, Pohit G. A large deflection model for the pull-in analysis of electrostatically actuated microcantilever beams. Journal of Sound and Vibration. 2009;**322**:969-986

[25] Pratiher B. Stability and bifurcation analysis of an electrostatically controlled highly deformable microcantilever-based resonator. Nonlinear Dynamics. 2014;**78**(3):1781-1800

[26] Nayfeh AH, Mook DT. Nonlinear Oscillations. New York: Wiley-VCH; 1995

[27] Cartmell MP. Introduction to Linear, Parametric and Nonlinear Vibrations. London: Chapman and Hall; 1990

[28] Nayfeh AH, Balachandran B. Applied Nonlinear Dynamics: Analytical. Computational and Experimental Methods: Wiley; 1995

[29] Harsha CS, Prasanth CSR, Pratiher B. Prediction of pull-in phenomena and structural stability analysis of an electrostatically actuated microswitch. Acta Mechanica. 2016; **227**(9):2577-2594

[30] Pratiher B. Tuning the nonlinear behaviour of resonant MEMS sensors actuated electrically. Procedia Engineering. 2012;**47**:9-12

[31] Harsha CS, Prasanth CSR, Pratiher B. Modeling and non-linear responses of MEMS capacitive accelerometer. MATEC Web of Conferences. 2014;**16**:04003

[32] Harsha CS, Prasanth CSR, Pratiher B. Electrostatic pull-in analysis of a nonuniform micro-resonator undergoing large elastic deflection. Journal of Mechanical Engineering Sciences. 2018;**232**:3337-3350

[33] Harsha CS, Prasanth CSR, Pratiher B. Effect of squeeze film damping and AC actuation voltage on pull-in phenomenon of electrostatically actuated microswitch. Procedia Engineering. 2016;**144**:891-899

Chapter 14

Nonlinear Oxygen Transport with Poiseuille Hemodynamic Flow in a Micro-Channel

Terry E. Moschandreou and Keith C. Afas

Abstract

In a recent paper by the authors, a well-known governing nonlinear PDE used to model oxygen transport was formulated in a generalized coordinate system where the Laplacian was expressed in metric tensor form. A reduction of the PDE to a simpler problem, subject to specific integrability conditions, was shown, and in the present work, a novel approximate analytical solution is obtained in terms of the degenerate Weierstrass P function using a compatibility relation through the factorization of the reduced almost linear ode and subject to similar boundary conditions for a microfluidic channel used in recent work by the authors. A specific form of the initial equation which was reduced has been used by Nair and coworkers describing the intraluminal problem of oxygen transport in large capillaries or arterioles and more recent work by the corresponding author describing the release of adenosine triphosphate (ATP) in micro-channels. In the present problem, a channel with a central core, rich in red blood cells, and with a thin plasma region near the boundary wall, free of RBCs is considered.

Keywords: almost linear ODE, Poiseuille flow, oxygen transport, Painlevé analysis

1. Introduction

Various biophysical phenomena are modeled using nonlinear differential equations. Such is the case of a model used by Nair et al. [1–3] to describe oxygen transport in large capillaries [1–3]. This model incorporates two regions of blood flow. One is a core region of the blood with RBCs present, and in this core, oxygen dissociates into blood oxyhemoglobin. The velocity of blood in the core region is a function of the plasma velocity and rate of oxygen dissociating. The second region is a thin strip of flowing plasma with no RBCs at the wall of the micro-fluidic channel. The appropriate Robin condition and no-flux conditions are incorporated at and around a permeable membrane with oxygen transport through the membrane. Since there are two distinct regions of flow of liquid, RBCs with plasma and plasma alone, it is necessary to match the rate of change of partial pressure of oxygen at the common boundary of the liquid in each of the two regions. This kind of model has been used previously by Moschandreou et al. [4] studying the influence of tissue metabolism and capillary oxygen supply on arteriolar oxygen transport. In that study a numerical approach was used to solve the governing equations. In the present work we seek an analytical solution for a different highly nonlinear problem

 IntechOpen

in a micro-fluidic channel. Flows that are fully developed at inlet were studied by Ng [5] who studied oscillatory dispersion in a tube with chemical species undergoing linear reversible and irreversible reactions at the tube wall. Relative importance to the present work is that fully developed flow occurs at inlet of channel with boundary conditions specified on the wall. No inlet mass transport is specified at inlet of channel similar to [5] but unlike [6], for example. Using Painlevé analysis [7], certain nonlinear second-order ordinary differential equations (ODE) can be factorized and solved. We consider the model of Nair et al. [1–3], where the nonlinear PDE reduced in [8], to an "almost linear" second-order ode is considered. It is well-known that the Weierstrass P function in its series form is problematic to use in computational work due to very slow convergence of numerical methods. It is the aim of the present work to show that a degenerate form of the special function can be used as in [9] in the reduction of the nonlinear PDE in [1–3]. A thorough and recent review of oxygen control with microfluidics has been carried out in [10] and all of its references within. In this work we see how the microscale can be leveraged for oxygen control of RBCs.

2. General tensorial mass transport

Regardless of the kinematics of a surface (dynamic or stationary), all surfaces, $S = \partial\Omega$, enclosing a solid volume, Ω, obey the following intuitive conservation relation for an enclosed observable mass, m_o of some arbitrary substance:

$$\frac{d}{dt}m_o + \int_S \mathbf{j}_o \cdot \mathbf{dS} = \Sigma, \tag{1}$$

where $\mathbf{j}_o$ is the flux of the observable out or into the surface and Σ represents the net increase or decrease in the observable's mass.

The relation states intuitively that *any change of the observable's mass within the solid, plus all observable mass entering or leaving the boundary, should represent the net change in the mass of the object.*

Any mass transport can be derived from the above relation converted into the differential form. We first recognize that the observable's mass can be represented through the observable's density:

$$m_o = \int_\Omega \rho_o \quad d\Omega,$$

In addition, we can make the same statement about the net equilibrium constant. Suppose there is a local equilibrium density, σ, such that

$$\Sigma = \int_\Omega \sigma \quad d\Omega.$$

We then can obtain a full integral form of the conservation relation:

$$\frac{d}{dt}\int_\Omega \rho_o \quad d\Omega + \int_S \mathbf{j}_o \cdot \mathbf{dS} - \int_\Omega \sigma \quad d\Omega = 0. \tag{2}$$

In general, it can be shown that for a general surface that is moving, the time derivative of a volume integral defined over the dynamic volume enclosed by the surface can be summarized as [11]

$$\frac{d}{dt}\int_{\Omega}\psi \quad d\Omega = \int_{\Omega}\frac{\partial}{\partial t}\psi \quad d\Omega + \int_{S}\tilde{C}\psi \quad dS, \tag{3}$$

where $\tilde{C}$ is defined as the normal projection of the surface's perpendicular speed, also named, the *surface velocity*. This is encapsulated (using Einstein summation convention) by the tensorial equation:

$$\tilde{C} = \mathbf{V}_{\perp}\cdot\mathbf{N} = V_{\perp}^{i}N_{i},$$

where $\mathbf{N} = \mathbf{Z}^{i}N_{i}$ is the unit normal to the surface and $\mathbf{Z}^{i}$ is the contravariant basis for the coordinate space. Thus, we can simplify our conservation relation:

$$\int_{\Omega}\frac{\partial}{\partial t}\rho_{o} \quad d\Omega + \int_{S}\tilde{C}\rho_{o} \quad dS + \int_{S}\mathbf{j}_{o}\cdot\mathbf{dS} - \int_{\Omega}\sigma \quad d\Omega = 0,$$

We can also simplify the vector surface element, $\mathbf{dS} = \mathbf{N}dS$, and convert the flux term into a tensorial formation, by recognizing $j_{o} = (j_{o})^{i}\mathbf{Z}_{i}$:

$$\int_{\Omega}\frac{\partial}{\partial t}\rho_{o} \quad d\Omega + \int_{S}\tilde{C}\rho_{o} \quad dS + \int_{S}(j_{o})^{i}N_{i} \quad dS - \int_{\Omega}\sigma \quad d\Omega = 0.$$

Finally, we will use the definition of the surface velocity and combine the two surface integral terms into one:

$$\int_{\Omega}\frac{\partial}{\partial t}\rho_{o} \quad d\Omega + \int_{S}(V_{\perp}^{i}\rho_{o} + j_{o}^{i})N_{i} \quad dS - \int_{\Omega}\sigma \quad d\Omega = 0.$$

We can use Gauss' divergence theorem on the surface integral term and unite all the terms under one volume integral:

$$\int_{\Omega}\frac{\partial}{\partial t}\rho_{o} + \nabla_{i}\left(V_{\perp}^{i}\rho_{o} + j_{o}^{i}\right) - \sigma \quad d\Omega = 0.$$

Using the localization theorem, the integrand must be zero inside the volume integral, and we obtain the differential form of the conservation relation:

$$\frac{\partial}{\partial t}\rho_{o} + \nabla_{i}\left(V_{\perp}^{i}\rho_{o} + j_{o}^{i}\right) - \sigma = 0. \tag{4}$$

We can simplify the equation, further by considering a particular form of the flux. In this, we consider advective flux (flux due to bulk movement of an observable's mass) and diffusive flux (flux due to a concentration gradient). Using advective formulas and Fick's first law, we obtain the flux to be

$$j_{o}^{i} = v_{o}^{i}\rho_{o} - D_{o}\nabla^{i}\rho_{o}, \tag{5}$$

where $\mathbf{v}_{o} = v_{o}^{i}\mathbf{Z}_{i}$ is the velocity of the observable within the volume. Substituting these into the differential conservation relation, we obtain

$$\frac{\partial}{\partial t}\rho_{o} + \nabla_{i}\left(V_{\perp}^{i}\rho_{o} + v_{o}^{i}\rho_{o} - D_{o}\nabla^{i}\rho_{o}\right) - \sigma = 0.$$

We simplify the equation, expanding the covariant derivative to obtain the final form of the conservation relation:

$$\frac{\partial}{\partial t}\rho_o + \nabla_i\left(\left(V_\perp^i + v_o^i\right)\rho_o\right) = \nabla_i\left(D_o \nabla^i \rho_o\right) + \sigma. \tag{6}$$

This form can also be put into an invariant tensorial form by utilizing the invariant time derivative operator $(\dot{\nabla})$ from the calculus of moving surfaces [11]:

$$\dot{\nabla}\rho_o + V_\perp^i \nabla_i \rho_o + \nabla_i\left(\left(V_\perp^i + v_o^i\right)\rho_o\right) = \nabla_i\left(D_o \nabla^i \rho_o\right) + \sigma. \tag{7}$$

This equation can also be put into a vector form:

$$\dot{\nabla}\rho_o + V_\perp \cdot \vec{\nabla}\rho_o + \vec{\nabla} \cdot \left(\left(V_\perp + v_o\right)\rho_o\right) = \vec{\nabla} \cdot \left(D_o \vec{\nabla}\rho_o\right) + \sigma. \tag{8}$$

2.1 Application to oxygen transport

We consider a biological application of the observable's mass transport equation to microfluidic arterial oxygen transport. In this case, $o = O_2$. For this, we are required to make a few assumptions:

(A1) We first tentatively assume that the arteriole's surface is stationary and **not**. This would necessarily imply $V_\perp^i \mathbf{Z}_i = \mathbf{0}$.

(A2) We also consider steady-state solutions, by assuming that the density is not dependent on time. This means that $\frac{\partial}{\partial t}\rho_{O_2} = 0$.

(A3) In addition, we restrict our studies to microfluidic environments which are in equilibrium. This would imply that the net local density change is zero, or $\sigma = 0$.

(A4) We also assume that the diffusion constant is a constant.

Using the above relations, we reduce the conservation relation to

$$\nabla_i\left(v_{O_2}^i \rho_{O_2}\right) = D_{O_2} \nabla_i \nabla^i \rho_{O_2}. \tag{9}$$

Both of the operators can be expanded using the Voss-Weyl formula [11] and restated in terms of partial derivatives with respect to the spatial coordinates, Z^i, and the spatial metric tensor, Z_{ij}:

$$\frac{1}{\sqrt{|Z_{jk}|}}\frac{\partial}{\partial Z^i}\left(\sqrt{|Z_{jk}|}v_{O_2}^i \rho_o\right) = \frac{1}{\sqrt{|Z_{jk}|}} D_{O_2} \frac{\partial}{\partial Z^i}\left(\sqrt{|Z_{jk}|}Z^{i\ell}\frac{\partial}{\partial Z^\ell}\rho_{O_2}\right). \tag{10}$$

We assume for a moment that the coordinate system chosen is some general axial coordinate system consisting of two arbitrary coordinates, (Z^1, Z^2), and a third coordinate corresponding to the standard z coordinate found in cylindrical and Euclidean coordinate systems.

This forms a three-dimensional coordinate system of (Z^1, Z^2, z). We first assume that the velocity of the observable is only along the z coordinate. This means that the term on the left is greatly simplified:

$$\frac{1}{\sqrt{|Z_{jk}|}}\frac{\partial}{\partial z}\left(\sqrt{|Z_{jk}|}\left(v_{O_2}\right)_z \rho_{O_2}\right) = \frac{1}{\sqrt{|Z_{jk}|}} D_{O_2} \frac{\partial}{\partial Z^i}\left(\sqrt{|Z_{jk}|}Z^{i\ell}\frac{\partial}{\partial Z^\ell}\rho_{O_2}\right).$$

We then assume that the velocity of the observable does not depend on the z-coordinate. This means that the particle moving along a streamline parallel to the length of the tube will not accelerate. This will produce the equation:

$$\frac{1}{\sqrt{|Z_{jk}|}}(v_{O_2})_z \frac{\partial}{\partial z}\left(\sqrt{|Z_{jk}|}\rho_{O_2}\right) = \frac{1}{\sqrt{|Z_{jk}|}} D_{O_2} \frac{\partial}{\partial Z^i}\left(\sqrt{|Z_{jk}|}Z^{i\ell} \frac{\partial}{\partial Z^\ell}\rho_{O_2}\right).$$

From here, if we assume the coordinates $(Z^1, Z^2) = (x, y)$, then we can simplify greatly to obtain the final form:

$$(v_{O_2})_z \frac{\partial \rho_{O_2}}{\partial z} = D_{O_2} \frac{\partial}{\partial Z^i}\left(Z^{i\ell} \frac{\partial \rho_{O_2}}{\partial Z^\ell}\right). \tag{11}$$

In addition, we assume that ρ_{O_2} does not depend on x. This produces the final equation of

$$(v_{O_2})_z \frac{\partial \rho_{O_2}}{\partial z} = D_{O_2} \frac{\partial^2 \rho_{O_2}}{\partial y^2}. \tag{12}$$

In addition, for all the above equations, since by Boyle's law, at a constant temperature, all instances of oxygen density can be equivalently replaced by oxygen pressure.

3. Governing equation for oxygen transport

As can be extrapolated from above, the general form of the nonlinear PDE for consideration defining oxygen transport in core region with Poiseuille hemodynamic flow is given by Eq. (13). The boundary conditions are shown in Section 10.1 of the Appendix, the velocity profile is shown in Section 10.2, and **Figure 1** shows the geometry of the problem:

$$\left[v_p(1 - H_T) + v_{RBC} H_T \frac{K_{RBC}}{K_p}\left(1 + \frac{[Hb_T]}{K_{RBC}} \frac{dSO_2}{dPO_2}\right)\right] \frac{\partial PO_2}{\partial z} = D_p \nabla^2 PO_2 \tag{13}$$

There is a core region of blood flow with RBCs and plasma and a cell-free region with only plasma flowing. In the plasma region near the wall, the governing equation is as in Eq. (13), without the second term in the square brackets. In general the geometry of the problem can be either a tube or a channel, and the Laplacian is generalized in [8]. In the present work, we confine the problem to a channel flow. The blood plasma velocity is v_p, and v_{RBC} is the velocity of RBCs together with plasma in the cell-rich region. The velocity of the RBCs is lower due to the slip

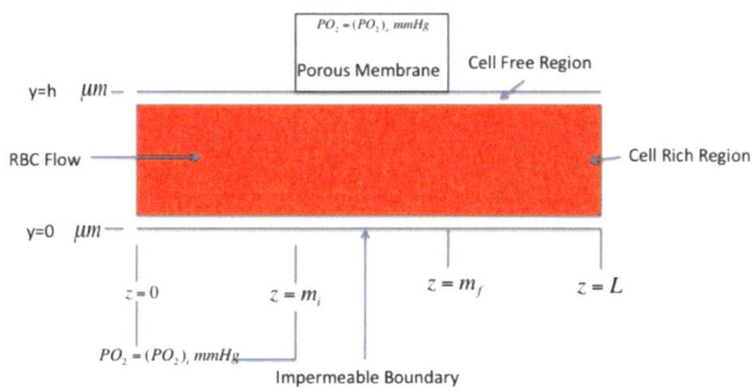

Figure 1.
Geometry of micro-channel with permeable membrane centered at the top of the channel.

between plasma and RBCs. The distribution of RBCs is such that the hematocrit is higher at the center of the channel and lower near the wall. The term $\frac{dSO_2}{dPO_2}$ is the slope of the oxyhemoglobin dissociation curve and is a highly nonlinear function of oxygen tension PO_2 [1–3]. The dissociation curve is approximated by the Hill equation [4, 6], where an empirical constant N is used and a constant P_{50} appears which is the oxygen tension that yields 50% oxygen saturation. $[Hb_T]$ is the total heme concentration, D_p is the given oxygen diffusion coefficient in plasma and has units of $\mu m^2/s$ and K_{RBC}, K_p are the solubilities of O_2 in the RBCs and plasma, respectively, and have units of $M/mmHg$. The values of the respective parameters are taken from [8].

In connection with [8], we let

$$\Sigma(PO_2) = \text{const} + \text{coeff}\,\frac{dSO_2}{dPO_2}.$$

This allows for a transformation of Eq. (13) into Eq. (14) where "const" and "coeff" are constants derived in [8].

We thus have a PDE of the form:

$$\Sigma(PO_2)\frac{\partial PO_2}{\partial z} = \frac{1}{v_p}\nabla^2 PO_2. \tag{14}$$

Choosing a separation form of PO_2

$$PO_2 = PO_2(\tilde{Z}, z), \tag{15}$$

where $\tilde{Z} = \left(\tilde{Z}^1, \tilde{Z}^2\right)$ indicates a semi-general coordinate system; we assume the following form of the solution:

$$\Sigma(PO_2(\tilde{Z}, z)) = P(\tilde{Z})L_1(z) + L_2(z). \tag{16}$$

Let $\tilde{Z}_{ij}$ be the metric tensor of the coordinate system composed of $\left(\tilde{Z}^1, \tilde{Z}^2\right)$. As derived in [8], we express the Laplacian of an arbitrary function, $\psi(\tilde{Z})$, in terms of the metric tensor in curvilinear coordinates, i.e.,

$$\nabla_i\nabla^i\psi(\tilde{Z}) = \nabla^2\psi(\tilde{Z}) = \frac{1}{\sqrt{|\tilde{Z}_{jk}|}}\frac{\partial}{\partial\tilde{Z}^i}\left(\sqrt{|\tilde{Z}_{jk}|}\tilde{Z}^{i\ell}\frac{\partial\psi}{\partial\tilde{Z}^\ell}\right). \tag{17}$$

Applying this definition to $\nabla^2\Sigma^{-1}(PL_1 + L_2)$ similar to [8], we show that

$$(PL_1 + L_2)\frac{\partial}{\partial z}\Sigma^{-1}(PL_1 + L_2) = \frac{1}{v_p}\left[L_1\frac{d\Sigma^{-1}}{d\Sigma}\nabla^2 P + L_1\tilde{Z}^{ij}(\nabla_i P)(\nabla_i P)\frac{d^2\Sigma^{-1}}{d\Sigma^2}\right].$$

Rearranging in terms of $\frac{d\Sigma^{-1}}{d\Sigma}$, we obtain

$$\left[(PL_1 + L_2)\left(P\frac{dL_1}{dz} + \frac{dL_2}{dz}\right)v_p - L_1\nabla^2 P\right]\frac{d\Sigma^{-1}}{d\Sigma} = \left[L_1\tilde{Z}^{ij}(\nabla_i P)(\nabla_i P)\right]\frac{d^2\Sigma^{-1}}{d\Sigma^2}.$$

Isolating Σ terms and allowing a zero separation constant, we obtain

$$(PL_1 + L_2)\left(P\frac{dL_1}{dz} + \frac{dL_2}{dz}\right)v_p - L_1\nabla^2 P = 0. \tag{18}$$

where $L_1 \neq 0$ and P must obey the metric condition

$$\tilde{Z}^{ij}(\nabla_i P)(\nabla_i P) \neq 0. \tag{19}$$

Also, Σ^{-1} must obey the conditions

$$\left\{\frac{d\Sigma^{-1}}{d\Sigma} \neq 0, \frac{d^2\Sigma^{-1}}{d\Sigma^2} = 0\right\}. \tag{20}$$

It has been shown in [8] that for $L_2(z) = 0$, Eq. (13) can be reduced to

$$\left\{\nabla^2 P + \mathbb{S}_2 v_p P^2 = 0, \frac{dL_1}{dz} + \mathbb{S}_2 = 0\right\}, \tag{21}$$

where $\mathbb{S}_2$ is a separation constant, $v_p\left(\tilde{Z}\right) = c - d\|\mathbf{r}\|^2$ is Poiseuille flow, and

$$PO_2 = A_1 P L_1 + A_2, \tag{22}$$

where A_1, A_2 are arbitrary constants.

In the present work, we consider the general case where the forms are chosen for L_1 and L_2:

$$\left\{L_1 = -\mathbb{S}_1\left(Cz + \frac{1}{3}m_i\right)^2, L_2 = C_1 z L_1\right\},$$

where C, C_1, and m_i are constants. C_1 is a free constant, C is to be determined using boundary conditions, and m_i is a fixed known constant. The reason for this choice of functions $L_1(z)$ and $L_2(z)$ will be made apparent in Sections 4 and 5. The constant m_i is chosen as in **Figure 1**, and $\mathbb{S}_1$ is a free constant.

4. Transformation of associated equation

The equation to solve is a PDE related to Eq. (18) and condition Eq. (20), for L_1 and L_2 defined above:

$$\begin{aligned} P_{rr} = &-\mathbb{S}_1\left(c - dr^2\right)\{2C(Cz + 1/3m_i)P^2 + \{2zC_1(Cz + 1/3m_i) \\ &+ C_1(Cz + 1/3m_i)^2 + 2zC_1(Cz + 1/3m_i)C\}P \\ &+ zC_1\left[C_1(Cz + 1/3m_i)^2 + 2zC_1(Cz + 1/3m_i)C\right]. \end{aligned} \tag{23}$$

Let

$$\xi = \xi(r) = \left(c - dr^2\right), \tag{24}$$

The transformed equation becomes

$$-2d\frac{\partial P}{\partial \xi}+4d(c-\xi)\frac{\partial^2 P}{\partial \xi^2}=-\mathbb{S}_1\xi\{2CMP^2+\left(2zC_1M+C_1M^2+2zC_1MC\right)P \\ +zC_1\left(C_1M^2+2zC_1MC\right),\} \tag{25}$$

where $M=Cz+\frac{1}{3}m_i$, $z=\frac{M-\frac{1}{3}m_i}{C}$, and $C_1=-(-\mathbb{S}_1)^{1/n}=-\epsilon$, small, for $n=5,4,3,2$.

5. General form of nonlinear equation

The following form of the nonlinear nonhomogeneous PDE, Eq. (25), is considered:

$$\frac{\partial^2 Y(\xi,z)}{\partial \xi^2}+F_1(\xi)\frac{\partial Y(\xi,z)}{\partial \xi}+F_2(\xi,z)Y^2(\xi,z)-(\epsilon/2)F_3(\xi,b_0(z))Y(\xi,z)+ \\ (1/2)G\left(\xi,z;\epsilon^2\right)=0, \tag{26}$$

where

$$F_1(\xi)=-(2c-2\xi)^{-1}, \tag{27}$$

$$F_2(\xi,z)=-1/4\,\frac{\mathbb{S}_1(2CM)\xi}{d(c-\xi)}, \tag{28}$$

and $G(\xi,z;\epsilon^2)$ involve a small parameter by choice of L_1 and L_2 and can be made arbitrarily small, whereas $F_3(\xi,b_0(z))$ can be made large due to choice of $b_0(z)$.
In light of work in [7], upon substitution of F_1 and F_2 for λ as defined in Eq. (33):

$$\lambda(\xi,z)=\frac{1}{\sqrt[5]{-1/4\,\frac{\mathbb{S}_1(2CM)\xi}{d(c-\xi)}}}\frac{1}{\sqrt[5]{2c-2\xi}}, \tag{29}$$

and

$$Y(\xi,z)=\lambda(\xi,z)(W(Z(\xi,z))-b_0(z))+\epsilon M\left(c-dt^2\right)\times \\ [z/3\ (9Cz+m_i)]t^{-2}\times F_3^{-1}(\xi,b_0(z)), \tag{30}$$

where Z is defined such that

$$\partial Z=\phi(\xi,z)\partial\xi, \tag{31}$$

where $F_3^{-1}=1/F_3$ is defined by Eq. (35), $\phi(\xi,z)$ is defined by Eq. (34), and $W(Z)$ is defined by Eq. (32). It follows that substitution of Eq. (30) for Y into Eq. (26) will cancel part of the coefficient of the Y term leaving the first term on the right side of Eq. (30). (The second term on the right side of Eq. (30) is the simplification of the last term on the right-hand side of Eq. (25) except the F_3^{-1} term). Next, ε^2 terms will cancel in Eq. (26) leaving a form of the equation which is homogeneous similar to that in [7] (where we have assumed that $\sup_z b_0(z)$ is large for z far down the channel), i.e.:

$$\frac{\partial^2 Y(\xi,z)}{\partial \xi^2}+F_1(\xi)\frac{\partial Y(\xi,z)}{\partial \xi}+F_2(\xi,z)Y^2(\xi,z)+F_3(\xi,z)Y(\xi,z)=0.$$

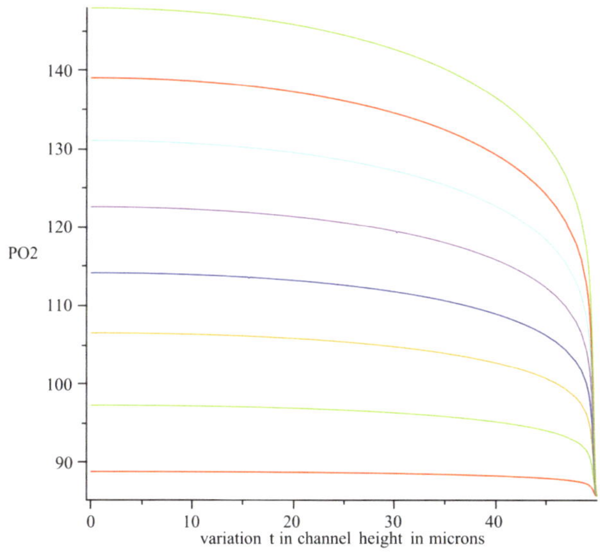

Figure 2.
Oxygen tension PO_2 [mmHg] versus channel height in microns at axial distance $z = m_i = -7000$ (at top) through -7950 microns (bottom) in intervals of 95 microns for lower human hematocrit.

It can be verified that as $\mathbb{S}_1 \to 0$ and $\epsilon \to 0$, the solution obtained from Eq. (26) is a linear function in ξ and independent of z, and one possible solution is a decreasing function on the height of the channel, which is intuitive as the PO_2 should drop at the wall of the channel (see **Figures 1** and **2**). In an approximating sense, though, the other terms will contribute in Eq. (26) as shown below.

The functions $\lambda(\xi,z)$,$\phi(\xi,z)$ and function $b_0(z)$ chosen to be large in magnitude in the supremum sense for all z values are selected so that the transformed equation, Eq.(26), through Eqs. (29)–(31) is written as one of the Painlevé classifications of the second-order differential equations [7].

There are two independent canonical forms for this equation [7], one of which is

$$\frac{d^2W(Z)}{dZ^2} - 6W^2(Z) + 6b_0^2 = 0. \tag{32}$$

According to Estevez et al. [7], the functions must be of the form

$$\lambda(\xi,z) = F_2^{-1/5}(\xi,z)e^{-\frac{2}{5}\int F_1(\xi)d\xi}, \tag{33}$$

and

$$\phi(\xi,z)^2 = -\frac{\lambda(\xi,z)F_2(\xi,z)}{6}. \tag{34}$$

The functions F_1, F_2 and F_3 should satisfy the compatibility relation [7]:

$$F_3(\xi,b_0(z)) = 2b_0(z)F_2(\xi,z)\lambda(\xi,z) - \frac{\frac{\partial^2}{\partial\xi^2}\lambda(\xi,z)}{\lambda(\xi,z)} - \frac{F_1(\xi)\ \frac{\partial}{\partial\xi}\lambda(\xi,z)}{\lambda(\xi,z).} \tag{35}$$

The previous equation, Eq.(35), after substitution of Eqs. (27)–(29) reduces considerably to the following for F_3:

$$F3(\xi, b_0(z)) = \left(-50\,(c-\xi)\xi^2\right)^{-1}\times \left\{12c - 7\xi + 25b_0(z)\,\xi^2\sqrt[5]{2}\left(-\frac{S1\xi CM}{d(c-\xi)}\right)^{4/5}(2c-2\xi)^{4/5}\right\}. \tag{36}$$

The two forms of the solution for $F3$ (one being the coefficient of P simplified in Eq. (25) and the other Eq. (36)) are equated to each other and compared:

$$-(1/50)\left(12c - 7\xi + 25b_{00}zM2^{2/5}\xi^2\left(-\frac{\mathbb{S}_1\xi CM}{d(c-\xi)}\right)^{4/5}(2c-2\xi)^{4/5}(-\mathbb{S}_1)^{1/5}(CM)^{-4/5}\right)\times$$
$$(c-\xi)^{-1}\xi^{-2} = M\frac{\mathbb{S}_1\xi}{c-\xi}\frac{z(2+3C)+m_i}{2} \sim M\frac{\mathbb{S}_1\xi}{c-\xi}\frac{z(2+3C)}{2}.$$

For z large $m_i/2$ is dropped from the total expression. The $\xi^{4/5}$ term is scaled by multiplying by a factor of 4 to obtain ξ approximately (see **Figure 3**).
Here $b_{00} = (1/8)(2+3C)/\left(2\left[(-S1)^{1/5}\right]\right)$

$$b_0(z) = b_{00}z\frac{M(-S1)^{1/5}}{(CM)^{4/5}}. \tag{37}$$

Since z can be large downstream, even for some small $\mathbb{S}_1$, then the term $12c - 7\xi$ drops out. This results in equality of the two forms of F3 presented. A final substitution

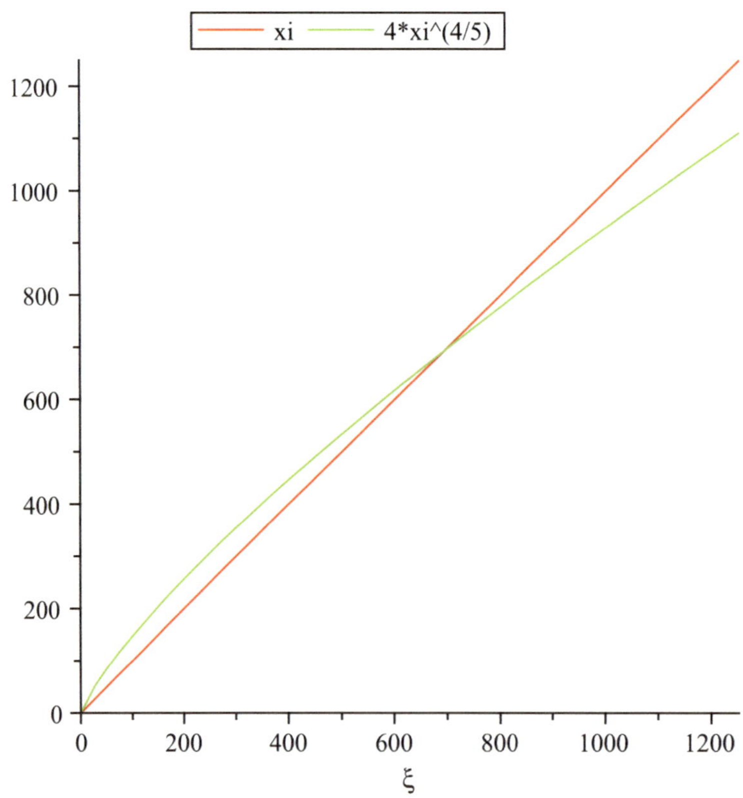

Figure 3.
ξ versus $4\xi^{4/5}$.

for z with a large number multiplied with itself maintains the equality (i.e., $z \to \alpha z$, for α large and positive). The solution is valid for z downstream, and we exclude the interval $[0, m_i]$ in **Figure 1** with no inlet $PO2$ value specified as a boundary condition at $z = 0$ as this would give erroneous results in the upstream region.

The final form of the general solution to Eq. (25), downstream (defined on the interval $[m_i, m_f]$ in **Figure 1**) using Eq. (30) is

$$\begin{aligned} P(r,z) = 2CM\{N(r)\epsilon \times [z(9Cz + m_i)/6C] \times F_3^{-1}(\xi, b_0(z)) + \\ \frac{(2CM)^{-1}\frac{\pi^2}{4}\left(-\sin\left(\frac{\pi}{2}K(r,z) + Q\right)^{-2} + \frac{1}{3}\right)\cdot 2^{4/5}}{M_1(r,z)} - \\ 10^4 z(-S1)^{1/5}\left[(2+3C)/16C^{4/5}\right]M^{1/5}\ (2CM)^{-1}\frac{1}{M_1(r,z)}\}, \end{aligned} \tag{38}$$

where

$$K(r,z) = 0.29(c - dr^2)\sqrt{\frac{-(CM)\mathbb{S}_1(c - dr^2)}{\left(-(2CM)\mathbb{S}_1(c - dr^2)\right)^{1/5}}}, \tag{39}$$

$$M_1(r,z) = \left(-(2CM)\mathbb{S}_1(c - dr^2)\right)^{1/5}, \tag{40}$$

$$N(r) = \frac{c - dr^2}{r^2}, \tag{41}$$

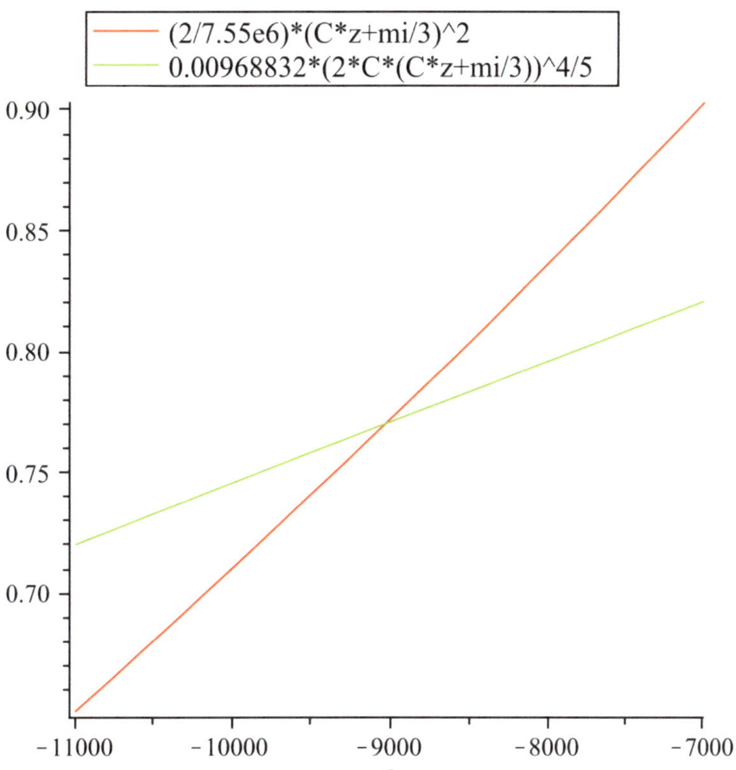

Figure 4.
2CM versus 0.01040 ∗ $(2CM)^{9/5}$.

C and Q are constants and $\epsilon = (-\mathbb{S}_1)^{1/n}$, small, for n = 5, 4, 3, 2.

As an approximation, if we divide Eq. (26) by z, using Eq. (30), for large z downstream in the channel, it can be seen that the following equation emerges:

$$\frac{\partial^2}{\partial \xi^2}(\lambda W) + F_1(\xi)\frac{\partial}{\partial \xi}(\lambda W) = F_2(\xi)\lambda^2 W^2 - F_3(\xi)(\lambda W).$$

Now the z part appearing in λ is $1/(2CM)^{1/5}$. Multiplying the equation above by $(2CM)^{6/5}$ will result in the left side of the equation to consist of two $2CM$ terms, and the right-hand side of the equation to have two $2(CM)^{9/5}$ terms where $F_3(\xi)$ has a $(2CM)^{4/5}$ term from Eq. (36). In Eq. (38) the $\frac{\sin(\pi/2K(r)+Q)^{-2}}{M_1(r,z)}$ term will oscillate to zero as z approaches minus infinity.

A best line of fit can be made by scaling the term $2(CM)^{9/5}$. Multiplying this by approximately 0.01040 results in a best approximation as shown in **Figure 4**. This allows us to cancel almost all z dependence in the equation except oscillating term and get a homogeneous "almost" z-independent equation as in [7]. Hence, the PDE of Eq. (26) can be reduced to an ode in ξ of similar form but homogeneous as z approaches minus infinity.

6. Boundary and matching conditions at wall and core-plasma interface

We employ the Robin boundary condition Eq. (58) in Appendix and derivative matching conditions at the interface of plasma and RBC core regions, respectively, at $z = m_i$ and $z = m_f$, shown in **Figure 1**, to determine constants Q and C as well as two additional constants for linear solution in plasma layer. The solution to the linear part of Eq. (13) defined in the plasma layer, i.e.,

$$v_p \frac{\partial PO_2(plasma)}{\partial z} = D_p \nabla^2 PO_2(plasma), \tag{42}$$

is

$$PO_2(plasma) = CP_1\, CylinderU\left(1/2c\sqrt{-\frac{\mathbb{S}_1}{D_p}\frac{1}{\sqrt{d}}}, r\sqrt[4]{-4\frac{\mathbb{S}_1 d}{D_p}}\right) + CP_2\, CylinderV\left(1/2c\sqrt{-\frac{\mathbb{S}_1}{D_p}\frac{1}{\sqrt{d}}}, r\sqrt[4]{-4\frac{\mathbb{S}_1 d}{D_p}}\right),$$

where CylinderU and CylinderV are parabolic cylinder functions and CP_1 and CP_2 are constants to be determined. $\mathbb{S}_1$ is a separation constant for Eq. (42). The following boundary matching condition is utilized at the interface of the plasma layer (1 micron in height) and RBC core region (49 microns in height):

$$\frac{\partial PO_2(plasma)}{\partial r} = \frac{\partial PO_2(core)}{\partial r}, \tag{43}$$

at $z = m_i$ and $z = m_f$, respectively (see **Figure 1**). Four equations in four unknowns were solved in Maple 18, where two constants are from each of the two solutions in plasma and core regions, respectively. The total of eight constants was

Constants	C	Q	CP_1	CP_2
High hematocrit = 0.45	−0.1009687192	6.261616672×10^7	1.111333619	1.542724257
Low hematocrit t = 0.15	−0.09824747318	2063.831966	1.111333619	1.542724258

Table 1.
System of constants for associated system of four unknowns solved in maple.

determined and is shown in **Table 1** for low and high hematocrit. The value of D_p is obtained from Table 1 in [6] and $c = 1250$ and $d = 0.5$. For high hematocrit as shown in the Appendix, the solution incorporates different values of c and d. Also we let $\mathbb{S}_1 = -0.911 \times 10^{-8}$ in the core region and $\mathbb{S}_1 = -0.211 \times 10^{-3}$ the in plasma layer. Here there are two different separation constants for two different regions.

The fourth equation used was that the change in flux at the wall at the edge of the plasma layer far downstream $(z < < m_f)$ is zero. In addition the no-flux condition shown in **Figure 1** (i.e., impermeable membrane) at $r = 0$ was satisfied exactly for all z for the core region solution. This is the fifth boundary condition. There is no inlet PO_2 specified at $z = 0$. The oxygen tension in core is $PO_2(\text{core}) = 33.07440049 + 3.63804058210^{-7}P$ which is in the form of Eq. (22) and gives a PO_2 of approximately 150 mmHg at $r = 0, z = m_i = -7000$ microns.

7. Weierstrass elliptic function

The Weierstrass P function is defined as

$$W(Z) = \frac{1}{Z^2} + \sum_w \frac{1}{(Z-w)^2} - \frac{1}{w^2}. \tag{44}$$

As in [9], we consider the following expression:

$$\eta = \exp\,(u\pi i/\omega), \tag{45}$$

and derive a function of η which behaves like the Weierstrass P function at $\eta = 0$. The development of η in the neighborhood of $u = 0$ is

$$\eta = 1 + \frac{u\pi i}{\omega} + \frac{1}{2!}\left(\frac{u\pi i}{\omega}\right)^2 + \,\dots, \tag{46}$$

or

$$\eta - 1 = \frac{u\pi i}{\omega}\left[1 + \frac{1}{2!}\frac{u\pi i}{\omega} + \frac{1}{3!}\left(\frac{u\pi i}{\omega}\right)^2\right] \tag{47}$$

Observe that the function

$$(\eta - 1)^2 = -\frac{u^2\pi^2}{\omega^2}\left[1 + \frac{u\pi i}{\omega} + \frac{7}{12}\left(\frac{u\pi i}{\omega}\right)^2 + \,\dots\right], \tag{48}$$

is zero of the second order at all the points $u = 2\mu\omega(\mu = 0, \pm 1, \pm 2, \pm 3, ..)$.

Consider a function $J(\eta)$ such that $J(\eta) \neq 0$ for $\eta = 1$, the function

$$\frac{J(\eta)}{(\eta-1)^2}, \tag{49}$$

is infinite of the second order for all values $u = 0$ and $u = 2\mu\omega$.
This behavior at these points is the same as the Weierstrass P function.
We write $J(\eta) = a + b\eta + c\eta^2$ where a, b, c are constants.
Since $\eta^2 = e^{\frac{2u\pi i}{\omega}}$, we have

$$\frac{J(\eta)}{(\eta-1)^2} = \frac{a + b\left[1 + \frac{u\pi i}{\omega} + \frac{1}{2!}\left(\frac{u\pi i}{\omega}\right)^2 + \dots\right] + c\left[1 + \frac{2u\pi i}{\omega} + \frac{1}{2!}\left(\frac{2u\pi i}{\omega}\right)^2\right]}{-\frac{u^2\pi^2}{\omega^2}\left[1 + \frac{u\pi i}{\omega} + \frac{1}{3}\left(\frac{u\pi i}{\omega}\right)^2 + \dots\right]}$$
$$-\frac{\omega^2}{\pi^2}\frac{\left[a + b\left(1 + \frac{u\pi i}{\omega} + \frac{1}{2}\left(\frac{u\pi i}{\omega}\right)^2 + \dots\right) + c\left[1 + \frac{2u\pi i}{\omega} + \dots\right]\right]}{u^2}\left[1 - \frac{u\pi i}{\omega} + u^2\right]. \tag{50}$$

As in [9] the Weierstrass P function is shown to have the following degenerate form which is used in Section 5:

$$\frac{J(\eta)}{(\eta-1)^2} = (\pi/2\omega)^2\left[\sin^{-2}(u\pi/2\omega) - 1/3\right]. \tag{51}$$

8. Results and discussion

The present work shows that near the inlet of the permeable membrane (see **Figure 1**), there is a significant drop at the wall of the channel in PO_2 as compared to downstream values as shown in **Figures 2** and **5**. It is worthy to note that the structure of the solution obtained in terms of degenerate Weierstrass P function forms a near constant solution across the height of a micro-fluidic channel downstream at end of the membrane region. This is seen in both **Figures 2** and **5**, and in **Figure 5**, the contours flatten out as the flow of blood proceeds far downstream. The release of ATP has been shown to be caused by a change in oxygen saturation [6]. It is concluded that there is a significant decrease in oxygen tension to the right

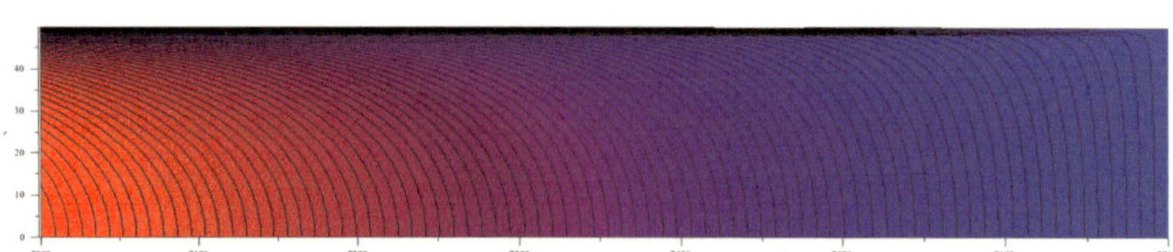

Figure 5.
Contour plot for channel from $z_0 = -7000$ to $z = -7900$ microns. Vertical axis is r variation across the channel. Horizontal axis is z variation.

of the permeable membrane downstream. It is in this region that there is a significant concentration of ATP released. It is important to mention that a rapid decrease in oxygen saturation is one means to produce ATP; it is also accomplished by means of applying shear stress on the RBC as shown in [12].

9. Conclusion

A well-known governing nonlinear PDE used to model oxygen transport was formulated in a recent paper in a generalized coordinate system where the Laplacian is expressed in metric tensor form. A reduction of the PDE to simpler problem subject to specific integrability conditions was shown there. A reduced almost linear ode was derived, and in the present paper, a solution has been obtained using a well-known factorization method for the second-order ode where a compatibility equation has been used in equating it to a specific form of the original differential equation. Approximate oxygen tension profiles have been determined downstream in a micro-channel in the vicinity of a permeable membrane with an oxygen supply on the other side of the membrane. Although it is expected that ATP will be released as blood flows past the permeable membrane downstream, it has been shown mathematically that this is the case and increases in hematocrit produce more ATP. Future work remains to apply tensor equations for a moving arterial surface as generalized at the start of the present work.

Funding

This research did not receive any specific grant from funding agencies in the public, commercial, or not-for-profit sectors.

Appendix

A.1 Cell free

For cell-free region as shown in **Figure 1**, we have that

$$v_p \frac{\partial PO_2}{\partial z} = D_p \nabla^2 PO_2, \tag{52}$$

where v_p is the velocity of plasma, given at the end of the Appendix.

A.2 Core region

In this case there is a central core region where oxygen dissociates to form HbO_2. Therefore, the velocity in core region, i.e., $v_{\ell O_2} = f\left(v_p\left(q^i\right), \frac{dSO_2}{dPO_2}\right)$.
The model consists of the following partial differential equation:

$$\left[v_p(1-H_T) + v_{RBC}H_T \frac{K_{RBC}}{K_p}\left(1 + \frac{[Hb_T]}{K_{RBC}}\frac{dSO_2}{dPO_2}\right)\right]\frac{\partial PO_2}{\partial z} = D_p \nabla^2 PO_2. \tag{53}$$

A.3 Boundary conditions

The boundary conditions are defined as follows:

$$PO_2(r, m_i) = (PO_2)_i, \quad r \in (0, h), \tag{54}$$

1. Pre-membrane

$$\left.\frac{\partial PO_2}{\partial r}\right|_{r=h} = 0, z \in (0, m_i). \tag{55}$$

2. Bottom of membrane region

$$\left.\frac{\partial PO_2}{\partial r}\right|_{r=0} = 0, z \in (0, L). \tag{56}$$

3. Post-membrane

$$\left.\frac{\partial PO_2}{\partial r}\right|_{r=h} = 0, z \in \left(m_f, L\right). \tag{57}$$

The only region not covered by the above three flux equations is the region of PO_2 occupying $z \in \left(m_i, m_f\right)$ at $r = h$ the membrane. This region is governed by a Robin condition specified in [8].

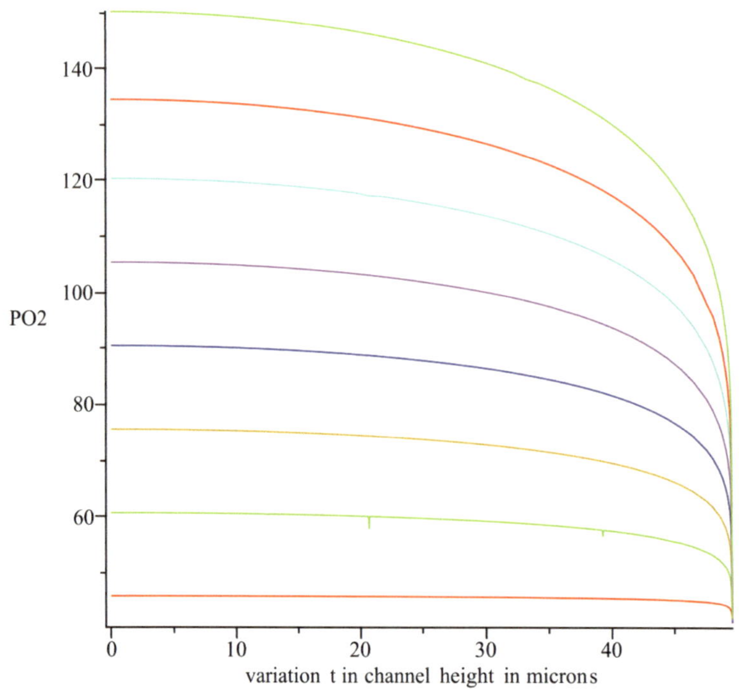

Figure 6.
Oxygen tension PO_2 versus channel height in microns at axial distance $z = m_i = -7000$ (at top) through $z = m_f = -7700$ microns (at the bottom) for higher human hematocrit in intervals of approximately 95 microns.

The boundary condition is

$$\left.\frac{\partial PO_2}{\partial r}\right|_{r=h} = \frac{D_m K_m}{D_p K_p}\left(\frac{(PO_2)_o - PO_2(h,z)}{\tau}\right), z \in (m_i, m_f), \tag{58}$$

where $(PO_2)_o$ is the PO_2 level on the other side of the membrane.

All of these conditions act on the entire system, and constants appearing are found in Table 1 in [6].

A.4 Velocity profile function

The following velocity profile is used in channel:

$$v_p(\hat{\mathbf{x}}) = \frac{3(130)\cdot 10^6}{4w\left(h^3 + \left(\mu_p/\mu_c\right)y_i^3\right)}\begin{cases}(h^2 - \hat{\mathbf{x}}^2), & y_i \le |\hat{\mathbf{x}}| \le h \\ (h^2 - y_i^2) + \mu_p/\mu_c(y_i^2 - \hat{\mathbf{x}}^2) & 0 \le |\hat{\mathbf{x}}| \le y_i\end{cases} \tag{59}$$

Relative apparent viscosity, $\mu_p/\mu_c \sim 1$, for low discharge hematocrit. The previous results (**Figures 2** and **5**) were based on this data. **Figure 6** is based on a discharge hematocrit of approximately 40% higher with relative apparent viscosity equal to 1.7. It is apparent from the graph that in comparison to **Figure 2** with lower discharge hematocrit that there is an increase in the drop of PO_2 profiles downstream with a resulting higher production of ATP.

Author details

Terry E. Moschandreou[1]* and Keith C. Afas[2]

1 Department of Applied Mathematics, Faculty of Science, Western University, London, Ontario, Canada

2 Medical Biophysics, Faculty of Medical Science, Western University, London, Ontario, Canada

*Address all correspondence to: tmoschan@uwo.ca

References

[1] Nair PK, Huang NS, Olson JS. A simple model for prediction of oxygen transport rates by flowing blood in large capillaries. Microvascular Research. 1990;**39**:203-211

[2] Nair PK, Hellums JD, Olson JS. Prediction of oxygen transport rates in blood flowing in large capillaries. Microvascular Research. 1989;**36**: 269-285

[3] Nair PK. Simulation of Oxygen Transport in Capillaries. Rice University; 1988

[4] Moschandreou TE, Ellis CG, Goldman D. Influence of tissue metabolism and capillary oxygen supply on arteriolar oxygen transport: A computational model. Mathematical Biosciences. 2011;**232**(1):1-10

[5] Ng C-O. Dispersion in steady and oscillatory flows through a tube with reversible and irreversible wall reactions. Proceedings of the Royal Society A. 2006;**462**:481-515

[6] Sove RJ, Ghonaim N, Goldman D, Ellis CG. A computational model of a microfluidic device to measure the dynamics of oxygen-dependent ATP release from erythrocytes. PLoS One. 2013;**8**(11):1-9

[7] Estevez PG, Kuru S, Negro J, Nieto LM. Factorization of a class of almost linear second-order differential equations. Journal of Physics A: Mathematical and Theoretical. 2007;**40**: 9819-9824

[8] Afas KC, Moschandreou TE. Analytic multiplicative separation and existence investigation of non-linear oxygen transport with Poiseuille hemodynamic flow in a semi-generalized co-ordinate system. International Journal of Differential Equations and Applications. 2015;**14**:281-311

[9] Hancock H. Lectures on the Theory of Elliptic Functions. Dover; 2004

[10] Brennan MD, Rexius-Hall ML, Elgass LJ, Eddington DT. Oxygen control with microfluidics. Lab on a Chip. 2014;**22**:1-14

[11] Grinfeld P. Introduction to Tensor Analysis and the Calculus of Moving Surfaces. Springer; 2010

[12] Wan J, Ristenpart WD, Stone HA. Dynamics of shear-induced ATP release from red blood cells. PNAS;**105**(43): 16432-16437